AF614753

Methods in Molecular Biology

Series Editor
John M. Walker
School of Life and Medical Sciences
University of Hertfordshire
Hatfield, Hertfordshire, AL10 9AB, UK

For further volumes:
http://www.springer.com/series/7651

SiRNA Delivery Methods

Methods and Protocols

Edited by

Kato Shum and John Rossi

Beckman Research Institute of City of Hope, Duarte, CA, USA

Editors
Kato Shum
Beckman Research Institute of City of Hope
Duarte, CA, USA

John Rossi
Beckman Research Institute of City of Hope
Duarte, CA, USA

ISSN 1064-3745 ISSN 1940-6029 (electronic)
Methods in Molecular Biology
ISBN 978-1-4939-3111-8 ISBN 978-1-4939-3112-5 (eBook)
DOI 10.1007/978-1-4939-3112-5

Library of Congress Control Number: 2015947262

Springer New York Heidelberg Dordrecht London

Printed on acid-free paper

Humana Press is a brand of Springer
Springer Science+Business Media LLC New York is part of Springer Science+Business Media (www.springer.com)

Preface

RNA interference (RNAi) is a cellular process for gene silencing based on recognition and subsequent degradation of specific mRNA sequences. The process was initially observed in plants and was later demonstrated in the nematode worm by Andrew Fire and Craig Mello. This groundbreaking discovery won Fire and Mello the Nobel Prize in physiology or medicine in 2006. The RNAi process can be triggered by chemically synthesized small interfering RNAs (SiRNAs) or vector-expressed short hairpin RNAs (shRNAs). SiRNAs are short double-stranded RNA molecules (20–25 bp) with two well-defined overhanging nucleotides at the 3′ end, and shRNAs are longer double-stranded RNA molecules (~70 bp) that contain a tight hairpin turn.

Since RNAi harnesses a natural pathway that can turn off gene expression at the post-transcriptional level even before protein translation occurs, the knowledge that SiRNA can affect gene expression has had a tremendous impact on basic and applied research, and RNAi is currently one of the most promising new approaches for disease therapy. RNAi-based therapeutics possess several advantages over small molecules and biologics, and one major advantage is that SiRNA/shRNA therapeutics can be optimally designed for any target, including targets that are considered "undruggable." Currently, there are approximately 22 different SiRNA/shRNA therapeutics undergoing rigorous clinical testing for diseases ranging from genetic disorders to human immunodeficiency virus infections and cancers. For example, Patisiran (ALN-TTR02), led by Alnylam® Pharmaceuticals, is currently in phase III clinical trials and represents the most advanced RNAi-based therapeutic. Patisiran silences *TTR* mRNA transcripts to reduce accumulation of amyloid deposits in the liver for the treatment of transthyretin (TTR)-mediated amyloidosis. If successful, Patisiran will be a significant milestone for the entire field of RNAi therapeutics and will provide a model for future trial design, FDA regulatory approval, and strategies for RNAi commercialization.

Although much is known about the mechanisms of RNAi, there are a number of challenges that applications of this gene-silencing technology need to overcome, such as cytoplasmic delivery of SiRNA/shRNA. Delivering SiRNA/shRNA molecules to the right tissue and cell type to treat disease is extremely difficult. Getting SiRNAs across the cell membrane and into the cytoplasm where they can effectively silence their target mRNAs poses significant hurdles. Investigators in academia and biotech/pharmaceutical industry have made intensive efforts to understand the molecular mechanism of RNAi and develop more advanced RNAi delivery formulations. The two main strategies are delivery of SiRNAs by chemical formulation (nonviral delivery) or delivery of shRNA-encoding genes by engineered viruses that will ultimately generate SiRNAs by transcription in the target cells (viral delivery). Viral vectors are one of the major vehicles in RNAi-based therapeutics that have advantages for difficult-to-transfect cells and for stable shRNA expression; however, concerns of potent immunogenicity, insertional mutagenesis, and biohazards of viral vectors may present a variety of potential problems to the patient. Nonviral methods could offer certain advantages over viral methods, and various innovative nonviral vectors have been vigorously developed to provide a safer and more efficient delivery system. Nevertheless, few of these vectors have so far been developed clinically owing to their low delivery efficiency relative to viral vectors.

In *SiRNA Delivery: Methods and Protocols*, we present a collection of cutting-edge strategies in delivery of SiRNA/shRNA developed or refined in the past few years, produced by a team of internationally renowned authors. This book is divided into two parts. The larger, first part covers a variety of nonviral SiRNA delivery formulations, including modified nucleotides and backbones (Chapter 1), peptides (Chapters 2–5), lipid-based nanoparticles (Chapters 6–9), exosomes (Chapter 10), dendrimers (Chapter 11), chitosans (Chapter 12), carbon nanotubes (Chapter 13), recombinant yeasts (Chapter 14), oligonucleotides (Chapters 15–17), and antibodies (Chapter 18). The smaller, second part covers three viral vehicles for shRNA delivery: lentiviruses (Chapter 19), adenoviruses (Chapter 20), and adeno-associated viruses (Chapters 21 and 22). Each protocol contains a brief introduction to their respective topic, list of the necessary materials and reagents, step-by-step readily reproducible laboratory protocols, and problem-solving tips. Lastly, Chapter 23 will discuss and compare various delivery strategies employed to solve the problem of SiRNA/shRNA delivery.

Finally, it has been an honor to work with each of the authors in assembling this compilation of protocols and procedures. We congratulate them all on their achievements marking their methods available. We are grateful to John M. Walker, the series editor, who gave us this opportunity and guided us patiently through the entire process. We hope you will find *SiRNA Delivery: Methods and Protocols* a useful reference easily used at the lab bench.

Duarte, CA, USA

Kato Shum
John Rossi

Contents

Contributors

DONG SUNG AN • *Hematology–Oncology, The Department of Medicine, David Geffen School of Medicine at UCLA, Los Angeles, CA, USA; School of Nursing, University of California Los Angeles, Los Angeles, CA, USA; UCLA AIDS Institute, Los Angeles, CA, USA*

SAMIR EL ANDALOUSSI • *Department of Physiology, Anatomy and Genetics, University of Oxford, Oxford, UK; Department of Laboratory Medicine, Karolinska Institutet, Huddinge, Sweden*

PATRICK ARBUTHNOT • *Wits/SAMRC Antiviral Gene Therapy Research Unit, School of Pathology, Health Sciences Faculty, Johannesburg, South Africa*

STEPHANIE E. BARRETT • *Pharmaceutical Sciences and Clinical Supply, Merck Research Laboratories, West Point, PA, USA*

ULRICH BRINKMANN • *Large Molecule Research, Roche Pharma Research and Early Development, Roche Innovation, Penzberg, Germany*

XIAOJUN CAI • *Shanghai East Hospital, The Institute for Biomedical Engineering & Nano Science (iNANO), Tongji University School of Medicine, Shanghai, China*

SILVIA CATUOGNO • *Istituto di Endocrinologia ed Oncologia Sperimentale, CNR, Naples, Italy*

JIANJUN CHENG • *Department of Materials Science and Engineering, University of Illinois at Urbana–Champaign, Urbana, IL, USA*

CAROL CROWTHER • *Wits/SAMRC Antiviral Gene Therapy Research Unit, School of Pathology, Health Sciences Faculty, Johannesburg, South Africa*

SUVARNA DASH-WAGH • *Department of Neuroscience, Karolinska Institutet, Stockholm, Sweden*

JUSTIN P. DASSIE • *Department of Internal Medicine, University of Iowa, Iowa City, IA, USA*

DAVID D. DICKEY • *Department of Internal Medicine, University of Iowa, Iowa City, IA, USA*

HAIQING DONG • *Shanghai East Hospital, The Institute for Biomedical Engineering & Nano Science (iNANO), Tongji University School of Medicine, Shanghai, China*

STEVEN F. DOWDY • *Department of Cellular & Molecular Medicine, UCSD School of Medicine, La Jolla, CA, USA*

BRETT D. DUFOUR • *Division of Neuroscience, Oregon National Primate Research Center, Beaverton, OR, USA; Department of Behavioral Neuroscience, Oregon Health and Science University, Portland, OR, USA*

MITSUHIRO ENOMOTO • *Department of Orthopaedic and Spinal Surgery, Graduate School, Tokyo Medical and Dental University, Tokyo, Japan; Hyperbaric Medical Center, Tokyo Medical and Dental University, Tokyo, Japan*

CARLA L. ESPOSITO • *Istituto di Endocrinologia ed Oncologia Sperimentale, CNR, Naples, Italy*

KARIEM EZZAT • *Department of Physiology, Anatomy and Genetics, University of Oxford, Oxford, UK*

Vittorio de Franciscis • *Istituto di Endocrinologia ed Oncologia Sperimentale, CNR, Naples, Italy*

Paloma H. Giangrande • *Department of Internal Medicine, University of Iowa, Iowa City, IA, USA; Department of Radiation Oncology, University of Iowa, Iowa City, IA, USA*

Jordan J. Green • *Department of Biomedical Engineering, Johns Hopkins University School of Medicine, Baltimore, MD, USA; Translational Tissue Engineering Center, and Institute for Nanobiotechnology, Johns Hopkins University School of Medicine, Baltimore, MD, USA; Department of Material Science and Engineering, Johns Hopkins University School of Medicine, Baltimore, MD, USA; Department of Ophthalmology, Johns Hopkins University School of Medicine, Baltimore, MD, USA; Department of Neuroscience, Johns Hopkins University School of Medicine, Baltimore, MD, USA*

Michael Grote • *Large Molecule Research, Roche Pharma Research and Early Development, Roche Innovation, Penzberg, Germany*

Erin N. Guidry • *Discovery Chemistry, Merck Research Laboratories, Kenilworth, NJ, USA*

Tomoya Hada • *Faculty of Pharmaceutical Sciences, Hokkaido University, Sapporo, Japan*

Alexander S. Hamil • *Department of Cellular & Molecular Medicine, UCSD School of Medicine, La Jolla, CA, USA*

Hideyoshi Harashima • *Laboratory of Innovative Nanomedicine, Faculty of Pharmaceutical Sciences, Hokkaido University, Hokkaido, Japan*

Takashi Hirai • *Department of Orthopaedic and Spinal Surgery, Graduate School, Tokyo Medical and Dental University, Tokyo, Japan; School of Dentistry, Oral Biology, Oral Biology and Medicine, University of California Los Angeles, Los Angeles, CA, USA*

Dewan Md Sakib Hossain • *Department of Cancer Immunotherapeutics & Tumor Immunology, Beckman Research Institute of City of Hope, Duarte, CA, USA*

Hidetoshi Kaburagi • *Department of Orthopaedic and Spinal Surgery, Graduate School, Tokyo Medical and Dental University, Tokyo, Japan*

Kazunori Kataoka • *Department of Materials Engineering, Graduate School of Engineering, The University of Tokyo, Tokyo, Japan; Center for Disease Biology and Integrative Medicine, Graduate School of Medicine, The University of Tokyo, Tokyo, Japan; Department of Bioengineering, Graduate School of Engineering, The University of Tokyo, Tokyo, Japan; Kawasaki Institute of Industry Promotion, Kawasaki, Japan*

Hyun Jin Kim • *Department of Bioengineering, University of California, Berkeley, CA, USA; Department of Materials Engineering, Graduate School of Engineering, The University of Tokyo, Tokyo, Japan*

Michinori Kohara • *Department of Microbiology and Cell Biology, Tokyo Metropolitan Institute of Medical Science, Tokyo, Japan*

James Koropatnick • *Department of Pathology, Surgery, and Oncology, University of Western Ontario, London, ON, Canada; London Regional Cancer Program, Lawson Health Research Institute, London, ON, Canada*

Marcin Kortylewski • *Department of Cancer Immunotherapeutics & Tumor Immunology, Beckman Research Institute of City of Hope, Duarte, CA, USA*

Kristen L. Kozielski • *Department of Biomedical Engineering, Johns Hopkins University School of Medicine, Baltimore, MD, USA; Translational Tissue Engineering Center, Johns Hopkins University School of Medicine, Baltimore, MD, USA; Institute for Nanobiotechnology, Johns Hopkins University School of Medicine, Baltimore, MD, USA*

Ülo Langel • *Department of Neurochemistry, Stockholm University, Stockholm, Sweden*

YONGYONG LI • *Shanghai East Hospital, The Institute for Biomedical Engineering & Nano Science (iNANO), Tongji University School of Medicine, Shanghai, China*
XIAOXUAN LIU • *Aix-Marseille Université, CNRS, Marseille Cedex, France*
ZHONGTIAN LIU • *College of Animal Science & Technology, Northwest A&F University, Shaanxi, China*
KLAUS MAYER • *Roche Pharma Research & Early Development, Large Molecule Research, Roche Innovation, Penzberg, Germany*
JODI L. MCBRIDE • *Division of Neuroscience, Oregon National Primate Research Center, Beaverton, OR, USA; Department of Behavioral Neuroscience, Oregon Health and Science University, Portland, OR, USA; Department of Neurology, Oregon Health and Science University, Portland, OR, USA*
WEI-PING MIN • *Department of Pathology, Surgery, and Oncology, University of Western Ontario, London, ON, Canada; London Regional Cancer Program, Lawson Health Research Institute, London, ON, Canada; Institute of Immunotherapy, Nanchang University and Jiangxi Academy of Medical Sciences, Nanchang, China*
KANJIRO MIYATA • *Center for Disease Biology and Integrative Medicine, Graduate School of Medicine, The University of Tokyo, Tokyo, Japan*
DAYSON MOREIRA • *Department of Cancer Immunotherapeutics & Tumor Immunology, Beckman Research Institute of City of Hope, Duarte, CA, USA*
BETTY MOWA • *Wits/SAMRC Antiviral Gene Therapy Research Unit, School of Pathology, Health Sciences Faculty, Johannesburg, South Africa*
SERGEY NECHAEV • *Department of Cancer Immunotherapeutics & Tumor Immunology, Beckman Research Institute of City of Hope, Duarte, CA, USA; Irell & Manella Graduate School of Biological Science, Beckman Research Institute at City of Hope, Duarte, CA, USA*
LING PENG • *Aix-Marseille Université, CNRS, Marseille Cedex, France*
VÉRONIQUE PRÉAT • *Louvain Drug Research Institute, Advanced Drug Delivery and Biomaterials, Université Catholique de Louvain, Brussels, Belgium*
HÉLOÏSE RAGELLE • *Louvain Drug Research Institute, Advanced Drug Delivery and Biomaterials, Université Catholique de Louvain, Brussels, Belgium; David H. Koch Institute for Integrative Cancer Research, Massachusetts Institute of Technology, Cambridge, MA, USA*
TIANBIN REN • *Shanghai East Hospital, The Institute for Biomedical Engineering & Nano Science (iNANO), Tongji University School of Medicine, Shanghai, China*
THOMAS C. ROBERTS • *Department of Molecular and Experimental Medicine, The Scripps Research Institute, La Jolla, CA, USA; Department of Physiology, Anatomy and Genetics, University of Oxford, Oxford, UK; Development, Aging and Regeneration Program, Sanford Burnham Prebys Medical Discovery Institute, La Jolla, CA, USA*
YU SAKURAI • *Faculty of Pharmaceutical Sciences, Hokkaido University, Sapporo, Japan*
YUSUKE SATO • *Laboratory of Innovative Nanomedicine, Faculty of Pharmaceutical Sciences, Hokkaido University, Hokkaido, Japan*
SAKI SHIMIZU • *School of Nursing, University of California, Los Angeles, Los Angeles, CA, USA; UCLA AIDS Institute, Los Angeles, CA, USA*
KING SUN SIU • *Department of Pathology, Surgery, and Oncology, University of Western Ontario, London, ON, Canada; London Regional Cancer Program, Lawson Health Research Institute, London, ON, Canada*
GABRIEL SKOGBERG • *Department of Rheumatology and Inflammation Research, Sahlgrenska Academy, University of Gothenburg, Gothenburg, Sweden*

Luisa Statello • *Department of Rheumatology and Inflammation Research, Sahlgrenska Academy, University of Gothenburg, Gothenburg, Sweden*
Piotr Swiderski • *Department of Molecular Medicine, Beckman Research Institute of City of Hope, Duarte, CA, USA*
Min Tang • *Shanghai East Hospital, The Institute for Biomedical Engineering & Nano Science (iNANO), Tongji University School of Medicine, Shanghai, China*
Esbjörn Telemo • *Department of Rheumatology and Inflammation Research, Sahlgrenska Academy, University of Gothenburg, Gothenburg, Sweden*
Gregory S. Thomas • *Department of Internal Medicine, University of Iowa, IA, USA*
Irmgard S. Thorey • *Roche Pharma Research & Early Development, Large Molecule Research, Roche Innovation, Penzberg, Germany*
Mats Ulfendahl • *Department of Neuroscience, Karolinska Institutet, Stockholm, Sweden*
Hadi Valadi • *Department of Rheumatology and Inflammation Research, Sahlgrenska Academy, University of Gothenburg, Gothenburg, Sweden*
Gaëlle Vandermeulen • *Louvain Drug Research Institute, Advanced Drug Delivery and Biomaterials, Université Catholique de Louvain, Brussels, Belgium*
Kevin Vanvarenberg • *Louvain Drug Research Institute, Advanced Drug Delivery and Biomaterials, Université Catholique de Louvain, Brussels, Belgium*
Jessica Wahlgren • *Department of Rheumatology and Inflammation Research, Sahlgrenska Academy, University of Gothenburg, Gothenburg, Sweden*
Marc S. Weinberg • *Department of Molecular and Experimental Medicine, The Scripps Research Institute, La Jolla, CA, USA; Antiviral Gene Therapy Research Unit, Department of Molecular Medicine and Haematology, University of the Witwatersrand Medical School, Johannesburg, South Africa; HIV Pathogenesis Research Unit, Department of Molecular Medicine and Haematology, University of the Witwatersrand Medical School, Johannesburg, South Africa*
Kun Xu • *College of Animal Science & Technology, Northwest A&F University, Shaanxi, China*
Swati Seth Yadav • *School of Nursing, University of California, Los Angeles, Los Angeles, CA, USA; UCLA AIDS Institute, Los Angeles, CA, USA*
Lichen Yin • *Institute of Functional Nano & Soft Materials (FUNSOM) & Collaborative Innovation Center of Suzhou Nano Science and Technology, Soochow University, Suzhou, China*
Takanori Yokota • *Department of Neurology and Neurological Science, Graduate School, Tokyo Medical and Dental University, Tokyo, Japan*
Long Zhang • *College of Animal Science & Technology, Northwest A&F University, Shaanxi, China; Frontier Institute of Science and Technology, Xi'an Jiaotong University, Shaanxi, China*
Tingting Zhang • *College of Animal Science & Technology, Northwest A&F University, Shaanxi, China; Research Institute of Applied Biology, Shaanxi University, Shaanxi, China*
Qifang Zhang • *Department of Cancer Immunotherapeutics & Tumor Immunology, Beckman Research Institute of City of Hope, Duarte, CA, USA*
Zhiying Zhang • *College of Animal Science & Technology, Northwest A&F University, Shaanxi, China*

YUJUAN ZHANG • *Institute of Immunotherapy, Nanchang University and Jiangxi Academy of Medical Sciences, Nanchang, China*
MENG ZHENG • *Department of Materials Engineering, Graduate School of Engineering, The University of Tokyo, Tokyo, Japan*
NAN ZHENG • *Department of Materials Science and Engineering, University of Illinois at Urbana–Champaign, Urbana, IL, USA*
XIUFEN ZHENG • *Department of Pathology, Surgery, and Oncology, University of Western Ontario, London, ON, Canada; London Regional Cancer Program, Lawson Health Research Institute, London, ON, Canada*
HAIYAN ZHU • *Shanghai East Hospital, The Institute for Biomedical Engineering & Nano Science (iNANO), Tongji University School of Medicine, Shanghai, China*

Chapter 1

Synthesis and Conjugation of Small Interfering Ribonucleic Neutral SiRNNs

Alexander S. Hamil and Steven F. Dowdy

Abstract

Due to their high potency (EC50 ~1 pM) and exquisite target selectivity for all expressed mRNAs, small interfering RNA (siRNA)-induced RNAi responses hold significant promise as a therapeutic modality. However, the size and high negative charge of siRNAs render them unable to enter cells without assistance from a delivery agent. Most current methods of siRNA delivery rely on encasing siRNA molecules in large nanoparticles or cationic liposomes. However, these approaches suffer from a number of problems, including a poor diffusion coefficient, cytotoxicity, and poor pharmacokinetics. To address the problem of siRNA in vivo delivery, we developed monomeric neutral RNAi prodrugs, termed siRibonucleic neutrals (siRNNs), that directly neutralize the phosphate backbone negative charge by synthesis with bioreversible phosphotriester groups that are enzymatically cleaved off in the cytoplasm of cells. Here we describe the synthesis and purification of siRNN conjugates that induce in vivo target gene knockdown following systemic delivery into mice.

Key words siRNN prodrug, RNAi delivery, siRNN delivery, Bioreversible phosphotriester, GalNAc

1 Introduction

siRNA-induced RNAi responses have become a standard methodology used in the study of gene function in vitro, but the biophysical properties of siRNAs have limited their use in vivo. siRNAs can be synthesized in a scalable manner, allowing for the rapid production of siRNAs directed against any target mRNA [1]. However, for all the promise as a therapeutic, siRNAs have several significant negative attributes that limit their usefulness. The high negative charge of siRNA phosphodiester backbone and the 14,000 Dalton (Da) size prevent siRNA molecules from crossing the cellular membrane [2]. These attributes also make siRNAs pharmacokinetically unfavorable, as naked siRNAs are removed from the bloodstream by the kidneys within minutes of injection into a mouse [3]. Moreover, naked native charged siRNAs are rapidly degraded in blood by serum nucleases.

Kato Shum and John Rossi (eds.), *SiRNA Delivery Methods: Methods and Protocols*, Methods in Molecular Biology, vol. 1364, DOI 10.1007/978-1-4939-3112-5_1, © Springer Science+Business Media New York 2016

Here, we demonstrate the synthesis of monomeric neutral RNAi prodrugs, termed siRibonucleic neutrals (siRNNs). Due to neutralization of the negatively charged phosphodiester with bioreversible phosphotriester protecting groups, siRNNs show high stability against serum nucleases and can be used for in vivo delivery. Inside of cells, the S-acyl-2-thioethyl (SATE) phosphotriester group used in siRNNs is specifically converted into a charged native phosphodiester linkage by cytoplasmic thioesterases, resulting in a native wild type charged siRNA that is loaded into Ago2 to induce robust RNAi responses [4]. RNN oligonucleotides can be synthesized on standard oligonucleotide synthesizers using commercially available phosphoramidites and RNN phosphoramidites that can be synthesized on site [5]. Similar to RNA, RNN oligonucleotides are purified by reversed-phase high-performance liquid chromatography (RP-HPLC). Purified RNN oligonucleotides are then analyzed by mass spectrometry and urea denaturing gel electrophoresis. RNNs show both high coupling and purification yields on par with, if not better than RNA oligonucleotides. Finally, RNN oligonucleotides containing reactive phosphotriester groups are conjugated to Tris-GalNAc, a hepatocyte-specific targeting domain, and duplexed to complementary RNN oligonucleotides to form siRNN conjugates ready for animal injection [6].

2 Materials

2.1 Synthesis and Purification of RNN Oligonucleotides

1. Phosphotriester phosphoramidites: Cannot contain a 2′-OH and must have ultramild exocyclic amino group protection (either phenoxyacetyl (PAC) or 4-isopropylphenoxyacetyl (iPr-PAC)) (*see* **Note 1**). Recommended RNA phosphoramidites are: 2′-OMe-5′-O-DMTr-PAC-A-CE (Glen Research, Cat. # 10-3601-XX), 2′-Fluoro-5′-O-DMTr-PAC-C-CE (R. I. Chemicals, Cat. # P1500-01), 2′-OMe-5′-O-DMTr-iPr-PAC-G-CE (Glen Research, Cat. # 10-3621-XX), and 2′-Fluoro-5′-O-DMTr-U-CE (R. I. Chemicals, Cat. # F-1015).
2. RNN phosphoramidites (synthesis methods described in Meade et al. [5]) are: 2′-OMe-5′-O-DMTr-PAC-A-*t*Bu-SATE, 2′-fluoro-5′-O-DMTr-PAC-C-*t*Bu-SATE, 2′-OMe-5′-O-DMTr-iPr-PAC-G-*t*Bu-SATE, 2′-fluoro-5′-O-DMTr-U-*t*Bu-SATE, aldehyde A-SATE phosphoramidites of the above, and dT-Q-CPG 500 (Glen Research, Cat. # 21-2230-XX) (*see* **Note 2**).
3. Oligonucleotide synthesis chemicals are the following: Activator: 5-Benzylthio-1H-tetrazole (BTT) in Acetonitrile (Glen Research, Cat.# 30-3170-XX) (*see* **Note 3**).
4. Deblocking reagent: 3 % TCA (w/v).
5. Cap A: Ultra Mild Cap A THF/Pyridine/Pac_2O (Glen Research, 40-4210-XX) (*see* **Note 4**).

6. Cap B: 16 % MeIm in THF.
7. Acetonitrile, anhydrous.
8. Oxidizing reagent: 0.02 M I_2O in THF/Pyridine/H_2O (70/20/10).
9. CPG cleavage solution: 10 % diisopropylamine (DIA), 90 % anhydrous methanol. Prepare immediately before use.
10. HPLC Buffer A: 50 mM triethylamine acetate (TEAA) in ultrapure water.
11. HPLC Buffer B: 90 % acetonitrile (ACN) in ultrapure water.
12. Stericup-GP, 0.22 μm, polyethersulfone, 1000 mL.
13. Matrix for MALDI-TOF analysis of RNNs: 10 mg/mL 2′,4′,6′-Trihydroxyacetophenone (THAP) and 50 mg/mL diammonium citrate in 50 % acetonitrile and deionized water (*see* **Note 5**).
14. DMT removal buffer: 80 % glacial acetic acid, 10 % ACN, 10 % ultrapure water.
15. Oligonucleotide synthesizer.
16. Centrifugal evaporator.
17. RP-HPLC.
18. MALDI-TOF.
19. Lyophilizer.

2.2 Analysis of RNN Oligonucleotides

1. Illustra NAP-10 Columns, Sephadex G-25 DNA Grade.
2. 40 % acrylamide/bis solution, 19:1.
3. 10× Tris-borate-EDTA (TBE).
4. 10 % Ammonium persulfate in deionized water.
5. *N*,*N*,*N*′,*N*′-Tetramethylethylenediamine (TEMED).
6. Gel running buffer: 1× TBE in deionized water.
7. 100× bromophenol blue (BPB) solution: 0.2 g Tris base, 0.2 g BPB dissolved in 9.9 mL of deionized water.
8. 2× Denaturing gel loading buffer (10 mL): 9.8 mL of formamide, 100 μL of 0.5 M EDTA, pH 8.0, and 100 μL of 100× BPB solution.
9. Methylene blue stain: 250 μL saturated 35 mg/L methylene blue solution and 19.75 mL of deionized water.

2.3 siRNN Conjugation

1. Aniline.
2. GalNAc-HyNic.
3. Ethanol (200 proof).

3 Methods

3.1 Synthesis and Purification of RNN Oligonucleotides

1. Synthesize RNN oligonucleotide on standard solid-phase oligonucleotide synthesizer. Use oligonucleotide synthesis chemicals, dT-Q-CPG solid support, and phosphoramidites listed in Subheading 2.1. Use 6-min coupling time for all RNN phosphoramidites. Synthesize RNN oligonucleotides DMT-On for RP-HPLC purification (*see* **Note 6**).
2. After completion of RNN oligonucleotide synthesis, remove column from synthesizer and transfer resin into a 2 mL screw-cap microcentrifuge tube. Add 1 mL of CPG cleavage solution to tube containing CPG-oligonucleotide. Incubate for 4 h at room temperature on a rotisserie rotator.
3. After 4 h, pellet CPG by brief centrifugation in a mini centrifuge (5 s at 2000 × *g*). Transfer crude RNN oligonucleotide supernatant to a new 1.5 mL microcentrifuge tube. Wash CPG with 0.5 mL of anhydrous methanol and pool with crude RNN oligonucleotide solution.
4. Dry crude RNN oligonucleotide by centrifugal evaporation (*see* **Note 7**).
5. Once dry, dissolve crude RNN oligonucleotide pellet in 50 % ACN by briefly heating at 65 °C and vortexing.
6. Inject into RP-HPLC column for purification with the following protocol: flow rate: 1 mL/min, gradient: 0–2 min (100 % Buffer A/0 % Buffer B), 2–42 min (0 % Buffer B to 60 % Buffer B) (*see* **Notes 8** and **9**) (Fig. 1a).
7. Collect 500 μL fractions across DMT-On peak.
8. Analyze HPLC fractions for presence of full-length RNN oligonucleotide on MALDI-TOF using THAP matrix (Fig. 1b).
9. Freeze fractions containing full-length RNN oligonucleotide and lyophilize until dry (*see* **Note 10**).
10. Once dry, dissolve each fraction in 100 μL of 50 % ACN in ultrapure water and pool with other fractions containing full-length RNN oligonucleotide. Freeze and lyophilize dry pooled RNN oligonucleotide.
11. To remove DMT group, add 200 μL of DMT-removal buffer to lyophilized RNN oligonucleotide pellet. Incubate solution for 1 h at 65 °C. After 1 h, freeze and lyophilize solution.
12. For RNN oligonucleotide desalting, equilibrate NAP-10 column with 15 mL of 20 % ACN in ultrapure water. Resuspend RNN oligonucleotide in 1 mL of 20 % ACN in ultrapure water. Incubate RNN oligonucleotide solution at 65 °C for several minutes and vortex periodically to bring oligonucleotide into solution. Once in solution, cool to room temperature and then

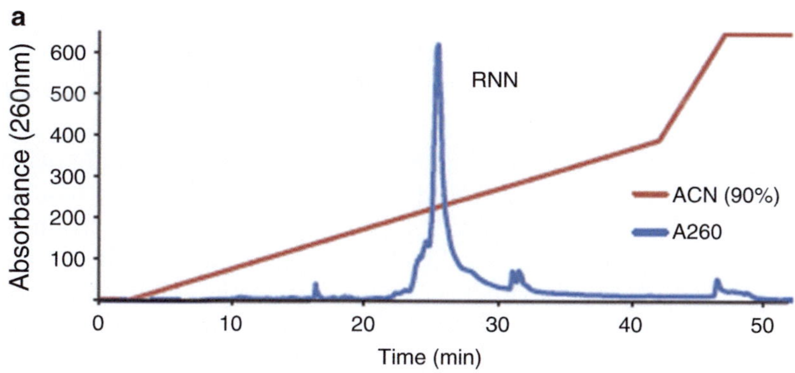

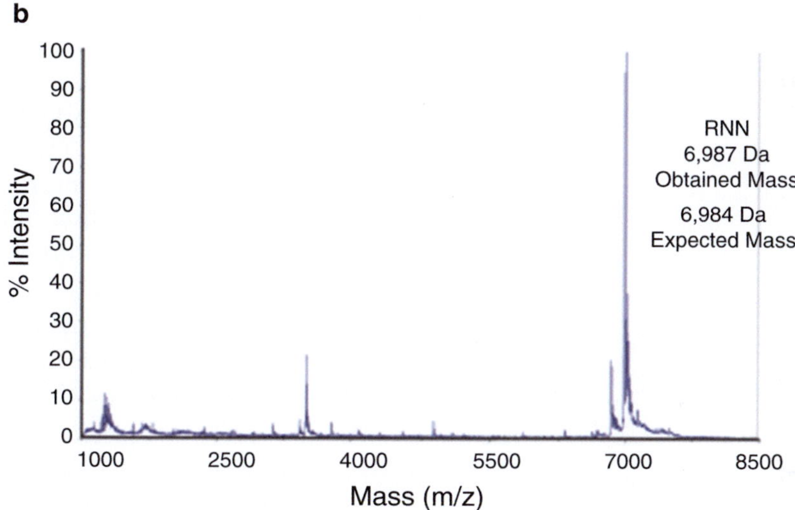

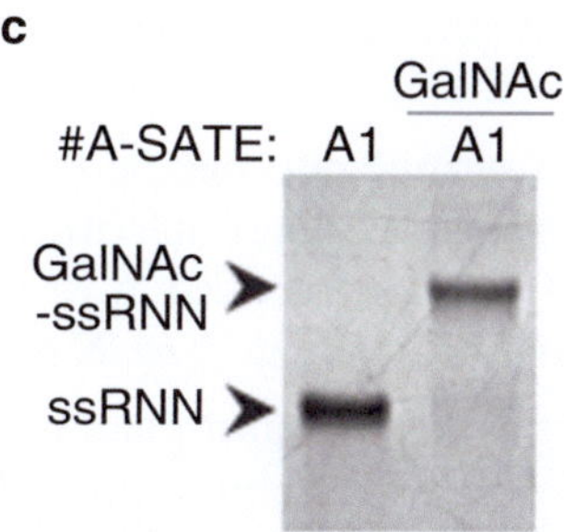

Fig. 1 Synthesis, purification, and conjugation of RNN oligonucleotides. (**a**) Representative RP-HPLC chromatogram of a crude RNN oligonucleotide. Gradient begins at 100 % 50 mM TEAA in H_2O and ends at 90 % ACN. Fractions are collected across the main peak and analyzed by MALDI-TOF mass spectrometry. Fractions containing full-length product are pooled, frozen, and lyophilized for further processing. (**b**) MALDI-TOF analysis of crude RNN oligonucleotide using THAP matrix. (**c**) Urea denaturing gel analysis of an RNN oligonucleotide containing one reactive A-SATE phosphotriester conjugated to Tris-GalNAc. *Left lane*: Unconjugated RNN oligonucleotide. *Right lane*: GalNAc-conjugated RNN oligonucleotide

add to equilibrated NAP-10 column. To elute RNN oligonucleotide add 1.4 mL of 20 % ACN in ultrapure water to the NAP-10 column and capture eluent. Freeze and lyophilize eluent (*see* **Note 11**).

13. Resuspend lyophilized RNN oligonucleotide pellet in 200 μL of 50 % ACN in water. If necessary, briefly incubate at 65 °C and vortex to bring oligonucleotide into solution. Measure absorbance of RNN oligonucleotide at 260 nm. Use the calculated extinction coefficient for your oligonucleotide and the measured A260 to determine oligonucleotide concentration. Dilute oligonucleotide to desired concentration with 50 % ACN in ultrapure water and store at −20 °C.

3.2 Analysis of RNN Oligonucleotides

1. Mix urea denaturing gel solution components in a 50 mL conical centrifugation tube. Add 10.5 g of urea, microwave for 9 s, and vortex until urea is fully dissolved. Add 200 μL of 10 % ammonium persulfate, mix, add 20 μL of TEMED, and cast two gels within 7.25 cm × 10 cm × 1.5 mm gel cassettes. Immediately insert 10-well gel combs without introducing air bubbles. Allow gels to sit at room temperature for 10–15 min to ensure complete polymerization before use.
2. Aliquot 1 nmol of RNN oligonucleotide to be analyzed in a fresh microcentrifuge tube. Dry sample by centrifugal evaporation (without heat) or by lyophilization (*see* **Note 12**).
3. Once oligonucleotide samples are completely dry, dissolve pellets in 3–5 μL of 2× denaturing gel loading buffer. Incubate samples in a 65 °C heating block for 1 min and cool on ice.
4. Place gels in running apparatus and fill with gel running buffer. Remove the comb gently and electrophorese for 10 min at 200 V constant to preheat the gel prior to gel loading.
5. Prior to gel loading, use a syringe and needle to clean out the wells of the gel. Load prepared oligonucleotide solution to each well and run at 200 V constant for 1 h or until bromophenol blue dye front runs off of the gel (*see* **Note 13**).
6. Remove gel from running apparatus and glass plates. Place in dish containing 20 mL of methylene blue stain and rock gently at room temperature until desired level of staining is accomplished (*see* **Notes 14** and **15**).
7. Analyze purified RNN oligonucleotides on MALDI-TOF using THAP matrix to ensure that the final product has the expected mass (Fig. 1b).

3.3 Conjugation of GalNAc-siRNN

1. Set up the following conjugation reaction: 120 μM RNN oligonucleotide (containing 1× A-SATE phosphotriester), 600 μM GalNAc-HyNic, 100 mM aniline in 50 % ACN, and ultrapure water (*see* **Note 16**).

2. Incubate reaction for 1 h at room temperature (*see* **Note 17**).
3. Freeze and lyophilize conjugation solution until dry.
4. Once dry, add 1 mL of ethanol to pellet. Incubate sample in a 65 °C heating block for 5 min with intermittent vortexing to break up pellet.
5. Incubate solution at −80 °C for 30 min.
6. Centrifuge sample for 5 min at 10,000 × *g* to pellet insoluble material.
7. Remove and discard supernatant (*see* **Note 18**).
8. Rinse pellet by adding 1 mL of −20 °C ethanol to pellet and vortexing.
9. Centrifuge sample for 5 min at 10,000 × *g*. Remove and discard supernatant.
10. Dry GalNAc-RNN pellet by centrifugal evaporation.
11. Once dry, resuspend pellet in 50 % ACN and measure absorbance of RNN oligonucleotide at 260 nm. Use the oligonucleotide extinction coefficient and the measured A260 to determine conjugate concentration (Fig. 1c).
12. To make GalNAc-siRNNs, aliquot an amount of GalNAc-RNN conjugate, add an equimolar amount of the complementary RNN oligonucleotide, heat at 65 °C for 5 min, vortex, and incubate at room temperature for 15 min to allow for RNN oligonucleotide and conjugated RNN oligonucleotide to duplex.
13. Freeze GalNAc-siRNN sample and lyophilize dry (*see* **Note 19**).
14. Once dry, resuspend GalNAc-siRNN in water or saline. Sample is now ready for injection into animal models or to treat primary hepatocytes.

4 Notes

1. RNN oligonucleotide synthesis is incompatible with phosphoramidites that contain 2′-OH (following deprotection). Presence of 2′-OH groups will also lead to strand scission at locations of phosphotriester nucleotides.
2. Any solid support that is susceptible to cleavage by 10 % diisopropylamine in methanol may be used for RNN oligonucleotide synthesis.
3. BTT activator may form crystals on the ends of oligonucleotide synthesizer activator injection lines. If this occurs, increase activator priming time or number of primes to remove obstructing crystal and ensure proper activator injection.

4. Ultra Mild Cap A must be used to avoid exchange of iPr-PAC or PAC protecting groups with acetyl groups.
5. 3-Hydroxypicolinic acid (3-HPA) matrix is generally recommended for MALDI-TOF analysis of oligonucleotides >3500 Da. However, THAP matrix is better suited for the analysis of fully 2′-OMe/F-modified oligonucleotides.
6. If synthesis of a phosphorothioate oligonucleotide is desired, RNN phosphoramidites are compatible with sulfurization during oligonucleotide synthesis.
7. Do not heat during centrifugal evaporation as this may lead to SATE phosphotriester reversal.
8. Buffer A should be filtered through a 0.22 μm filter prior to use.
9. Buffer B can also be made with 50 mM TEAA. However, presence of TEAA in Buffer B does not generally confer a purification advantage and will necessitate more than 2 lyophilizations to fully remove all TEAA from the collected fractions.
10. Do not pool fractions prior to the first lyophilization as this can lead to concentration of TEAA and reversal of SATE phosphotriesters.
11. RNN oligonucleotides that contain hydrophobic phosphotriesters (*t*Bu-SATE) may be more difficult to solubilize. For these oligonucleotides it is advisable to add 400 μL of 50 % ACN in ultrapure water to lyophilized oligonucleotide pellet, incubate solution at 65 °C for several minutes with periodic vortexing, and then add 600 μL of ultrapure water to the solution to bring the final oligonucleotide solution to 1 mL of 20 % ACN in ultrapure water.
12. The acetonitrile present in RNN oligonucleotide stocks will cause gel loading errors if the sample is not fully dried.
13. Urea will constantly diffuse out of the gel and into the wells that leads to poor resolution of samples. It is recommended that wells be flushed out with a syringe and needle, the flushed out wells loaded with sample, and then the process repeated until all samples have been loaded. The gel should be loaded and run promptly, as urea will continue to diffuse out of the gel and compress samples loaded into wells.
14. Neutralizing phosphotriesters decrease methylene blue staining of oligonucleotides. This will cause lighter staining of RNN oligonucleotides relative to RNA oligonucleotides despite equimolar loading.
15. 5–10 min of incubation will generally be sufficient to stain 21-mer RNN oligonucleotides. Afterwards, destain gel in deionized water on a rocker at room temperature. Examine destained gel for the presence of fast-migrating truncation

products that cannot be seen above background stain prior to destaining.

16. If conjugating to an RNN oligonucleotide containing more than 1× A-SATE phosphotriester maintain a 1:5 molar ratio of A-SATE to GalNAc-HyNic. For example, if conjugating an RNN oligonucleotide containing 3× A-SATEs the reaction should contain 120 μM RNN oligonucleotide and 1.8 mM GalNAc-HyNic.
17. For reaction volumes > 100 μL, incubate on a rotisserie rotator.
18. Supernatant contains unreacted GalNAc-HyNic and residual aniline. If desired, supernatant can be retained for re-purification of unreacted GalNAc-HyNic.
19. To examine conjugation status of the RNN oligonucleotide at any time during this procedure, lyophilize 1 nmol of GalNAc-conjugated RNN oligonucleotide and 1 nmol of unconjugated RNN oligonucleotide until dry. Run unconjugated and GalNAc-conjugated RNN oligonucleotide side by side on a urea denaturing gel and stain with methylene blue as described in Subheading 3.2 (Fig. 1c).

References

1. Beaucage SL (1993) Oligodeoxyribonucleotides synthesis. Phosphoramidite approach. Methods Mol Biol 20:33–61
2. Rettig GR, Behlke MA (2012) Progress toward in vivo use of siRNAs-II. Mol Ther 20: 483–512
3. Merkel OM, Librizzi D, Pfestroff A, Schurrat T, Behe M, Kissel T (2009) In vivo SPECT and real-time gamma camera imaging of biodistribution and pharmacokinetics of siRNA delivery using an optimized radiolabeling and purification procedure. Bioconjug Chem 20:174–182
4. Lefebvre I, Perigaud C, Pompon A, Aubertin AM, Girardet JL, Kirn A, Gosselin G, Imbach JL (1995) Mononucleoside phosphotriester derivatives with S-acyl-2-thioethyl bioreversible phosphate-protecting groups: intracellular delivery of 3′-azido-2′,3′-dideoxythymidine 5′-monophosphate. J Med Chem 38:3941–3950
5. Meade BR, Gogoi K, Hamil AS, Palm-Apergi C, van den Berg A, Hagopian JC, Springer AD, Eguchi A, Kacsinta AD, Dowdy CF, Presente A, Lonn P, Kaulich M, Yoshioka N, Gros E, Cui XS, Dowdy SF (2014) Efficient delivery of RNAi prodrugs containing reversible charge-neutralizing phosphotriester backbone modifications. Nat Biotechnol 32:1256–1261
6. Sliedregt LA, Rensen PC, Rump ET, van Santbrink PJ, Bijsterbosch MK, Valentijn AR, van der Marel GA, van Boom JH, van Berkel TJ, Biessen EA (1999) Design and synthesis of novel amphiphilic dendritic galactosides for selective targeting of liposomes to the hepatic asialoglycoprotein receptor. J Med Chem 42:609–618

Chapter 2

Liver-Targeted SiRNA Delivery Using Biodegradable Poly(amide) Polymer Conjugates

Stephanie E. Barrett and Erin N. Guidry

Abstract

The realization of polymer conjugate-based RNA delivery as a clinical modality requires the development and optimization of novel formulations. Although many literature examples of polymer conjugate-based SiRNA delivery systems exist, the protocols described herein represent a robust and facile way of screening any poly(amine)-based polymer system for SiRNA delivery. In this chapter, we describe the synthetic methods used to prepare poly(amide) polymers using a controlled polymerization method, as well as the preparation of the resulting targeted SiRNA polymer conjugates. In addition, detailed methods are provided for the characterization of the biodegradable poly(peptides) as well as the polymer conjugate that ensues.

Key words Biodegradable polymers, Polymer conjugates, SiRNA delivery, Targeted delivery, SiRNA protocols, Polyamide

1 Introduction

The key to successful clinical application of RNA interference requires optimization and development of improved delivery vehicles. The delivery vehicle must protect SiRNA from degradation while in circulation, deliver the genetic material to the target cell of interest, and assist the SiRNA with endosomal escape (Fig. 1) [1–7]. Several modalities have emerged, showing progress towards this goal, including liposomes, lipid nanoparticles, and polymer conjugates, each offering its own advantages and challenges. In this chapter, we focus on the use of biodegradable polymer conjugates as delivery vehicles for the targeted delivery of SiRNA to hepatocellular tissue. Polymer conjugates offer several advantages over other modalities, including their proven effectiveness as a targeted-delivery strategy, their small sizes for improved tissue penetration (diameters <50 nm), and their high degree of chemical control over the structure of the conjugate [3, 8–10].

Kato Shum and John Rossi (eds.), *SiRNA Delivery Methods: Methods and Protocols*, Methods in Molecular Biology, vol. 1364, DOI 10.1007/978-1-4939-3112-5_2, © Springer Science+Business Media New York 2016

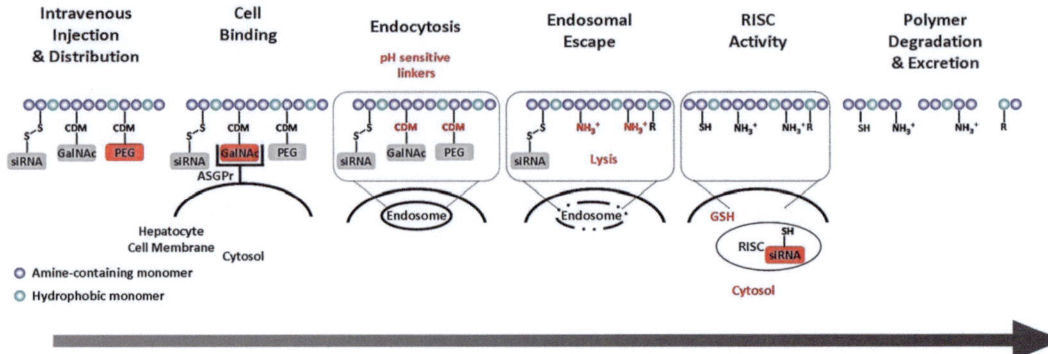

Fig. 1 The process of in vivo siRNA delivery. Components highlighted in *red* are thought to be the most significant constituent for that particular step

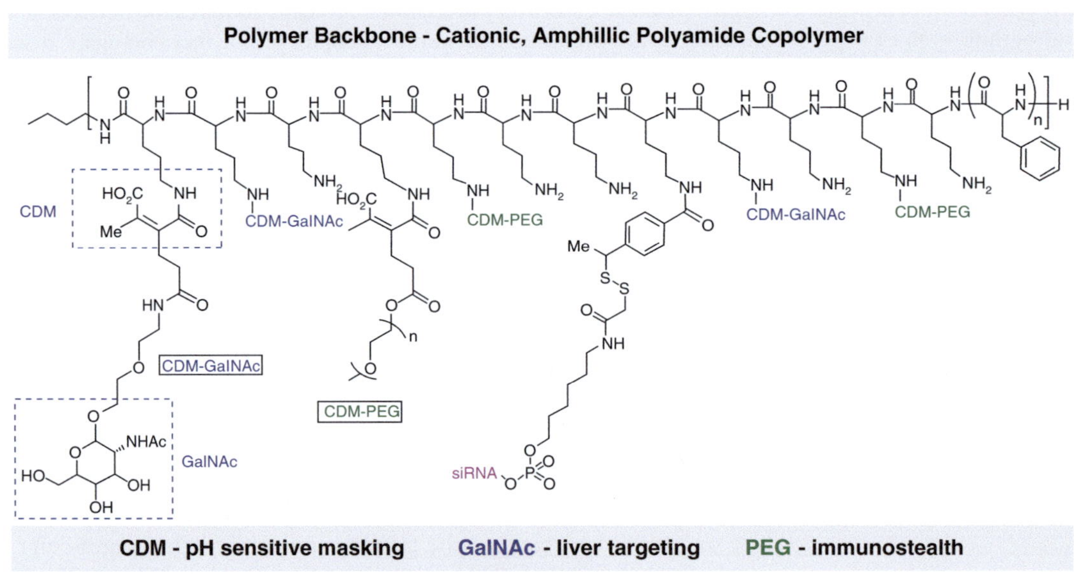

Fig. 2 Graphical representation of siRNA polymer conjugates

Recently, our group [11–14] and others [15, 16] have reported pH responsive polymer conjugate delivery systems for the delivery of SiRNA to liver hepatocytes, resulting in in vivo mRNA knockdown in hepatocellular tissue (Fig. 2). The design employed in these pH responsive polymer conjugates consists of a cationic or cationic-amphiphilic polymer in which the cationic functionality of the polymer is reversibly functionalized, masking the cationic character of the polymer conjugate in systemic circulation until it reaches the acidic environment of the endosome. Additionally, the targeting ligand (*N*-acetylgalactosamine, GalNAc), which promotes uptake into parenchymal liver cells using the asialoglycoprotein receptor (ASGPr), is incorporated through these pH-sensitive masking groups (carboxydimethyl maleic anhydride or CDM).

In this chapter, we describe the step-by-step synthesis and characterization of biodegradable poly(amide) polymer conjugates, which have been specifically designed for the targeted delivery of SiRNA to hepatocyte cells. Moreover, the protocols for preparing polymer conjugates for SiRNA delivery described in this chapter can be used for any polymer containing pendant amine groups, including homopolymers, block copolymers, random copolymers, and branched polymers.

2 Materials

2.1 Boc-L-Ornithine N-Carboxyanhydride NCA Monomer Synthesis

1. *N*-Boc-L-ornithine, triphosgene, anhydrous tetrahydrofuran (THF), isopropylacetate (IPAc), hexanes, L-phenylalanine, silica gel (*see* **Note 1**), NMR solvents, water, and acetonitrile (ACN).

2.2 Poly(amide) Synthesis

1. *N*-butylamine, anhydrous *N,N*-dimethylacetamide (DMA), dichloromethane (DCM), trifluoroacetic acid (TFA), and GPC elution solvent: 10 mM LiCl in *N,N*-dimethylformamide (DMF).

2.3 CDM-PEG Synthesis

1. 2-(2-Carboxyethyl)-3-methylmaleic anhydride (CDM), dichloromethane (DCM), *N,N*-dimethylformamide (DMF), oxalyl chloride, pyridine, ammonium chloride, 5 M HCl, water, trifluoroacetic acid (TFA), acetonitrile (ACN), and polyethylene glycol monomethyl ether.

2.4 SATA-Modified siRNA Synthesis

1. DMSOm sodium bicarbonate and *N*-succinimidyl-S-acetylthioacetate (SATA).
2. SiRNA (*see* **Note 2**).
3. 0.1 M Sodium bicarbonate: Sodium bicarbonate (840 mg) was dissolved in endonuclease-free water (100 mL). The solution was stored at room temperature until use (*see* **Note 3**).

2.5 Polymer Conjugate Synthesis

1. 100 mM Tris–HCl pH 9.0: Trizma base (12.1 g) was dissolved in endonuclease-free water (1 L). The solution was pH adjusted to 9.0 by dropwise addition of a 1.0 M HCl solution.
2. SATA-modified oligonucleotide solution (35 mg/mL): 15 mg of oligonucleotide was dissolved in endonuclease-free water (428 μL). The solution was stored at 5 °C until use (*see* **Note 4**).
3. Polymer solution (30 mg/mL): 61 mg of polymer (41 wt%, with TFA as the major component, 25 mg actual polymer) was dissolved in DMSO (835 μL) in a 40 mL glass vial. The solution was stored at 5 °C until use (*see* **Note 4**).

4. SMPT solution (1 mg/100 μL): 5 mg of SMPT (4-succinimidyloxycarbonyl-alpha-methyl-alpha(2-pyridyldithio)toluene) was dissolved in 500 μL DMSO. The solution was stored at 5 °C until use (*see* **Note 4**).
5. CDM masking reagents: Weigh out 51.5 mg CDM-GalNac (0.109 mmol, 50 equivalents) and 222 mg of CDM-PEG (0.327 mmol, 150 equivalents) as solids into a 40 mL vial.

2.6 Polymer Conjugate Analytical Characterization

1. 1 M dithiothreitol (DTT).
2. 100 mM Tris–HCl, pH 8.4 (*see* **Note 5**).
3. GPC elution solvent: 100 mM Tris–HCl, pH 8.4, 2.0 M sodium chloride. The solution was stored at room temperature (*see* **Note 5**).
4. HPLC chromatography solutions: 0.1 % TFA in water: Dissolve trifluoroacetic acid (4 mL) in water (4 L). 0.1 % TFA in 70:30 methanol (MeOH):acetonitrile (ACN): Dissolve 4 mL trifluoroacetic acid in 2.8 L MeOH and 1.2 L ACN (*see* **Note 5**).

2.7 Equipment and Supplies

1. Chemical fume hood.
2. Glassware oven.
3. Round-bottom flasks (various sizes, for monomer and polymer preparation).
4. Glass bottles (various sizes, for buffer preparation).
5. Glass vials (40 mL, 8 mL).
6. HPLC vials.
7. Karl-Fisher Titrator.
8. Nitrogen/vacuum manifold.
9. Magnetic stir plate.
10. Gel-permeation chromatography system: Waters Alliance 2695 separation module equipped with a Waters 2414 refractive index detector.
11. TOSOH TSKgel Alpha 3000 column.
12. GE Healthsciences Superdex 75HR 10/300.
13. High-pressure liquid chromatography system.
14. C18 guard column (for HPLC).
15. Thermo betasil phenyl 50 μm, 50 × 4.6 mm.
16. Centrifuge dialysis tubes for polymer conjugate purification. Amicon Ultra-15 Centrifugal Filter Unit with Ultracel-3 Membrane (part number UFC900324). Filter units were dialyzed with 0.1 M sodium hydroxide (1 cycle, 15 mL) followed by rinsing with endonuclease-free water immediately prior to use to remove the glycerol preservative per the manufacturer's instructions.

17. Centrifuge dialysis tubes for determination of free CDM-GalNAc and CDM-PEG. Millipore Amicon Ultra-0.5 Centrifugal Filter Unit with Ultracel-10 membrane (part number UFC501008).
18. Dry-ice acetone bath.
19. Ice-water bath.
20. Lyophilizer (vacuum manifold may be used as an alternative).
21. Shakes.

3 Methods

3.1 Boc-L-Ornithine N-Carboxyanhydride Monomer Synthesis

1. Charge an oven dried 1 L round-bottom flask with a magnetic stir bar and allow it to cool to room temperature under a stream of nitrogen.
2. Dissolve boc-L-ornithine (40.0 g, 150 mmol) in THF (400 mL, water content = 40 ppm, *see* **Note 6**) in round-bottom flask equipped with condenser and overhead stirrer.
3. Add triphosgene (17.8 g, 60 mmol) to the slurry and stir at 25 °C.
4. After 20 min, heat the mixture to 50 °C (*see* **Note 7**).
5. After 3.5 h, cool the mixture to 10 °C and carefully quench with ice water.
6. Extract the solution with isopropylacetate (IPAc) (3 × 400 mL) and wash the organic layer with ice cold water (2 × 200 mL).
7. Filter the separated organic layer through a silica pad (200 g, prewetted with THF), wash with THF (600 mL), and concentrate via rotary evaporation.
8. Dilute the crude product with IPAc (200 mL, water content = 60 ppm).
9. Concentrate the product by rotary evaporation until solid precipitate forms.
10. Age the solution for 0.5 h at 25 °C.
11. Add hexanes (350 mL) over 1 h.
12. Filter the product and wash with IPAc/Hexanes (1:2 v/v) (300 mL).
13. Dry the solid under vacuum overnight at 25 °C.

3.2 L-Phenylalanine N-Carboxyanhydride Monomer Synthesis

1. Charge an oven dried round-bottom 1 L flask with a magnetic stir bar and allow it to cool to room temperature under a stream of nitrogen.
2. Add phenylalanine (50.0 g, 303 mmol) to the flask.

3. Add anhydrous THF (600 mL, 0.5 M) to the flask to give a suspension.
4. Heat the suspension to 50 °C.
5. Add triphosgene (35.9 g, 121 mmol, 0.4 equivalents) and stir the suspension until the reaction goes clear (~30 min).
6. Cool the reaction mixture to ambient temperature and then concentrate by rotary evaporation to an oil.
7. Slowly pour the oil into 3 L of hexanes with rapid stirring to yield a white precipitate.
8. Cap the resulting suspension using aluminum foil and place in the freezer for 3 h.
9. Filter the precipitate via vacuum filtration while maintaining an inert environment and blowing nitrogen into the top of the filter.
10. Rinse the white solid with hexanes (3 × 20 mL) to obtain the product.
11. Collect the solid from the filter and dry overnight under vacuum.

3.3 Monomer Characterization

3.3.1 Confirmation of N-Carboxyanhydride Monomer Synthesis by ^1H NMR Spectroscopy

1. Add 1 mg of monomer into a small vial and dissolve in 1 mL of $CDCl_3$.
2. Using a pasture pipette, transfer the sample into a clean, dry ^1H NMR tube without forming bubbles and immediately measure in a 500 MHz ^1H NMR spectrometer (*see* **Note 8**).
3. Confirm the synthesis of the monomer from the appearance of proton peaks for the N-carboxyanhydride at the chiral carbon (*see* **Note 9**).

3.3.2 Confirmation of N-Carboxyanhydride Synthesis by High-Performance Liquid Chromatography

1. Add 1 mg of monomer into a small vial and dissolve in 10 mL of 1:1 v/v ACN:H_2O.
2. Transfer 1 mL of this solution into an HPLC vial.
3. Immediately measure in Agilent 1100 HPLC (*see* **Notes 10** and **11**).

3.4 Polymer Synthesis

1. Charge an oven dried round-bottom flask, flask #1, with a magnetic stir bar and allow it to cool to room temperature under a stream of nitrogen.
2. Add L-boc-ornithine-N-carboxyanhydride monomer (250 mg, 0.97 mol) to the flask and purge with an atmosphere of nitrogen using a nitrogen/vacuum manifold.
3. Add 2 mL anhydrous DMA.
4. Charge a second oven dried flask (20 mL), flask #2, with a magnetic stir bar and allow it to cool to room temperature under a stream of nitrogen.

5. In this second flask, add 0.5 mL of *n*-butyl amine and 5 mL of DMA, keep under nitrogen.
6. Add the *n*-butylamine solution in DMA (32 μL, 0.032 mol, 0.03 equivalents relative to monomer) from flask #2 to flask #1.
7. Place flask #1 under vacuum (<10^{-6} mmHg) and stir at room temperature overnight using a magnetic stir plate (*see* **Note 12**).
8. Release vacuum using nitrogen and add L-phenylalanine-N-carboxyanhydride monomer (43 mg, 0.225 mol) to the reaction flask #1.
9. Place the reaction flask #1 under vacuum (<10^{-6} mmHg) again for an additional 8 h (*see* **Note 9**).
10. Remove solvent (DMA) under reduced pressure.
11. Take crude, concentrated protected polymer solution and add it dropwise into rapidly stirring water (20 times the crude polymer solution volume) to cause polymer precipitation.
12. Filter the solid polymer, place it in a vacuum flask, freeze the polymer solid using a dry-ice:acetone bath, and place under vacuum for 48 h to dry the polymer solid.
13. Store protected poly(amide) at −20 °C.

3.4.1 Confirmation of Block Co-polymer Synthesis by ^1H NMR Spectroscopy

1. Add 5 mg of *boc*-protected polymer into a small vial and dissolve in 1 mL of DMSO-d_6.
2. Using a pasture pipette, transfer 1 mL of the sample into a clean, dry ^1H NMR tube without forming bubbles and measure in a 500 MHz ^1H NMR spectrometer.
3. Confirm the synthesis of the polymer from the appearance of broadened proton peaks for the protected block co-polymer.

3.4.2 Molecular Weight Analysis of the Protected Polymer by Gel Permeation Chromatography

1. Add 100 μL of the crude reaction mixture of protected block co-polymer to 400 μL of diluent (DMF with 10 mM LiCl).
2. Transfer the solution into an HPLC vial.
3. Immediately measure by GPC.
4. The molecular weight and molecular weight distribution are calculated by comparing them to poly(styrene) standards using Waters software.

3.5 Polymer Deprotection

1. Add the boc-protected polymer solution to a round-bottom flask.
2. Add dichloromethane at a concentration of 35 mg polymer/mL (*see* **Note 13**).
3. Stir the solution at room temperature under nitrogen.
4. Add trifluoroacetic acid to the solution (1:1 dichloromethane: trifluoroacetic acid by volume).

5. Stir at room temperature for 20 min.
6. Evaporate solvent (DCM/TFA) under reduced pressure to afford solid, deprotected polyamide (*see* **Note 14**).
7. Store protected poly(amide) at −20 °C.

3.5.1 Confirmation of Deprotection of Block Co-polymer by ^1H NMR Spectroscopy

1. Add 5 mg of deprotected polymer into a small vial and dissolve in 1 mL of DMSO-d_6.
2. Using a pasture pipette, transfer 1 mL of the sample into a clean, dry ^1H NMR tube without forming bubbles, and measure in 500 MHz ^1H NMR spectrometer.
3. Confirm the synthesis of the polymer from the appearance of broadened proton peaks for the block co-polymer (*see* **Note 15**).

3.5.2 Determination of Ornithine:Phenylalanine Ratio in the Poly(amide) Block Copolymer

1. Use the spectra obtained from the ^1H NMR of the deprotected poly(amide) block copolymer.
2. The calculation of the average ratio of ornithine:phenylalanine is completed by examining the signals between 1.4 and 1.8 ppm (equivalent to 4 protons, a) for ornithine and the signals from 7 to 7.4 ppm for the phenylalanine (equivalent to 5 protons, b) (*see* Eq. 1).
3. Determine the number of hydrogens associated with ornithine by integrating from 1.4 to 1.8 ppm/number of protons associated with that signal (4 protons).
4. Determine the number of hydrogens associated with phenylalanine by integrating from 7 to 7.4 ppm/number of protons associated with that signal (5 protons).
5. Determine the ratio of ornithine:phenylalanine by dividing the number of hydrogens associated with ornithine by the number of hydrogens associated with phenylalanine.
6. To determine the average number of phenylalanine monomers per chain, take the number average molecular weight (Mn) obtained from GPC and divide it by (molecular weight of the boc-protected ornithine repeat unit × NMR ratio of ornithine to phenylalanine + the molecular weight of the phenylalanine repeat unit) (147 g/mol) (*see* Eq. 2).
7. Take the ornithine:phenylalanine ratio that was determined in **step 5**, and multiply it by the average number of phenylalanine in a chain to obtain the average number of ornithine monomers in a chain (*see* Eq. 3):

$$\text{Ratio of ORN:PHE} = \frac{\text{Integration value for "a"}/4\text{ protons}}{\text{Integration value for "b"}/5\text{ protons}} \quad (1)$$

$$\#\text{ of PHE repeat units} = \frac{\text{Mn obtained from GPC}}{\left(\text{MW}_{\text{ORN repeat unit}}\left(\text{ratio of ORN : PHE}\right) + \text{MW}_{\text{PHE repeat unit}}\right)} \quad (2)$$

$$\# \text{of ORN repeat units} = \text{Ratio of ORN} : \text{PHE} \times \# \text{PHE units} \quad (3)$$

3.6 CDM-PEG Synthesis

1. Weigh out CDM (10 g) into a 500 mL round-bottom flask and add dichloromethane (250 mL) to dissolve solids.
2. Add DMF (0.84 mL).
3. Chill reaction vessel to 0 °C using an ice bath and add oxalyl chloride (7.3 mL) over 40 min.
4. Remove the ice bath and stir overnight at room temperature.
5. Blow nitrogen through the reaction mixture to remove excess HCl, venting through 10 N NaOH to trap HCl, for 2.5 h.
6. Remove the reaction solvent under reduced pressure using a rotary evaporator to give a crude oil (CDM-chloride).
7. Dissolve crude CDM-chloride in dichloromethane (250 mL) and transfer the solution to a 1 L round-bottom flask.
8. Chill to 0 °C using an ice batch.
9. Dissolve polyethylene glycol monomethyl ether (28.3 g) in dichloromethane (60 mL) and pyridine (4.38 mL).
10. Add the polyethylene glycol monomethyl ether:pyridine solution to CDM-chloride solution over 70 min.
11. Continue to stir the reaction at 0 °C for 1 h.
12. To quench the reaction, add 22 % aqueous ammonium chloride (180 mL).
13. Adjust the pH of the reaction mixture to 1.98 with 300 μL of 5 N HCl.
14. Pour the biphasic reaction mixture into a separatory funnel (1 L) and collect the organic layer.
15. Remove the solvent under reduced pressure using a rotary evaporator.
16. Purify the crude oil using reverse phase chromatography (C18 column, water:ACN with 0.1 % TFA, gradient of 5–100 % ACN over 30 min).

3.7 SATA-Modified siRNA Synthesis

1. Weigh out SiRNA (50 mg) into an 8-mL vial equipped with a magnetic stir bar. Add 0.1 M sodium bicarbonate buffer (1 mL, 50 mg/mL) to dissolve the SiRNA solid and cool the vial containing the SiRNA solution to 0–5 °C in an ice water bath.
2. In a separate vial, weigh out N-succinimidyl S-acetylthioacetate (SATA, 8.2 mg, 10 equivalents) and add DMSO (80 μL) to dissolve the solid SATA.
3. Add the SATA solution to the SiRNA solution slowly over 1 min.
4. Stir the reaction mixture for 2 h at 0–5 °C.

5. Using a pipet, transfer the crude SATA-SiRNA solution to the centrifugal dialysis filter unit and dilute to a volume of 15 mL with water.
6. Using a centrifuge operating at 1,200 × *g*, concentrate the SATA-SiRNA solution to a volume of ~3 mL. Discard permeate. Dilute the SATA-SiRNA solution retentate to a volume of 15 mL with water. Repeat this procedure four additional times.
7. After the final dialysis cycle, transfer the purified SATA-SiRNA solution to a plastic falcon tube. Freeze the aqueous SATA-SiRNA solution in a dry-ice/acetone bath. Place the frozen SATA-SiRNA solution in a lyophilization container and freeze-dry to give a white solid.

3.8 Polymer Conjugate Synthesis

1. To a 40 mL vial containing polymer solution in DMSO, add SMPT solution (75 μL) and shake at RT for 1 h at 300 rpm (*see* **Note 16**).
2. Dilute polymer-SMPT solution with pH 9 100 mM Tris buffer with 5 % glucose (9.85 mL) (*see* **Note 17**).
3. To the polymer-SMPT solution, add the SiRNA solution (143 μL) and shake at RT for 1 h at 300 rpm (*see* **Note 18**).
4. Using a pipet, transfer the polymer-SMPT-SiRNA solution to the 40 mL vial containing solid CDM-GalNAc and CDM-PEG. Shake at RT at 300 rpm for 10 min to dissolve the solids (*see* **Note 19**).
5. Using a pipet, transfer the polymer conjugate to the centrifugal dialysis filter unit and dilute to a volume of 15 mL with 100 mM Tris buffer, pH 9, with 5 % glucose.
6. Using a centrifuge operating at 1,200 × *g*, concentrate the polymer conjugate to a volume of ~3 mL. Discard permeate. Dilute the polymer conjugate retentate to a volume of 15 mL with pH 9 100 mM Tris buffer with 5 % glucose. Repeat this procedure two additional times.
7. After the final dialysis cycle, concentrate the polymer conjugate to a volume of ~5 mL (~0.4–2.0 mg/mL SiRNA concentration). Transfer polymer conjugate to a plastic falcon tube and store at −20 °C until use (*see* **Note 20**).

3.9 Polymer Conjugate Analysis: siRNA Concentration

1. Prepare SiRNA stock solutions at concentrations of 0.5 mg/mL and 1.0 mg/mL in 100 mM Tris–HCl pH 8.4.
2. Create an SiRNA calibration curve: Inject 2 μL of each stock solution onto an HPLC equipped with an SEC column (GE Healthsciences Superdex 75HR 10/300 column) with a mobile phase composed of 100 mM Tris with 2 M NaCl, pH 8.4 (UV at 260 nm). Integrate the AUC of each SiRNA peak and plot AUC area versus concentration to generate a calibration curve.

3. Determine SiRNA concentration of a polymer conjugate: Into an HPLC vial, dispense 10 μL of polymer conjugate solution, and dilute with 100 mM Tris, pH 8.4 to an approximate SiRNA concentration of 0.1 mg/mL. Add 1 M dithiothreitol (DTT) solution in water (10 μL). Age the sample for 30 min. Inject 2 μL of DTT-treated polymer conjugate solution onto an HPLC equipped with an SEC column (GE Healthsciences Superdex 75HR 10/300 column) with a mobile phase composed of 100 mM Tris with 2 M NaCl, pH 8.4 (UV at 260 nm). Integrate the AUC of the SiRNA peak. Use the previously generated SiRNA calibration curve to determine the SiRNA concentration of the sample.

3.10 Polymer Conjugate Analysis: Conjugation Efficiency

1. Determine the unbound SiRNA concentration of a polymer conjugate: Into an HPLC vial, dispense 10 μL of polymer conjugate solution and dilute with 100 mM Tris, pH 8.4 to an approximate SiRNA concentration of 0.1 mg/mL. Inject 2 μL of diluted polymer conjugate solution onto an HPLC equipped with an SEC column (GE Healthsciences Superdex 75HR 10/300 column) with a mobile phase composed of 100 mM Tris with 2 M NaCl, pH 8.4 (UV at 260 nm).
2. Integrate the AUC of the SiRNA peak. Use the previously generated SiRNA calibration curve to determine the SiRNA concentration of the sample.
3. Conjugation efficiency: Subtract the unbound SiRNA from the total SiRNA (*see* **Note 21**).

3.11 Polymer Conjugate Analysis: Masking Efficiency

1. Prepare CDM-GalNAc stock solution at concentrations of 10 mg/mL and 5 mg/mL in water.
2. Prepare CDM-PEG stock solutions at concentrations of 44 mg/mL and 22 mg/mL in water.
3. Create a CDM-GalNAc and CDM-PEG Calibration Curve: Inject 10 μL of each stock solution onto a reverse phase HPLC (C18 guard column + thermo betasil phenyl 50 × 4.6 mm, 5 μm) with a mobile phases composed of 0.1 % TFA in water and 0.1 % TFA in 70/30 methanol:acetonitrile (UV at 260 nm). Integrate the AUC of each SiRNA peak and plot AUC area versus concentration to generate a calibration curve.
4. Determine total CDM-GalNAc and CDM-PEG concentration: Into an HPLC vial, dispense polymer conjugate solution. Inject 10 μL of undiluted polymer conjugate solution onto a reverse-phase HPLC (C18 guard column + thermo betasil phenyl 50 × 4.6 mm, 5 μm) with a mobile phases composed of 0.1 % TFA in water and 0.1 % TFA in 70/30 methanol:acetonitrile (UV at 260 nm). Integrate the AUC of the CDM-GalNAc and CDM-PEG peaks. Use the previously generated CDM-GalNAc and CDM-PEG calibration curves to determine the CDM-GalNAc and CDM-PEG concentration in the sample.

5. Determine Free CDM-GalNAc and CDM-PEG Concentration: Place polymer conjugate (0.5 mL) in the centrifugal filter unit and using a centrifuge (19,700 × *g*, 10 min), filter the sample, collecting approximately 300 μL of permeate sample. Remove the permeate from the collection portion of the centrifugal filter unit and inject 10 μL of the undiluted permeate solution onto a reverse-phase HPLC (C18 guard column + thermo betasil phenyl 50 × 4.6 mm, 5 μm) with a mobile phase composed of 0.1 % TFA in water and 0.1 % TFA in 70/30 methanol:acetonitrile (UV at 260 nm). Integrate the AUC of the CDM-GalNAc and CDM-PEG peaks. Use the previously generated CDM-GalNAc and CDM-PEG calibration curves to determine the CDM-GalNAc and CDM-PEG concentration in the sample.
6. Estimate polymer concentration of a polymer conjugate: After measuring the SiRNA concentration for a given polymer conjugate sample (Subheading 3.9), the amount of polymer in the sample may be estimated by multiplying the SiRNA concentration by the SiRNA/polymer ratio.
7. Estimate the weight percent of ornithine in the polymer sample: The number of ORN repeat units was determined by ^{1}H NMR (Subheading 3.5.2). The MW of the ornithine (ORN) repeat unit is 114 g/mol and the MW of the phenylalanine (PHE) repeat unit is 147 g/mol (*see* Eq. 4).
8. Estimate ornithine concentration (moles of amine) per mL polymer conjugate: *See* Eq. 5 below. The concentration of polymer conjugate was determined in **step 5** (mg/mL) and the weight percent of ornithine was determined in **step 6**. The MW of the ornithine (ORN) repeat unit is 114 g/mol (*see* Eq. 5).
9. Determine masking efficiency: Subtract the free CDM-GalNAc/CDM-PEG from the total CDM-GalNAc/CDM-PEG. This is the amount of bound CDM-GalNAc/CDM-PEG in the sample (moles/mL). Use this value along with the ornithine per mL polymer conjugate calculated in **step 7** (*see* Eq. 6, *see* **Note 22**):

$$\text{Weight percent ornithine} = \frac{\#\text{ of ORN repeat units} \times \text{MW}_{\text{ORN repeat unit}}}{\#\text{ of ORN repeat units} \times \text{MW}_{\text{ORN repeat unit}} + \#\text{ of PHE repeat units} \times \text{MW}_{\text{PHE repeat unit}}} \quad (4)$$

$$\text{Concentration (Moles) of ornithine per mL polymer conjugate} = \left[\text{Polymer conjugate}\left(\frac{\text{mg}}{\text{mL}}\right)\right] \times \text{Weight percent ornithine} \times \left(1/\left(\text{MW}_{\text{ORN repeat unit}} \times 1000\right)\right) \quad (5)$$

$$\text{Masking efficiency}(\%) = [\text{Bound CDM}(\text{moles}/\text{mL})] / [\text{Ornithine per mL polymer conjugate}(\text{moles}/\text{mL})] \times 100 \quad (6)$$

4 Notes

1. Pre-packed silica columns were used for chromatography in conjunction with an ISCO Combiflash chromatography system; however manual chromatography and column packing are also acceptable.
2. Highly chemically modified SiRNA duplexes are recommended for the synthesis of SiRNA-polymer conjugates. Polymer conjugates do not offer the same level of protection from nucleases as particle-based SiRNA delivery vehicles and so improved efficacy is observed with highly chemically modified SiRNA. For the purposes of this protocol, the SiRNA targeting the gene Apolipoprotein B17 was utilized. Sequence and chemical modification pattern are shown below.

   5′-amil-iB-CUUUAACAAUUCCUGAAAUTsT-iB-3′
   3′-UsUGAAAUUGUUAAGGACUsUsUsA-5′

   AUGC—Ribose; amil = hexyl amino linker; iB—Inverted deoxy abasic; UC—2′Fluoro; AGT—2′Deoxy; AGU—2′OCH_3; UsA—phosphorothioate linkage
3. Hold solutions no more than 24 h.
4. These solutions should be prepared the day of conjugation.
5. Fresh solvent is prepared on an as needed basis, but not kept longer than 1 month.
6. The water content is measured using a Karl Fisher Titrator.
7. The reaction was monitored by HPLC and after 3.5 h the reaction is complete.
8. Avoid lag time in ^{1}H NMR measurement, as monomer may hydrolyze in the presence of water back to starting material and yield inaccurate results for percent conversion to product.
9. ^{1}H NMR confirmation of N-carboxyanhydride monomer synthesis by ^{1}H NMR spectroscopy (500 MHz $CDCl_3$, δ): 7.38 (m, 5H, Ar H), 6.96 (br. s, 1H Ar H), 5.13 (s, 2H; CH_2), 4.95 (br. s, 1H; CH), 4.35 (m, 1H; CH), 3.27 (q, 2H; CH_2), 2.00 (m, 1H; CH), 1.83 (m, 1H; CH), 1.72 (m, 2H; CH_2).
10. Monomer purity was analyzed using an Agilent 1100 HPLC equipped with an Ascentis Fused Core C18 column, 100 × 4.6 mm, 2.7 μm particle, with a gradient of 10–95 % MeCN/0.1 wt% H_3PO_4 in 6 min, hold 2 min, post 2 min,

1.8 mL/min, 10 min run, 2.0 UV 210 nm, 40 °C, sample 2.0 µL, 10 min run. Orn(Z) at 2.40 min. NCA-Orn(Z) at 3.40 min.

11. Avoid lag time in HPLC measurement, as monomer may hydrolyze in the presence of water back to starting material and yield inaccurate results for percent conversion to product.
12. Upon vacuum, bubbles will form in the solution, suggesting the release of CO_2.
13. A hazy solution is obtained.
14. The deprotected polymer is obtained as a trifluoroacetic acid salt after the solvent and volatile byproducts are removed by vacuum.
15. ^{1}H NMR confirmation of deprotection of block co-polymer by ^{1}H NMR spectroscopy (500 MHz $CDCl_3$, δ): 7.38 (m, 5H, Ar H), 6.96 (br. s, 1H Ar H), 5.13 (s, 2H; CH_2), 4.95 (br. s, 1H; CH), 4.35 (m, 1H; CH), 3.27 (q, 2H; CH_2), 2.00 (m, 1H; CH), 1.83 (m, 1H; CH), 1.72 (m, 2H; CH_2).
16. 5.4 equivalents of SMPT are used relative to SATA-SiRNA.
17. The target polymer concentration is ~2.5 mg/mL.
18. After addition of the SATA-SiRNA solution to the polymer solution, the reaction mixture may become a hazy suspension. Mild shaking of the reaction mixture is acceptable, but vigorous shaking or stirring may cause further aggregation of the suspended particles and is not recommended.
19. Addition of the hazy reaction mixture to the CDM solids should cause dissolution of the polymer conjugate due to masking of the amine functional groups with the CDM moieties.
20. Masking efficiencies of 40–65 % were typically obtained for polymer conjugate samples.
21. Conjugation efficiencies of 85–95 % were typically obtained for polymer conjugate samples.
22. Masking efficiencies of 40–65 % were typically obtained for polymer conjugate samples.

References

1. Whitehead KA, Langer R, Anderson DG (2009) Knocking down barriers: advances in siRNA delivery. Nat Rev Drug Discov 8:129–138
2. Stanton MG, Colletti SL (2011) Medicinal chemistry of siRNA delivery. J Med Chem 53:7887–7901
3. Convertine AJ, Benoit DSW, Duvall CL, Hoffman AS, Stayton PS (2009) Development of a novel endosomolytic diblock copolymer for siRNA delivery. J Control Release 133:221–229
4. Zimmermann TS, Lee ACH, Akinc A, Bramlage B, Bumcrot D, Fedoruk MN, Harborth J, Heyes JA, Jeffs LB, John M, Judge AD, Lam K, McClintock K, Nechev LV, Palmer LR, Racie T, Röhl I, Seiffert S, Shanmugam S, Sood V, Soutschek J,

Toudjarska I, Wheat AJ, Yaworski E, Zedalis W, Koteliansky V, Manoharan M, Vornlocher H-P, MacLachlan I (2006) RNAi-mediated gene silencing in non-human primates. Nature 441:111–114

5. Itaka K, Kanayama N, Nishiyama N, Jang W-D, Yamasaki Y, Nakamura K, Kawaguchi H, Kataoka K (2004) Supramolecular nanocarrier of siRNA from PEG-based block catiomer carrying diamine side chain with distinctive pKa directed to enhance intracellular gene silencing. J Am Chem Soc 126:13612–13613
6. Wang X-L, Nguyen T, Gillespie D, Jensen R, Lu Z-R (2008) A multifunctional and reversibly polymerizable carrier for efficient siRNA delivery. Biomaterials 29:15–22
7. Wang X-L, Ramusovic S, Nguyen T, Lu Z-R (2007) Novel polymerizable surfactants with pH-sensitive amphiphilicity and cell membrane disruption for efficient siRNA delivery. Bioconjug Chem 18:2169–2177
8. Rozema DB, Lewis DL, Wakefield DH, Wong SC, Klein JJ, Roesch PL, Bertin SL, Reppen TW, Chu Q, Blokhin AV, Hagstrom JE, Wolff JA (2007) Dynamic PolyConjugates for targeted in vivo delivery of siRNA to hepatocytes. Proc Natl Acad Sci U S A 104:12982–12987
9. Gaspar R, Duncan R (2009) Polymeric carriers: preclinical safety and the regulatory implications for design and development of polymer therapeutics. Adv Drug Deliv Rev 61:1220–1231
10. Matsumoto S, Christie RJ, Nishiyama N, Miyata K, Ishii A, Oba M, Koyama H, Yamasaki Y, Kataoka K (2008) Environment-responsive block copolymer micelles with a disulfide cross-linked core for enhanced siRNA delivery. Biomacromolecules 10:119–127
11. Barrett SE, Burke RS, Abrams MT, Bason C, Busuek M, Carlini E, Carr BA, Crocker LS, Fan H, Garbaccio RM, Guidry EN, Heo JH, Howell BJ, Kemp EA, Kowtoniuk RA, Latham AH, Leone AM, Lyman M, Parmar RG, Patel M, Pechenov SY, Pei T, Pudvah NT, Raab C, Riley S, Sepp-Lorenzino L, Smith S, Soli ED, Staskiewicz S, Stern M, Truong Q, Vavrek M, Waldman JH, Walsh ES, Williams JM, Young S, Colletti SL (2014) Development of a liver-targeted siRNA delivery platform with a broad therapeutic window utilizing biodegradable polypeptide-based polymer conjugates. J Control Release 183:124–137
12. Guidry EN, Farand J, Soheili A, Parish CA, Kevin NJ, Pipik B, Calati KB, Ikemoto N, Waldman JH, Latham AH, Howell BJ, Leone A, Garbaccio RM, Barrett SE, Parmar RG, Truong QT, Mao B, Davies IW, Colletti SL, Sepp-Lorenzino L (2014) Improving the in vivo therapeutic index of siRNA polymer conjugates through increasing pH responsiveness. Bioconjug Chem 25:296–307
13. Parmar RG, Busuek M, Walsh ES, Leander KR, Howell BJ, Sepp-Lorenzino L, Kemp E, Crocker LS, Leone A, Kochansky CJ, Carr BA, Garbaccio RM, Colletti SL, Wang W (2013) Endosomolytic bioreducible poly(amido amine disulfide) polymer conjugates for the in vivo systemic delivery of siRNA therapeutics. Bioconjug Chem 24:640–647
14. Parmar RG, Poslusney M, Busuek M, Williams JM, Garbaccio R, Leander K, Walsh E, Howell B, Sepp-Lorenzino L, Riley S, Patel M, Kemp E, Latham A, Leone A, Soli E, Burke RS, Carr B, Colletti SL, Wang W (2014) Novel endosomolytic poly(amido amine) polymer conjugates for systemic delivery of siRNA to hepatocytes in rodents and nonhuman primates. Bioconjug Chem 25:896–906
15. Wooddell CI, Rozema DB, Hossbach M, John M, Hamilton HL, Chu Q, Hegge JO, Klein JJ, Wakefield DH, Oropeza CE, Deckert J, Roehl I, Jahn-Hofmann K, Hadwiger P, Vornlocher H-P, McLachlan A, Lewis DL (2013) Hepatocyte-targeted RNAi therapeutics for the treatment of chronic hepatitis B virus infection. Mol Ther 21:973–985
16. Wong SC, Klein JJ, Hamilton HL, Chu Q, Frey CL, Trubetskoy VS, Hegge J, Wakefield D, Rozema DB, Lewis DL (2012) Co-injection of a targeted, reversibly masked endosomolytic polymer dramatically improves the efficacy of cholesterol-conjugated small interfering RNAs in vivo. Nucleic Acid Ther 22:380–390

Chapter 3

PepFect6 Mediated SiRNA Delivery into Organotypic Cultures

Suvarna Dash-Wagh, Ülo Langel, and Mats Ulfendahl

Abstract

Gene silencing by small interfering RNA (SiRNA) is an attractive therapeutic approach for pathological disorders that targets a specific gene. However, its applications are limited, as naked RNA is rapidly degraded by RNases and is inadequately internalized by the target cells in the body. Several viral and non-viral vectors have been described to improve the delivery of SiRNAs both in cultured cells as well as in vivo. Increasing evidence suggests that cell-penetrating peptides (CPPs) are an efficient, non-cytotoxic tool for intracellular delivery of SiRNA. Recently, a new peptide, PepFect6 (PF6), based system has been described for efficient SiRNA delivery in various cell types. PF6 is an amphipathic stearyl-TP10 peptide carrying a pH titratable trifluoromethylquinoline moiety that facilitate endosomal release. PF6 forms stable non-covalent complexes with SiRNA. Upon internalization, the complexes rapidly escape the endosomal compartment, resulting in robust RNA interference (RNAi) responses. This chapter describes a protocol to use the PF6-nanoparticle technology for SiRNA delivery into organotypic cultures of the inner ear i.e., cochlea. We also highlight different critical points in the peptide/SiRNA complex preparation, transfection and in analyzing the efficacy of PF6-SiRNA associated RNAi response.

Key words Cell-penetrating peptides (CPPs), PepFect6 (PF6), SiRNA, Inner ear, Cochlea, Organotypic cultures, Fluorescence recovery after photobleaching (FRAP)

1 Introduction

Small interfering RNAs (SiRNA) have emerged as a new class of therapeutics with a great potential for treating genetic disorders. However, when introduced into the tissue, naked SiRNAs in general are degraded by nucleases, which make the SiRNA delivery more complicated and challenging. In addition, unaided double-stranded RNA can induce innate immune response via interaction with RNA-binding proteins such as Toll-like receptors and protein kinase receptors [1]. The major hurdle in the development of RNAi therapies is the intracellular delivery of SiRNAs due to their negative charge and large size [2]. Various delivery strategies have been developed using both viral and nonviral vectors. In recent years, cell-penetrating peptides (CPPs) have emerged as potential

Kato Shum and John Rossi (eds.), *SiRNA Delivery Methods: Methods and Protocols*, Methods in Molecular Biology, vol. 1364, DOI 10.1007/978-1-4939-3112-5_3, © Springer Science+Business Media New York 2016

nonviral vectors for delivery of nucleic acids and other bioactive molecules, both in vitro and in vivo [3–5]. CPPs are short cationic and/or amphipathic peptides that can be covalently or non-covalently combined with negatively charged cargo, e.g., SiRNA. The major route of cellular entry for most of CPPs is via endocytosis. They are trafficked through the early endosomes to late endosomes, which are acidified to pH 5–6. Subsequently, the late endosomes fuse with lysosomes where the pH further drops to 4.5 and various degradative enzymes are present. The CPP cargo that fails to escape from these acidic vesicles is ultimately degraded [6]. Therefore, for efficient SiRNA delivery, CPP technology must be improved for higher endosomal escape.

Modification of the existing CPPs has helped to overcome this endosomal entrapment [3, 7, 8]. A novel CPP, called PepFect6 (PF6), has been recently developed that exhibits increased release of SiRNA molecules from the acidic endosomal compartments [3]. It is created by introducing a stearyl and trifluoromethyl quinoline (QN) moieties into Transportan 10 (TP10). Earlier studies show that stearylation of TP10 improves the activity and reduces cytotoxic effects of the peptide in serum, while addition of lysosomotropic agent chloroquine (CQ) derivative QN enhances the endosomal release of peptide/SiRNA complex [3, 8–10].

PF6 can be non-covalently complexed with SiRNAs. It forms homogenous, unimodal 70–100 nm nanoparticles with SiRNA, which are stable at 4 °C in water for at least 4 weeks. A molar excess of CPP over SiRNA neutralizes the negative charges of SiRNA and thus protects it from serum enzymes [11]. However, in the presence of serum, the nanoparticles are larger (125–200 nm) with wider distribution and are relatively less stable, especially at 37 °C [3, 11]. Nevertheless, the PF6-SiRNA nanoparticles are endocytosed within an hour by the cells, resulting in downregulation of the target gene without cytotoxicity or immunogenicity in vitro [3]. When injected systemically, PF6-SiRNA nanoparticles elicit an RNAi response without any acute toxicity to the vital organs [3, 10] suggesting promising perspectives for their future therapeutic applications without any risks of inflammation.

This chapter describes a protocol for the use of the non-covalent PF6 based intracellular delivery of SiRNA into the organotypic cultures of the cochlea. SiRNA targeting a gap junction protein, connexin 26, is used to test the functional efficacy of PF6-SiRNA complexes. Connexin 26 knockdown results in the interrupted intercellular communication among the supporting cells of the cochlea [12, 13]. Fluorescence recovery after photobleaching (FRAP) assay is a useful tool to study the gap junction functions. Rate of diffusion of fluorescent dye after FRAP displays functional effects of downregulation of connexin 26 expression by PF6-SiRNA complexes.

2 Materials

2.1 Cochlear Organotypic Cultures

1. Cell-Tak, used according to the manufacturer's protocol.
2. Phosphate-buffered saline (PBS).
3. 35 mm disposable dishes.
4. Forceps # 5 and #55 for fine dissections of cochleae.
5. 70 % ethanol.
   (a) Cochlear medium: Advanced reduced Dulbecco's modified Eagle's medium with phenol red (DMEM), 1 % N1 supplement, 1.65 % Glucose, 1 % Penicillin.
6. 4 well plate (3–4 explant/well; for RNA extraction) or Center-Well Organ Culture Dish for FRAP assay (1 explant/dish).

2.2 Peptide, Oligonucleotide and Transfection Reagents

1. PF6 is developed as a peptide-based carrier for SiRNA delivery by introducing modifications into the well-characterized TP10 peptide. Protocols for the synthesis and purification of PF6 are described by Andaloussi et al. [3, 9]. PF6 and derivatives are stable for at least 1 year when stored at −20 °C in a lyophilized form.
2. SiRNA targeting the hypoxanthine phosphoribosyltransferase (HPRT) 1 gene sense 5′-GCCAGACUUUGUUGGAUUU GAA ATT-3′ and antisense 5′-AAUUUCAAAUCCAACA AAGUCUGG CUU-3′ (biotin conjugated).
3. SiRNAs targeting the connexin 26 gene (NM_001004099) conjugated with biotin sense 5′-AAGUUCAUGAAGGG AGAGAUATT-3′- Cy5/biotin and antisense 5′-UAUCUCU CCCUUCAUGAACUU-3′.
4. Salt-free water.
5. Cochlear medium (as above).
6. 1.5 ml Eppendorf tubes.

2.3 Fluorescence Recovery After Photobleaching (FRAP) Assay

1. Cochlear cultures in the center-well culture dishes.
2. Calcein, AM.
3. PBS.
4. Laser confocal microscope (LSM 510, Zeiss).

3 Methods

3.1 Organotypic Cultures of the Cochleae

The dissection of the inner ear cochlea is described elsewhere [14]. In brief, the explants are cultured as follows:

1. Coat the dishes with Cell-Tak (135 μg/ml) for 1 h and wash with water, 3 times at room temperature.

2. Clean all instruments with 70 % ethanol or autoclave before dissection.
3. Decapitate the pups and dip the heads in 70 % alcohol for disinfection.
4. Split the skull in two and using #4 jeweler's forceps cut out the cochlea.
5. Immerse cochleae in PBS maintained on ice.
6. Dissect away the bony capsule using a #5 jeweler's forceps and transfer the soft tissue to PBS supplemented with 5.5 μl/ml of 30 % glucose.
7. Carefully identify the stria vascularis, spiral ganglion, and modiolus under high power and pull away from the explant one at time (*see* **Note 1**).
8. Place one explant per coated center-well dishes with 1 ml cochlear culture medium.
9. Check under microscope and adjust so the hair cells face upward (*see* **Note 2**).

3.2 Formation of PF6/siRNA Nanocomplexes and Delivery of PF6-siRNA Nanocomplexes into Organotypic Cochlear Cultures

The procedure for formation of PF6/SiRNA nanocomplexes constitutes an important factor in the successful transfection and should be followed carefully.

1. Take the vial containing the PF6 lyophilized powder out of the freezer and equilibrate for 30 min at room temperature without opening the vial, to avoid exposure to humidity. Dilute PF6 in salt-free water to a final storage concentration of 1 mM in a 1.5 ml tube (*see* **Note 3**). Aliquot the peptides in 20 μl aliquots to avoid several cycles of freeze-thawing, which will decrease the efficacy of the peptide. This stock solution is stable for about 2–3 months when stored at −20 °C.
2. When running an experiment, thaw one tube and add 180 μl of water to obtain a 100 μM peptide solution.
3. Mix gently by tapping the tube or by vortexing at low speed for a few seconds.
4. Prepare the nanocomplexes for experiments in duplicates in RNase-free environment (*see* **Note 4**). The final treatment volume is 1 ml per well, i.e., 2 ml for duplicate. Prepare the complexes in 1/10th of the final volume in salt-free water (*see* **Note 5**), i.e., 200 μl complexes. Do not exceed the volume required for four transfections, as this might cause aggregation.
5. Start by adding the appropriate volume of SiRNA and then water and finally the PF6 for each reaction (at a molar ratio of 1:20 or 1:40 of SiRNA over peptide) (*see* **Note 6**). Conduct a dose response assessment at each molar ratio (Fig. 1).

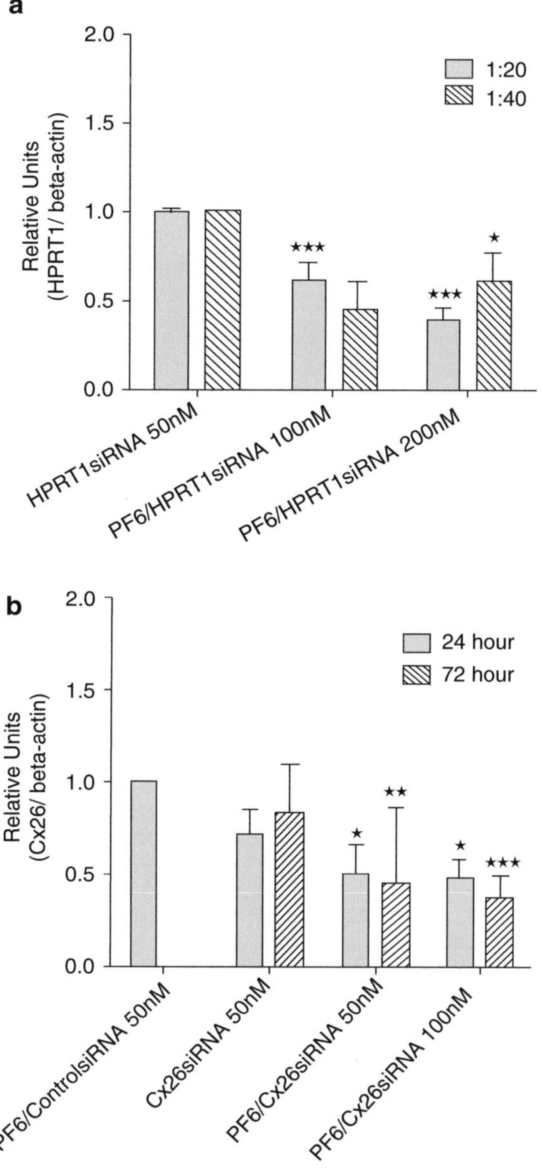

Fig. 1 PF6-mediated delivery of siRNA targeting of HPRT1 and Connexin (Cx) 26 genes in the organotypic cochlear cultures. (**a**) Efficiency of PF6 was analyzed by comparing the HPRT1 RNAi responses of PF6/HPRT1-siRNA and HPRT1-siRNA alone in the cultures by RT-qPCR following 24 h of incubation ($n=4$). (**b**) The cultures were incubated either with Cx26-siRNA alone (50 nM), or PF6/control-siRNA (50 nmol/L) or PF6/Cx26-siRNA nanoparticles in two different concentrations (50 and 100 nM). Quantitative analysis of transcript levels of Cx26, relative to beta-actin mRNA, was determined by RT-qPCR and compared with PF6/control-siRNA cultures after 24 ($n=6$) and 72 ($n=4$) hours. Actin was used as internal standard. *Error bars* indicate mean ± SEM, ***$P<0.001$, **$P<0.01$, *$P<0.05$ (Mann–Whitney *U*-test). *HPRT* hypoxanthine phosphoribosyl transferase, *RNAi* RNA interference, *RT-qPCR* reverse transcription-quantitative PCR, *siRNA* short interfering RNA. (Reproduced from ref. 15 with permission from Nature Publication Group)

Nevertheless, do not exceed the final concentration of PF6 over 2 μM in serum-free media and 5 μM in serum containing media to avoid peptide associated toxicity.

6. For control, prepare only SiRNA sample in similar manner (PF6 is replaced with water).
7. Incubate for the tubes for 30–60 min at RT.
8. After incubation, prepare a serial dilution from prepared complex solution in fresh tubes with 100 μl water (*see* **Notes 6** and 7).

3.3 FRAP Assay: Addressing the PF6-Based siRNA Delivery Efficacy

1. Check if all the cultures are attached and growing on the plate. Randomly divide the dishes into three groups; control, only SiRNA (50 nM) and PF6-SiRNA complex (SiRNA 100 nM).
2. Replace the media in wells with fresh media, 900 μl in each well.
3. Add 100 μl of prepared complex to each well and shake the plate slightly. Incubate the plates at 37 °C, 5 % CO_2 for 24 or 72 h (*see* **Note 8**).
4. Take out one plate at a time from the incubator, at the specific incubation time point. Add 2.5 μmol/L calcein to the medium, incubate for another 10 min at 37 °C.
5. Wash the culture thoroughly with warm PBS to remove free calcein and image immediately using a confocal laser scanning microscope equipped with a 40× water immersion objective at room temperature (*see* **Note 9**).
6. Select a cell in the outer sulcus that is surrounded by other cells from all sides. Acquire images of the size of 256 × 256 pixels, one every second. Photobleach the selected cell with 100 % laser power of 488 nm between image 15th and 16th and image further (Fig. 2). Analyze the cells with bleaching efficiencies larger than 20 % for FRAP as reported earlier in ref. [15]. The mean fluorescence intensity of the bleached region is corrected for the background variations.
7. Image only two or three cells are per culture, to avoid calcium responses (*see* **Note 10**). Analyze five to six different cultures per condition from three to four independent experiments for calculating the significant differences.

4 Notes

1. Dissection of cochleae should be done very carefully. The hair cells are very sensitive to mechanical damages. While dissecting, hold the organ of Corti only in the hook region or at the extreme distal part.

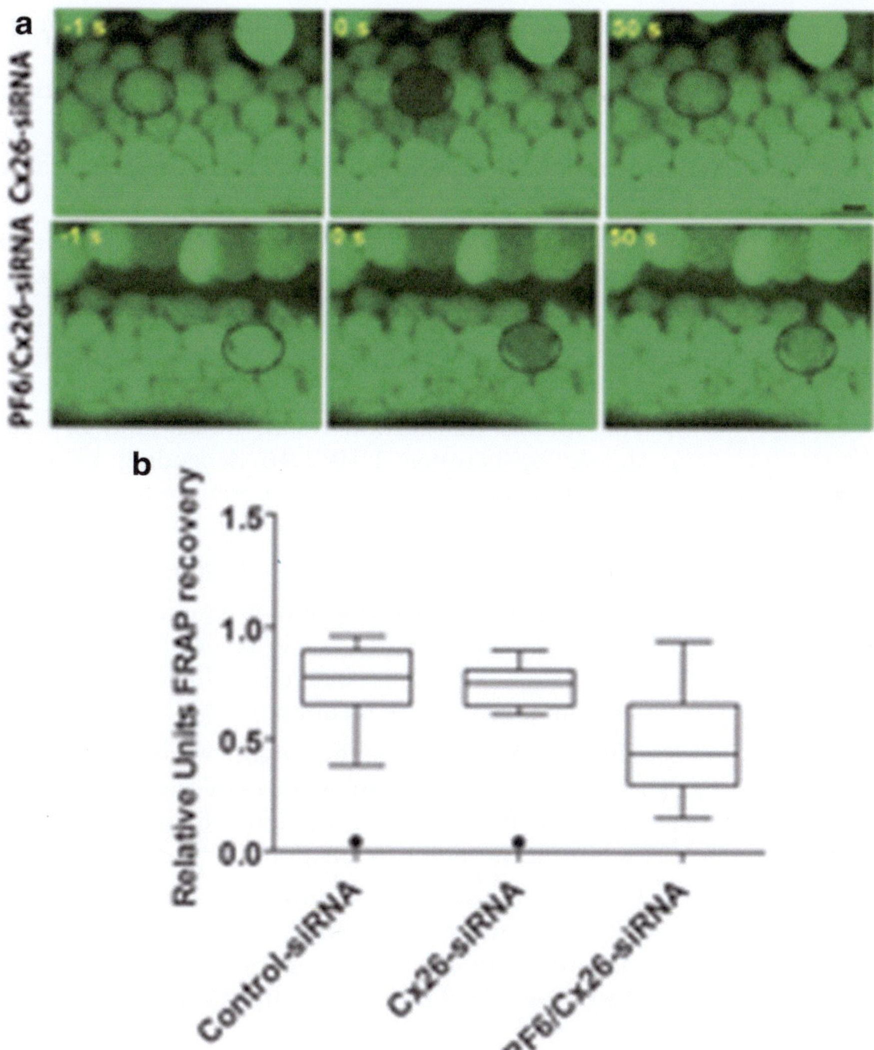

Fig. 2 Reduced gap junctional communication in PF6/Cx26-siRNA-treated cochlear cultures. (**a**) Representative images of the Cx26-siRNA-treated supporting cells (*upper panel*) and PF6/Cx26-siRNA-treated supporting cells (*lower panel*) before (−1 s, image 15), immediately (0 s, between image 15 and 16), and 50 s after photobleaching. The bleached cell (marked with a *circle*) recovered after 50 s in the Cx26-siRNA-treated cultures, while the PF6/Cx26-siRNA-treated cells showed reduced recovery. Bar = 5 μm. (**b**) The quantitative measurement of the fluorescence recovery revealed significantly reduced recovery of PF6/Cx26-siRNA-treated cells in comparison with the cells from the cochlear cultures treated with Cx26-siRNA alone or PF6/control-siRNA for 72 h. $^{**}P<0.01$, $^{*}P<0.05$ (Mann–Whitney *U*-test). *FRAP* fluorescence recovery after photobleaching, *siRNA* short interfering RNA. (Reproduced from ref. 15 with permission from Nature Publication Group)

2. Make sure that the hair cells in the cochlea faces upwards. The cochlea should be pushed down so it attaches the bottom of the plate and over time the cells can spread otherwise the cochlea will float away during the washing.
3. It is advisable to resuspend the peptide in salt-free water rather than buffers.
4. RNases can degrade the SiRNA so clean the laminar hood and tube stands with ethanol and RNase-free solutions.
5. Excess salt may hinder with the homogeneous complex formation. Hence, SiRNA and PF6 should be mixed in salt-free water.
6. For optimal transfection result, molar ratio of SiRNA: PF6 should be between 1:10–1:40 depending upon tissue/cell types and amount of SiRNA required for efficient downregulation of target gene expression.
7. It is advisable to include a dose–response curve for each SiRNA.
8. If the medium contains serum, it is suggested to incubate the cells with complexes for 4 h in serum-free medium and then add serum containing medium. However, cochlear culture medium is serum-free medium.
9. The cultures should be thoroughly washed with warm PBS in order to remove all possible traces of calcein from the solution, otherwise free calcein in solution will interfere with the FRAP occurring through gap junctions.
10. Calcein fluorescence is sensitive to the changes in the intracellular calcium concentration. Exposure to laser sometimes leads calcium influx. Hence, cells showing calcium response (grater than 5 %) should be excluded from analysis.

References

1. Hornung V, Guenthner-Biller M, Bourquin C, Ablasser A, Schlee M, Uematsu S, Noronha A, Manoharan M, Akira S, de Fougerolles A, Endres S, Hartmann G (2005) Sequence-specific potent induction of IFN-alpha by short interfering RNA in plasmacytoid dendritic cells through TLR7. Nat Med 11:263–270
2. Aagaard L, Rossi JJ (2007) RNAi therapeutics: principles, prospects and challenges. Adv Drug Deliv Rev 59:75–86
3. Andaloussi SE, Lehto T, Mager I, Rosenthal-Aizman K, Oprea II, Simonson OE, Sork H, Ezzat K, Copolovici DM, Kurrikoff K, Viola JR, Zaghloul EM, Sillard R, Johansson HJ, Said Hassane F, Guterstam P, Suhorutsenko J, Moreno PM, Oskolkov N, Halldin J, Tedebark U, Metspalu A, Lebleu B, Lehtio J, Smith CI, Langel U (2011) Design of a peptide-based vector, PepFect6, for efficient delivery of siRNA in cell culture and systemically in vivo. Nucleic Acids Res 39:3972–3987
4. Crombez L, Morris MC, Dufort S, Aldrian-Herrada G, Nguyen Q, Mc Master G, Coll JL, Heitz F, Divita G (2009) Targeting cyclin B1 through peptide-based delivery of siRNA prevents tumour growth. Nucleic Acids Res 37:4559–4569
5. Kim WJ, Christensen LV, Jo S, Yockman JW, Jeong JH, Kim YH, Kim SW (2006) Cholesteryl oligoarginine delivering vascular endothelial growth factor siRNA effectively inhibits tumor growth in colon adenocarcinoma. Mol Ther 14:343–350

6. Zhang B, Mallapragada S (2011) The mechanism of selective transfection mediated by pentablock copolymers; part II: nuclear entry and endosomal escape. Acta Biomater 7:1580–1587
7. Crombez L, Aldrian-Herrada G, Konate K, Nguyen QN, McMaster GK, Brasseur R, Heitz F, Divita G (2009) A new potent secondary amphipathic cell-penetrating peptide for siRNA delivery into mammalian cells. Mol Ther 17:95–103
8. Mae M, El Andaloussi S, Lundin P, Oskolkov N, Johansson HJ, Guterstam P, Langel U (2009) A stearylated CPP for delivery of splice correcting oligonucleotides using a non-covalent co-incubation strategy. J Control Release 134:221–227
9. Anko M, Majhenc J, Kogej K, Sillard R, Langel U, Anderluh G, Zorko M (2012) Influence of stearyl and trifluoromethylquinoline modifications of the cell penetrating peptide TP10 on its interaction with a lipid membrane. Biochim Biophys Acta 1818:915–924
10. Suhorutsenko J, Oskolkov N, Arukuusk P, Kurrikoff K, Eriste E, Copolovici DM, Langel U (2011) Cell-penetrating peptides, PepFects, show no evidence of toxicity and immunogenicity in vitro and in vivo. Bioconjug Chem 22:2255–2262
11. van Asbeck AH, Beyerle A, McNeill H, Bovee-Geurts PH, Lindberg S, Verdurmen WP, Hallbrink M, Langel U, Heidenreich O, Brock R (2013) Molecular parameters of siRNA—cell penetrating peptide nanocomplexes for efficient cellular delivery. ACS Nano 7:3797–3807
12. Maeda Y, Fukushima K, Nishizaki K, Smith RJ (2005) In vitro and in vivo suppression of GJB2 expression by RNA interference. Hum Mol Genet 14:1641–1650
13. Ortolano S, Di Pasquale G, Crispino G, Anselmi F, Mammano F, Chiorini JA (2008) Coordinated control of connexin 26 and connexin 30 at the regulatory and functional level in the inner ear. Proc Natl Acad Sci U S A 105:18776–18781
14. Parker M, Brugeaud A, Edge AS (2010) Primary culture and plasmid electroporation of the murine organ of Corti. J Vis Exp
15. Dash-Wagh S, Jacob S, Lindberg S, Fridberger A, Langel U, Ulfendahl M (2012) Intracellular delivery of short interfering RNA in rat organ of Corti using a cell-penetrating peptide PepFect6. Mol Ther Nucleic Acids 1, e61

Chapter 4

Highly Efficient SiRNA Delivery Mediated by Cationic Helical Polypeptides and Polypeptide-Based Nanosystems

Lichen Yin, Nan Zheng, and Jianjun Cheng

Abstract

RNA interference (RNAi) mediated by small interfering RNA (SiRNA) has recently emerged as a potent machinery in regulating gene expression at the post-translation step. Despite its efficiency and specificity, the biggest hurdle against its wide application is the safe and effective delivery of the SiRNA cargo into target cells. Here, we describe the highly effective SiRNA delivery mediated by the cationic helical polypeptides and polypeptide-based nanosystems both in vitro and in vivo via oral administration.

Key words Helical polypeptide, SiRNA delivery, Gene silencing, Oral delivery, Tumor necrosis factor α (TNF-α), Acute hepatic injury

1 Introduction

RNA interference (RNAi) mediated by small interfering RNA (SiRNA) is an important mechanism that regulates gene expression in eukaryotic cells via site-specific mRNA cleavage and degradation. Because of its efficiency and sequence specificity, the RNAi machinery affords an exciting modality for the treatment of various human diseases [1–4]. To realize the efficiency of the RNAi, an effective and safe delivery vector/method is required to deliver the SiRNA into target cells.

Nonviral delivery vectors, exemplified by cationic lipids and polymers, possess desired biocompatibility and minimal mutagenesis, and thus serve as desired candidates for SiRNA delivery. However, in comparison with their viral counterparts, nonviral vectors often suffer from low transfection efficiencies, mainly due to the various systemic barriers. For instance, SiRNA is liable to nuclease-assisted degradation; it can hardly traverse the cell membranes and the internalized SiRNA molecules tend to be entrapped by endosomes/lysosomes and ultimately get degraded.

Kato Shum and John Rossi (eds.), *SiRNA Delivery Methods: Methods and Protocols*, Methods in Molecular Biology, vol. 1364, DOI 10.1007/978-1-4939-3112-5_4, © Springer Science+Business Media New York 2016

Here, we report a newly developed nonviral gene delivery material, the cationic helical polypeptide [5, 6] and polypeptide-based supramolecular self-assembled nanoparticles (SSNPs) [7], which can effectively overcome the aforementioned barriers to mediate effective SiRNA delivery both in vitro and in vivo. Particularly, the stabilized helical structure allows the polypeptide to create pores on cell membranes and endosomal membranes, thereby facilitating the cellular internalization as well as endosomal escape of the SiRNA cargo. When the polypeptide was assembled with other rationally designed materials, the obtained SSNPs can target intestinal M cells and gut-associated macrophages (GAMs) after oral administration to greatly enhance the intestinal absorption of SiRNA against tumor necrosis factor-α (TNF-α) and mediate effective gene knockdown in macrophages, thus triggering systemic TNF-α silencing towards the treatment of acute inflammation. We herein describe the protocol for in vitro and in vivo SiRNA delivery mediated by the helical polypeptide (PVBLG-8) and PVBLG-8-containing SSNPs. We also demonstrate that the helical PVBLG-8, but not the random coiled analogue PVBDLG-8, can trigger effective knockdown of the luciferase gene in HeLa-Luc cells. Compared with the commercial reagent Lipofectamine 2000, the SSNPs mediated an approximately tenfold higher TNF-α knockdown efficiency in macrophages in vitro in terms of reduced SiRNA dose, and after oral administration, they prevented the lipopolysaccharide (LPS)-induced systemic TNF-α production to protect mice from inflammatory hepatic injury at the low dose of 200 μg SiRNA/kg. This finding also represents a 1–2 orders of magnitude improvement over existing delivery vehicles, which typically require dosing from 500 μg/kg to 50 mg/kg in mice via i.v. injection [8–10].

2 Materials

1. Oleyl-trimethyl chitosan chloride (OTMC, MW = 200 kDa, quaternization degree of 28.7 %, oleyl conjugation ratio of 20.3 %, structure shown in Fig 1), oleyl-PEG_{3400}-mannose (structure shown in Fig. 1.), and oleyl-PEG_{3400}-cysteamine (structure shown in Fig. 1.) were synthesized according to previously published protocols [7]. Poly(γ-4-((2-(piperidin-1-yl)ethyl)aminomethyl)-benzyl-L-glutamate) (PVBLG-8, degree of polymerization 200, structure shown in Fig. 1), a cationic helical polypeptide, was synthesized according to our published protocols [5, 6]. Poly(γ-4-((2-(piperidin-1-yl)ethyl)aminomethyl)-benzyl-D,L-glutamate) (PVBDLG-8), a random coiled analogue of PVBLG-8 was also synthesized according to our published protocols.

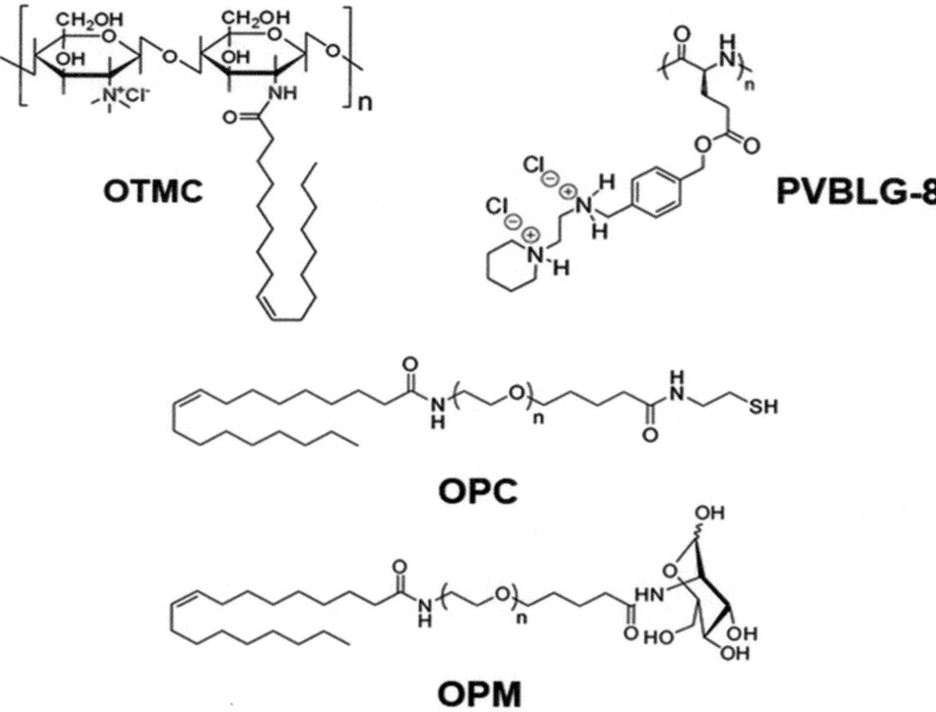

Fig. 1 Chemical structures of OTMC, OPC, OPM, and PVBLG-8

2. SiRNA duplex against mouse tumor necrosis factor-α (TNF-α) has the sequence of sense strand: 5′-GUCUCAGCCUCUUC UCAUUCCUGct-3′ and antisense strand: 5′-AGCAGGA AmUGmAGmAAmGAmGGmCUmGAmGAmCmAmU-3′, wherein the (m) pattern mN represents a 2′-*O*-methyl base. Sequences of the primers used were as follows. TNF-α F:CCACCACGCTCTTTCTGTCTACTG, TNF-α R: GGG CTACAGGCTTGTCACTCG.
3. SiRNA duplex against luciferase (Luc SiRNA) has the sequence of sense strand: 5′-CUUACGCUGAGUACUUCGAtt-3′ and antisense strand: 5′-UCGAAGUACUCAGCGUAAGtt-3′.
4. Weigh 2 mg lipopolysaccharide (LPS, from *E. coli* 0111:B4) in a biosafety cabinet, transfer to the 25-mL glass vial, and dissolve by 20 mL sterilized PBS to obtain a stock solution of 100 μg/mL. Aliquot to 1 mL/vial in 1.5-mL centrifuge tubes, and store at −20 °C. Thaw the solution at room temperature before use, take out 10 μL, transfer into a 25-mL glass vial, dilute with 9990 μL cell culture medium to obtain a solution of 100 ng/mL (for use in RAW 264.7 cells). Alternatively, take out 100 μL of the 100 μg/mL solution, transfer into a 25-mL glass vial, dilute with 9900 μL PBS to obtain a solution of 1 μg/mL (for use in mouse). Freshly prepare the solution directly before use.

5. Weigh 1 g D-galactosamine (D-GalN), transfer to the 25-mL glass vial, and dissolve by 10 mL sterilized PBS to obtain a solution of 100 mg/mL. Freshly prepare the solution directly before use.
6. Prepare all solutions using diethylpyrocarbonate (DEPC)-treated water and analytical grade reagents.
7. Pre-warm the cell culture media to 37 °C before use.
8. Sterilize all glass vials and centrifuge tubes before use.
9. Diligently follow all waste disposal regulations when disposing waste materials.

3 Methods

3.1 In Vitro siRNA Delivery by PVBLG-8

1. Weigh 10 mg PVBLG-8 and transfer to the 25-mL glass vial. Add 10 mL DEPC-treated water to make the 1 mg/mL solution. Add 1 M hydrogen chloride to adjust the pH to 6.5 (*see* **Note 1**). Aliquot to 1 mL/vial in 1.5-mL centrifuge tubes, and store at −20 °C (*see* **Note 2**). Thaw the solution at room temperature before use.
2. Take out 200 μL PVBLG-8 solution, transfer into a 1.5-mL centrifuge tube, add 800 μL NaCl solution (150 mM, containing 20 mM HEPES) to obtain a final concentration of 0.2 mg/mL.
3. Dissolve Luc SiRNA with DEPC water at 0.2 mg/mL. Aliquot to 50 μL/tube in 0.5-mL centrifuge tubes, and store at −80 °C. Thaw the solution in ice bath before use (*see* **Note 3**).
4. Transfer 20 μL SiRNA solution into a 1.5-mL centrifuge tube. Add PVBLG-8 solution into SiRNA at determined PVBLG-8/SiRNA weight ratios (e.g., 20, 100, and 200 μL PVBLG-8 equals to PVBLG-8/SiRNA weight ratios of 1, 5, and 10, respectively), and mix by gentle pipetting (*see* **Note 4**).
5. Vortex the mixture for 30 s at 1500 rpm/min and incubate the mixture at RT for 15 min to allow formation of complexes (*see* **Note 5**).
6. Culture HeLa-Luc cells (purchased from the American Type Culture Collection) in DMEM supplemented with 10 % horse serum, 1 % penicillin–streptomycin, 1 % L-glutamine according to the manufacturer's protocol (ATCC website). Passage the cells for more than 3 times while less than 20 times before transfection studies (*see* **Note 6**).
7. Harvest the cells and resuspend in the cell culture media (*see* **Note 7**) at 1×10^5 cells/mL. Add 0.1 mL of the cell suspension into each well of 96-well plates at the seeding density of 1×10^4 cells/well (*see* **Note 8**). Slightly shake (left and right for 10 times, then up and down for 10 times), and incubate at 37 °C for 24 h (*see* **Note 9**).

8. Aspirate the media, wash each well once with 500 μL PBS, then add 500 μL serum-free DMEM (or Opti-MEM) to each well.
9. Add different volume of freshly prepared PVBLG-8/SiRNA complexes to the media at the final SiRNA concentrations of 0.4, 0.8, 1.2, and 1.6 μg/mL, respectively (*see* **Note 10**).
10. Shake the plate slightly (left and right for 5 times, then up and down for 5 times), and incubate the cells at 37 °C for 4 h. As a control, the complexes formed between SiRNA and PVBDLG-8, a random coiled analogue of PVBLG-8 with diminished membrane activities, can be prepared and used to transfect HeLa-Luc cells with the same method as described for PVBLG-8.
11. Aspirate all the media in each well, wash the cells once with 500 μL PBS, and add 500 μL serum-containing DMEM/well. Culture the cells at 37 °C for another 20 h.
12. Remove the cell culture medium and immediately add 50 μL of the Bright-Glo luciferase reagent (Promega) to each well. Measure the luminescence intensity using a microplate reader according to the manufacturer's protocols.
13. Calculate the gene silencing efficiency of PVBLG-8 which was denoted as the percentage luminescence intensity of control cells that did not receive PVBLG-8/SiRNA complex treatment (*see* Fig. 2.).

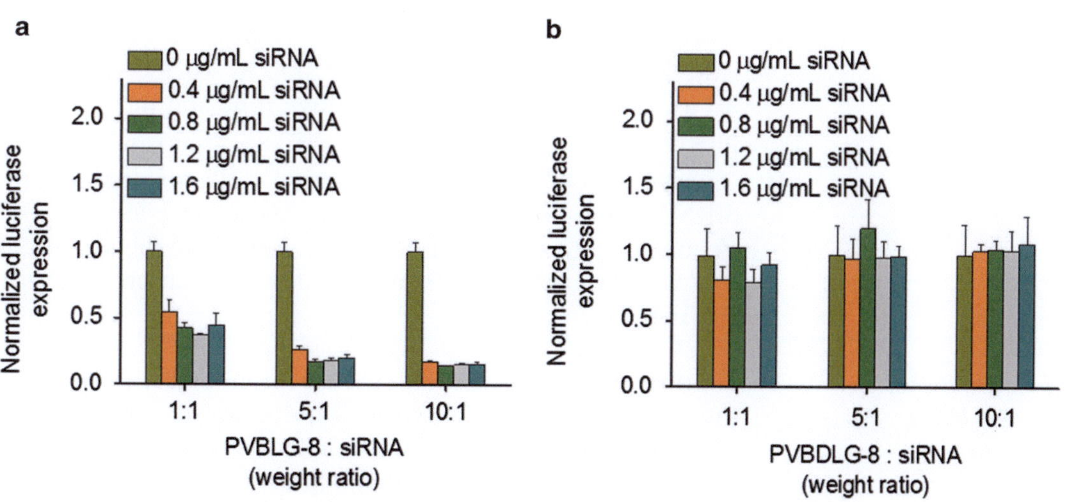

Fig. 2 (**a**) In vitro transfection of HeLa-Luc cells with luciferase siRNA at various PVBLG-8:siRNA weight ratios and siRNA concentrations. (**b**) In vitro transfection of HeLa-Luc cells with luciferase siRNA at various PVBDLG-8:siRNA weight ratios and siRNA concentrations (reproduced from ref. [6] with permission from Nature Publishing Group)

3.2 In Vitro TNF-α siRNA Delivery to Macrophages by Supramolecular Self-Assembled Nanoparticles (SSNPs)

3.2.1 Preparation of SSNPs

1. Weigh 20 mg OTMC and transfer to the 25-mL glass vial. Add 10 mL DEPC-treated water to make the 2 mg/mL solution.
2. Add 1 M hydrogen chloride to adjust the pH to 6.5 (*see* **Note 1**). Aliquot to 2 mL/vial in 7-mL vials, and store at −20 °C. Thaw the solution in 37 °C water bath before use (*see* **Note 11**).
3. Weigh 10 mg PVBLG-8 and transfer to the 25-mL glass vial. Add 10 mL DEPC-treated water to make the 1 mg/mL solution.
4. Add 1 M hydrogen chloride to adjust the pH to 6.5. Aliquot to 1 mL/vial in 1.5-mL centrifuge tubes, and store at −20 °C. Thaw the solution at room temperature before use (*see* **Note 2**).
5. Weigh 100 mg OPM or OPC and transfer to the 25-mL glass vial. Add 10 mL DEPC-treated water to make the 10 mg/mL solution. Aliquot to 1 mL/vial in 1.5-mL centrifuge tubes, and store at −20 °C. Thaw the solution in 37 °C water bath before use.
6. Weigh 10 mg sodium tripolyphosphate (TPP) and transfer to the 25-mL glass vial. Add 10 mL DEPC-treated water to make the 1 mg/mL solution. Aliquot to 1 mL/tube in 1.5-mL centrifuge tubes, and store at −20 °C. Thaw the solution at room temperature before use.
7. Dissolve SiRNA with DEPC water at 0.2 mg/mL. Aliquot to 50 μL/tube in 0.5-mL centrifuge tubes, and store at −80 °C. Thaw the solution in ice bath before use (*see* **Note 3**).
8. Add 20 μL SiRNA solution and 50 μL TPP solution into a 1.5-mL centrifuge tube and mix by gentle pipetting.
9. Add 200 μL OTMC, 80 μL PVBLG-8, 40 μL OPM, and 40 μL OPC into a 1.5-mL centrifuge tube and mix by gentle pipetting.
10. Transfer the OTMC/PVBLG-8/OPM/OPC mixture into the SiRNA/TPP mixture (*see* **Note 12**) and pipette for 10 times. Vortex for 30 s at 1500 rpm/min and incubate the mixture in 37 °C water bath for 30 min to allow formation of SSNPs (*see* Fig. 3a).

3.2.2 Transfection of siRNA to Mouse Macrophages

1. Culture RAW 264.7 cells (mouse monocyte macrophage, purchased from the American Type Culture Collection) in DMEM supplemented with 10 % fetal bovine serum (FBS) according to the manufacturer's protocol (ATCC website). Passage the cells for more than 3 times while less than 20 times before transfection studies (*see* **Note 13**).
2. Harvest the cells and resuspend in the cell culture media at 1×10^5 cells/mL. Add 0.5 mL of the cell suspension into each well of 24-well plates at the seeding density of 5×10^4 cells/well.

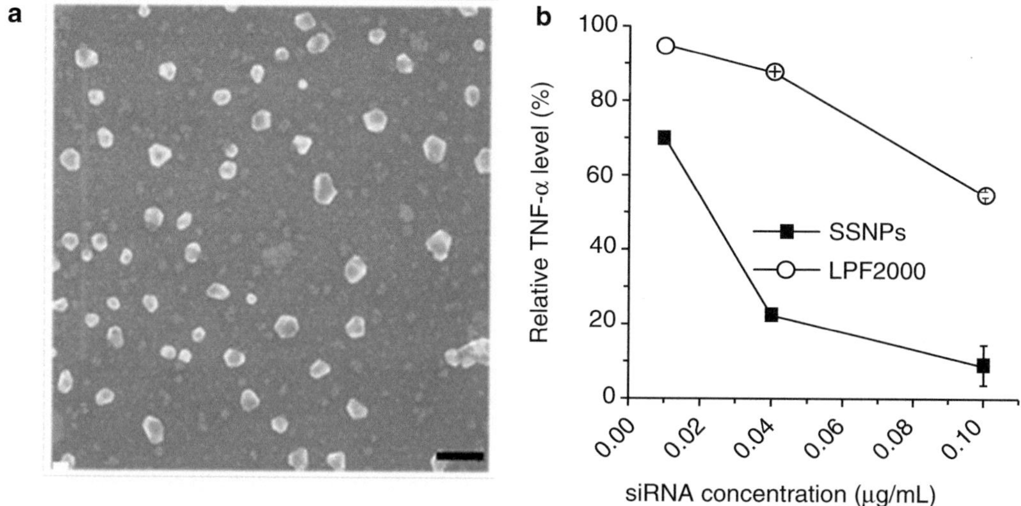

Fig. 3 (**a**) SEM image of SSNPs (bar = 200 nm). (**b**) TNF-α knockdown efficiencies of SSNPs and LPF2000/siRNA complexes at various siRNA doses ($n = 3$) (reproduced from ref. [7] with permission from Wiley)

Slightly shake (left and right for 10 times, then up and down for 10 times), and incubate at 37 °C for 24 h (*see* **Note 7**).

3. Aspirate the media (*see* **Note 14**), wash each well once with 500 μL PBS, then add 500 μL serum-free DMEM (or Opti-MEM) to each well. Add the freshly prepared SSNPs (5.4, 2.1, and 0.5 μL) to the media at the final concentrations of 0.1, 0.04 and 0.01 μg/mL, respectively (*see* **Note 10**).
4. Shake the plate slightly (left and right for 5 times, then up and down for 5 times), and incubate the cells at 37 °C for 4 h. Commercial reagent Lipofectamine 2000 (LPF2000) as an internal control was also used to transfect the RAW 264.7 cells according to the manufacturer's protocol.
5. Aspirate all the media in each well, wash the cells once with 500 μL PBS, and add 500 μL serum-containing DMEM/well. Culture the cells at 37 °C for another 20 h.
6. Aspirate the culture media, add 500 μL 100 ng/mL LPS solution, and incubate the cells at 37 °C for 3 h.
7. Take out 2 μL of the cell culture media and measure the extracellular TNF-α level by the mouse TNF-α ELISA kit according to the manufacturer's protocol (*see* **Note 15**).
8. Aspirate the cell culture media, wash the cells with 3× 500 μL PBS, and extract the total RNA using TRIzol according to the manufacturer's protocol (*see* **Note 16**).
9. Synthesize cDNA from 500-ng total RNA using the high capacity cDNA reverse transcription kit according to the manufacturer's protocol.

10. Measure the TNF-α mRNA level using an ABI PRISM 7900HT Real-Time PCR system according to the manufacturer's suggested protocols.
11. Measure the extracellular TNF-α level or the mRNA level of control cells that did not receive SSNPs treatment but were treated by the same method as described above.
12. Calculate the gene silencing efficiency of SSNPs which was denoted as the percentage of TNF-α or TNF-α mRNA levels of the control cells (*see* Fig. 3b).

3.3 Oral Delivery of TNF-α siRNA by Supramolecular Self-Assembled Nanoparticles (SSNPs) Against Hepatic Injury

1. Orally deliver SSNPs (430 μL/mouse, 200 μg SiRNA/kg) to male C57/BL-6 mice (20–22 g) via gavage.
2. Take 2 mL LPS solution (1 μg/mL) and 2 mL D-GalN solution (100 mg/mL), transfer to a 7-mL vial, and mix by gentle pipetting (*see* **Note 17**).
3. Inject the freshly prepared LPS/D-GalN solution (500 μL/mouse, equals to 12.5 μg LPS/kg and 1250 mg D-GalN/kg) intraperitoneally 24 h after oral gavage of SSNPs.
4. Anesthetize the mice with isoflurane 1.5 h after i.p. injection (*see* **Note 18**). Collect 0.5 mL by retro-orbital bleeding, put the blood in 1.5-mL centrifuge tube at RT for 10 min to let it coagulate, centrifuge at 10,500 × *g* for 5 min, and collect the serum.
5. Measure the serum TNF-α level using a mouse TNF-α ELISA kit according to the manufacturer's protocol.
6. Sacrifice the mice 5 h after i.p. injection of LPS/D-GalN. Harvest liver, spleen, and lung, wash with PBS, and homogenize with TRIzol reagent in ice (use the whole spleen and lung, while cut a small piece (0.2 g) of liver). Extract the RNA according to the manufacturer's protocol and measure the TNF-α mRNA level using real-time PCR as described for RAW 264.7 cells.
7. Measure the serum TNF-α level or the mRNA levels in each specific organ of control mice that received oral gavage of PBS (430 μL/animal) instead of SSNPs and were treated by the same method as described above.
8. Calculate the systemic TNF-α silencing efficiency of SSNPs which was represented as the percentage of TNF-αor TNF-α mRNA levels of the control animals (*see* Fig. 4).
9. In another experiment, gavage the SSNPs and i.p. inject the LPS/D-GalN in the same manner as **steps 1–8**.
10. Anesthetize the mice with isoflurane 5 h after i.p. injection. Collect 1 mL blood by retro-orbital bleeding, put the blood in 1.5-mL centrifuge tube at RT for 10 min to let it coagulate, centrifuge at 10,500 × *g* for 5 min, and collect the serum.

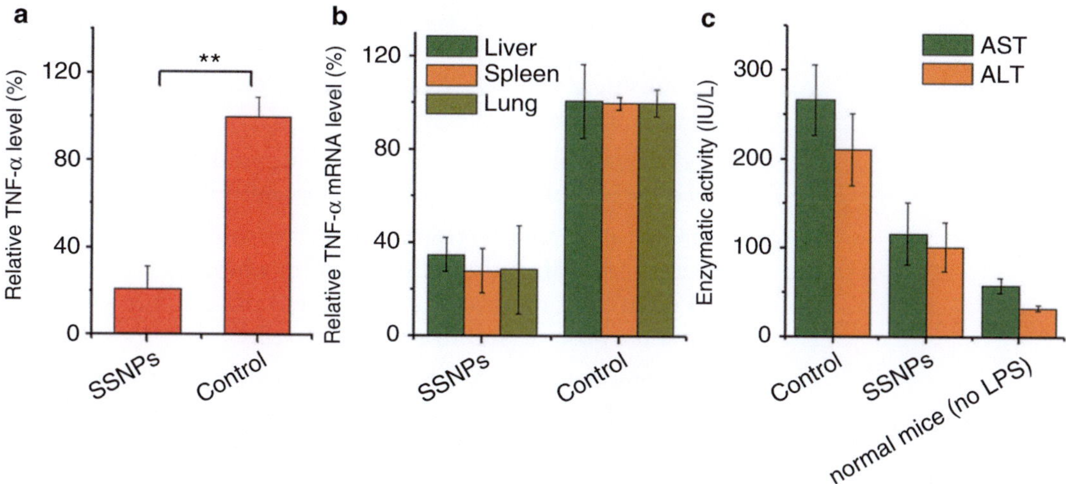

Fig. 4 (**a**) Serum TNF-α level of mice gavaged with SSNPs at 200 μg siRNA/kg ($n=6$). (**b**) Relative TNF-α mRNA levels in mouse liver, spleen, and lung 24 h after oral gavage of SSNPs ($n=3$). (**c**) Serum ALT and AST levels of mice 5 h after LPS/D-GalN stimulation ($n=4$) (reproduced from ref. [7] with permission from Wiley)

11. Measure the serum alanine transaminase (ALT) and aspartate aminotransferase (AST) levels using commercial kits according to the manufacturer's protocols. As a control, measure the serum AST/ALT levels of mice that received oral gavage of PBS (430 μL/animal) rather than SSNPs (*see* Fig. 4).

4 Notes

1. Using a pH paper to indicate the pH. pH values within the range of 6.2–6.8 are allowed.
2. Avoid repeated ice-thawing processes of PVBLG-8. Recommend less than five ice-thawing cycles for each preserved tube.
3. Avoid repeated ice-thawing processes of SiRNA. Recommend less than three ice-thawing cycles for each preserved tube.
4. Quickly add the PVBLG-8 solution into SiRNA at once and immediately mix for several times.
5. Incubation time can be longer than 15 min but less than 30 min.
6. Cell condition is critical to the success of transfection. Passage the cells at the confluence of 70–80 % and never let cells grow over-confluence. For each passage, keep the passage ratio higher than 1:3.
7. Culture the cells using antibiotic-containing media. However, use antibiotic-free media to seed cells. Also use antibiotic-free

media during and after transfection. Because PVBLG-8 mediates effective SiRNA delivery mainly via the pore-formation mechanism, it will also allow excessive diffusion of antibiotics into cells to hamper the cell viability.

8. For the convenient measurement of luciferase activity, seed the cells on white plates instead of transparent plates.
9. Usually do the transfection at the cell density of 70 %. Because we cannot visualize the cells grown on white plates, we usually seed some cells on another transparent plate at the same time for the purpose of monitoring cell density.
10. We recommend slowly add the complexes onto the surface of the culture media in small droplets. Do not touch the inner surface of each well. It will lead to inconsistence of the transfection results.
11. Fully thaw the solution at 37 °C until it becomes completely transparent. Thawing at room temperature sometimes cannot lead to incomplete dissolution because the amphiphilic polymer has the tendency to self-assemble.
12. Quickly add the OTMC/PVBLG-8/OPM/OPC mixture into the SiRNA/TPP at once and immediately mix for several times. Vortex vigorously to promote nanoparticle formation.
13. RAW 264.7 cells tend to be activated by the treatment of trypsin. It is therefore required to passage the cells using scrappers, which is also suggested by the ATCC. Scrape 3–5 times for each passage, and pipette 10–15 times to dissociate the cells. Avoid excessive scraping or pipetting that will also activate or damage the cells.
14. Do not let the cells stay in the dried condition for too long (usually less than 1 min). RAW 264.7 cells tend to detach from the plate when get dried.
15. The supernatant can be preserved at −20 °C if not assayed directly. However, avoid repeated ice-thawing process that would otherwise damage the TNF-α inside. It may need to dilute the sample with cell culture media if the TNF-α concentration exceeds the detection limit of the ELISA kit.
16. The homogenate can be preserved at −20 °C if not assayed directly and also avoid repeated ice-thawing processes.
17. i.p. injection of LPS/D-GalN establishes the acute hepatic injury model that is 100 % lethal to animals. This mixed solution needs to be freshly prepared and used within 20 min otherwise it would hamper its efficiency in inducing hepatic injury.
18. We recommend anesthetizing the animal before blood sampling because excessive struggling of the animal body may lead to inconsistent serum TNF-α levels.

References

1. Davis ME, Zuckerman JE, Choi CHJ, Seligson D, Tolcher A, Alabi CA, Yen Y, Heidel JD, Ribas A (2010) Evidence of RNAi in humans from systemically administered siRNA via targeted nanoparticles. Nature 464:1067–1070
2. Lee H, Lytton-Jean AKR, Chen Y, Love KT, Park AI, Karagiannis ED, Sehgal A, Querbes W, Zurenko CS, Jayaraman M, Peng CG, Charisse K, Borodovsky A, Manoharan M, Donahoe JS, Truelove J, Nahrendorf M, Langer R, Anderson DG (2012) Molecularly self-assembled nucleic acid nanoparticles for targeted in vivo siRNA delivery. Nat Nanotechnol 7:389–393
3. Woodrow KA, Cu Y, Booth CJ, Saucier-Sawyer JK, Wood MJ, Saltzman WM (2009) Intravaginal gene silencing using biodegradable polymer nanoparticles densely loaded with small-interfering RNA. Nat Mater 8:526–533
4. Dassie JP, Liu XY, Thomas GS, Whitaker RM, Thiel KW, Stockdale KR, Meyerholz DK, McCaffrey AP, McNamara JO, Giangrande PH (2009) Systemic administration of optimized aptamer-siRNA chimeras promotes regression of PSMA-expressing tumors. Nat Biotechnol 27:839–849
5. Gabrielson NP, Lu H, Yin LC, Li D, Wang F, Cheng JJ (2012) Reactive and bioactive cationic α-helical polypeptide template for nonviral gene delivery. Angew Chem Int Ed Engl 51:1143–1147
6. Gabrielson NP, Lu H, Yin LC, Kim KH, Cheng JJ (2012) A cell-penetrating helical polymer for siRNA delivery to mammalian cells. Mol Ther 20:1599–1609
7. Yin LC, Song ZY, Qu QH, Kim KH, Zheng N, Yao C, Chaudhury I, Tang HY, Gabrielson NP, Uckun FM, Cheng JJ (2013) Supramolecular self-assembled nanoparticles mediate oral delivery of therapeutic TNF-alpha siRNA against systemic inflammation. Angew Chem Int Ed Engl 52:5757–5761
8. Filleur S, Courtin A, Ait-Si-Ali S, Guglielmi J, Merle C, Harel-Bellan A, Clezardin P, Cabon F (2003) SiRNA-mediated inhibition of vascular endothelial growth factor severely limits tumor resistance to antiangiogenic thrombospondin-1 and slows tumor vascularization and growth. Cancer Res 63:3919–3922
9. McCaffrey AP, Meuse L, Pham TTT, Conklin DS, Hannon GJ, Kay MA (2002) Gene expression—RNA interference in adult mice. Nature 418:38–39
10. Soutschek J, Akinc A, Bramlage B, Charisse K, Constien R, Donoghue M, Elbashir S, Geick A, Hadwiger P, Harborth J, John M, Kesavan V, Lavine G, Pandey RK, Racie T, Rajeev KG, Rohl I, Toudjarska I, Wang G, Wuschko S, Bumcrot D, Koteliansky V, Limmer S, Manoharan M, Vornlocher HP (2004) Therapeutic silencing of an endogenous gene by systemic administration of modified siRNAs. Nature 432:173–178

Chapter 5

Disulfide-Bridged Cleavable PEGylation of Poly-L-Lysine for SiRNA Delivery

Min Tang, Haiqing Dong, Xiaojun Cai, Haiyan Zhu, Tianbin Ren, and Yongyong Li

Abstract

Engineered PEG-cleavable catiomers based on poly-L-lysine have been developed as nonviral gene vectors, which have been found to be one of important methods to balance "PEG dilemma." In this protocol, we aim at the standardization of the method and procedure of PEG-cleavable catiomers. Major steps including ring-opening polymerization (ROP) of ε-benzyloxycarbonyl-L-lysine *N*-carboxyanhydride (zLL-NCA) monomers to yield PEG-cleavable polylysine, examination on bio-stability and bio-efficacy of its gene complexes are described.

Key words Poly-L-lysine (PLL), Disulfide-bridged, Cleavable PEGylation, Gene delivery, Nonviral gene vector

1 Introduction

For nonviral gene vectors, poly-L-lysine (PLL) has been a regular cationic polypeptide that enables efficient gene package via electrostatic interaction. Effective gene capacity of PLL allows efficient gene condensation as well as the following gene transfection [1]. However, PLL/gene complexes show limited stability in the presence of biological relevant condition, such as serum, due to macrophage clearance by systemic administration [2, 3]. This is due to the existence of inherent aggregation effect induced by cationic surface of the complexes. Surface modification and functionalization is therefore essential to yield improved gene delivery system. Poly(ethylene glycol) (PEG)-shielding has become a leading coating strategy to prolong the in vivo circulation time of polyplexes (polymer/gene complex) in the cardiovascular system due to reduced macrophage clearance [4–7]. However, the shielding layer of PEG can not only cause the steric hindrance for cellular uptake, but also create a significant diffusion barrier to intracellular release of gene [8–10].

Kato Shum and John Rossi (eds.), *SiRNA Delivery Methods: Methods and Protocols*, Methods in Molecular Biology, vol. 1364, DOI 10.1007/978-1-4939-3112-5_5, © Springer Science+Business Media New York 2016

So the detachment of the PEG shell upon arrival at the target site is highly desired, and several chemical approaches have been developed to address the above issues [11–13]. Among them, introducing a disulfide bond (S–S) between the PEG segment and the PLL segment is a feasible route. During blood circulation, the S–S linkage remains fairly stable, and interaction with serum proteins is limited due to PEG-induced protection. However, the detachment of PEG layer takes place selectively via S–S cleavage as a result of the significant concentration gradient of glutathione (GSH) in tumor cells (four-fold higher than normal cells and hundred times higher in intracellular than that in extracellular) [14]. So far, there have been a number of works on disulfide-bridged cleavable PEGylation, but the standardization of this method to ensure consistent experimental results is unavailable, which is important for reliable evaluation as well as following clinical translation.

In this protocol, we aim at the standardization of the method and procedure of PEG-cleavable catiomers. Disulfide-bridged cleavable PEGylation of poly-L-lysine is employed as a model to document this method. Major steps including ring-opening polymerization (ROP) of ε-benzyloxycarbonyl-L-lysine *N*-carboxyanhydride (zLL-NCA) monomers to yield PEG cleavable polylysine, examination on bio-stability and bio-efficacy of its gene complexes are summarized and described. Some important data of this design are also presented to show how it can significantly improve the gene transfection efficiency compared to the control experiment with permanent linkage [15].

2 Materials

2.1 Synthesis of mPEG-SS-PLL

1. Dry tetrahydrofuran (THF), dichloromethane (DCM), *N,N*-dimethyl formamide (DMF) and *n*-hexane by refluxing over CaH_2. Distil or vacuum distil before use.
2. Poly(ethylene glycol) monomethyl ether (CH_3O-PEG of nominal Mw 1900).
3. Succinic anhydride (98 %).
4. 4-dimethylaminopyridine (DMAP).
5. Triethylamine (Et_3N, 99 %).
6. *N*-hydroxy succinimide (NHS).
7. *N,N′*-dicyclohexylcarbodiimide (DCC).
8. Cysteamine dihydrochloride (98 %).
9. ε-benzyloxycarbonyl-L-lysine.
10. Triphosgene (BTC, 99 %) (*see* **Note 1**).
11. Hydrogen bromide (HBr) 33 wt% solution in glacial acetic acid (*see* **Note 1**).

12. Trifluoroacetic acid (TFA) (*see* **Note 1**).
13. Dialysis membrane (MWCO: 1000 and 3500 Da).
14. Lyophilizer.
15. Rotary evaporator.
16. Vacuum pump.
17. Vacuum drying oven.
18. Magnetic stirrer with thermal and speed controller.

2.2 Effect of GSH on the Stability of mPEG-SS-PLL/Gene Complexes

1. 150 mM NaCl solution.
2. Glutathione (GSH, 99 %).
3. Microfiltration membrane (0.22 μm).
4. The reporter plasmids: pEGFP-C1 and pGL-3.
5. VEGF-siRNA.
6. 1 % (w/v) agarose gel: Dissolve 0.25 g agarose with 25 mL of TAE buffer, and heat the mixture till the agarose dissolute. Before the solidification of agarose, add 2.5 μL of 10 mg/mL EB to the solution and pour into plate.
7. 6× DNA loading buffer.
8. Nano-ZS 90 Nanosizer.
9. Electrophoresis apparatus.
10. Imago GelDoc system 2500.

2.3 Cell Culture

1. Cell lines: 293T cells, Hela cells and HepG2 cells cultured in full medium based on high-glucose DMEM at 37 °C in a humidified 5 % CO_2 atmosphere.
2. The full DMEM medium: high-glucose DMEM supplemented with 10 % fetal bovine serum (FBS) and 1 % penicillin-streptomycin (PS).
3. Dulbecco's Phosphate Buffered Saline, pH 7.4 (D-PBS).
4. 0.25 % trypsin–EDTA.
5. Dimethyl sulfoxide (DMSO).
6. MTT solution: Weigh 0.5 g MTT and dissolve in 100 mL of PBS, filter it by 0.22 μm microfiltration membrane, and store at 4 °C in the dark.
7. Shaking table.
8. Multiscan MK3 microplate reader.

2.4 In Vitro Transfection

1. Lysis buffer.
2. Bicinchoninic acid (BCA) protein assay kit.
3. Luciferase substrate.
4. Fluorospectrophotometer.

5. FCM buffer: PBS with 2 % FBS and 2 mM EDTA.
6. FCM stationary liquid: PBS with 2 % paraformaldehyde.
7. Flow cytometer.
8. Inverted fluorescence microscope.
9. Chemiluminescence apparatus.

2.5 Cellular Uptake

1. Label IT Tracker™ Cy3 reagent kit.
2. 5 M NaCl solution.
3. 100 % and 70 % Ethanol.
4. DEPC (diethyl pyrocarbonate) H_2O.
5. Sterile H_2O: ultrapure water was sterilized by autoclaving for 30 min at 120 °C.
6. Micro-spectrophotometer.
7. 4′,6-diamidino-2-phenylindole (DAPI).
8. 4 % paraformaldehyde.
9. Glycerin.
10. Confocal laser scanning microscopy.
11. Fluorescein isothiocyanate (FITC).
12. 0.4 % trypan blue.
13. Clathrin pathway inhibitor: 50 mM ammonia chloride solution and 10 μg/mL chlorpromazine.
14. Caveolin pathway inhibitor: 10 mM methyl-β-cyclodextrin (M-β-CD) and 200 μg/mL genistein.
15. Macropinocytosis pathway inhibitor: 100 μM Wortmannin.
16. ATP synthesis inhibitor: 10 mM sodium azide.
17. Actin polymerization inhibitor: 40 μg/mL colchicines.

3 Methods

3.1 Synthesis of mPEG-SS-PLL

Carry out all reactions under nitrogen atmosphere and all procedures at room temperature unless otherwise specified. In addition, dialysis was conducted against deionized water, followed by lyophilization.

1. Synthesis of mPEG-COOH: dissolve 2.5 mmol mPEG, 12.5 mmol succinic anhydride and 12.5 mmol DMAP in 30 mL dioxane, and add 12.5 mmol Et_3N dropwise. Continue the reaction overnight. After evaporating the dioxane under reduced pressure, dissolve the residual in DMF and dialyze (MWCO 1000 Da).
2. Synthesis of mPEG-SS-NH_2: to activate the carboxyl, mix 0.6 mmol DCC, 0.6 mmol NHS with 0.5 mmol mPEG-

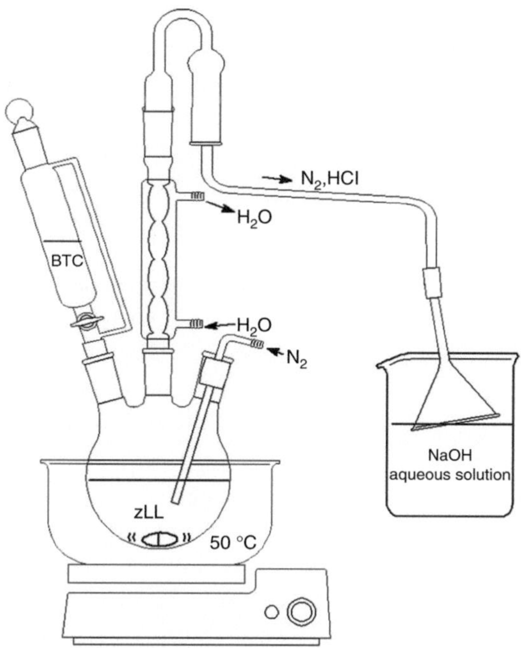

Fig. 1 Reaction device for synthesis of zLL-NCA. BTC was added dropwise into zLL under N_2 atmosphere. And the reaction would generate hydrogen chloride gas so that alkaline exhaust absorbing device was required

COOH in 30 mL DCM (30 mL) at 0 °C for 5 h. Then drop the mixture into 3.5 mmol cystamine (*see* **Note 2**) in DCM, and react for 24 h. After filtration to remove insoluble matter, precipitate the product in cold diethyl ether twice and dialyze (MWCO 1000 Da).

3. Synthesis of zLL-NCA (*see* **Note 3**): suspend 20 mmol zLL in dried 50 mL THF at 50 °C and add 9 mL THF of triphosgene in dropwise fashion (8.4 mmol) (Fig. 1). The reaction was not completed until the suspension turned into clear solution. After concentration by vacuum rotary evaporation treatment, pour the solution into excess dried n-hexane (*see* **Note 4**) to obtain crude crystals of zLL-NCA and recrystallize the zLL-NCA from dried THF/hexane (1:15, v/v) twice before being vacuum-dried (*see* **Note 5**).
4. Synthesis of mPEG-SS-PzLL: the copolymers were prepared by ring-opening polymerization (ROP) of zLL-NCA initiated by the amino-groups of mPEG-SS-NH_2. According to experiment designing, synthesize specific block polymers by varying the feed ratio of mPEG-SS-NH_2 and zLL-NCA or by regulation of reaction time. Conduct the reaction in DMF for 72 h and dialyze (MWCO of dialysis bag was chosen by the designed molecular weight of product).

5. Deprotection to get mPEG-SS-PLL: dissolve 160 mg mPEG-SS-PzLL in 10 mL of TFA and stir at 0 °C. Then add 2 equiv. of solution of 33 wt% HBr in HAc with respect to the benzyl carbamate (Z) groups and react for 1 h. After precipitating the product in cold diethyl ether twice, dissolve the sediment in DMF and dialyze (MWCO of dialysis bag was chosen by the designed molecular weight of product) (*see* **Note 6**).

3.2 Effect of GSH on the Stability of mPEG-SS-PLL/Gene Complexes

1. Dissolve mPEG-SS-PLL in 150 mM NaCl with a concentration of 2 mg/mL, and filter it using 0.22 μm microfiltration membrane. Then complex the mPEG-SS-PLL with DNA at various weight ratios. Vortex the complexes gently for 10 s and incubate them at 37 °C for 30 min.
2. Dilute the complexes with 1 mL of 150 mM NaCl, and measure the particle size of complexes. After yielding the particle with stable size, add GSH into the solutions to form 10 mM GSH final concentration (*see* **Note 7**), and then monitor the size change (Fig. 2).
3. To further evaluate the GSH-responsiveness of complexes with disulfide bonds, we also carried out the agarose gel retardation assay. Complex mPEG-SS-PLL and gene as **step 1**, then add GSH into the solutions to form 10 mM GSH final concentration.
4. After incubation with GSH for 0.5 h, load the solutions (10 μL) containing 0.5 μg pDNA (*see* **Note 8**) or siRNA (*see* **Note 9**) with 2 μL of 6× DNA Loading Buffer on 1 % (w/v) agarose gel containing ethidium bromide (EB), and carry the electrophoresis out in tris-acetate (TAE) running buffer at 100 V for 40 min (pDNA) or 15 min (siRNA). Then

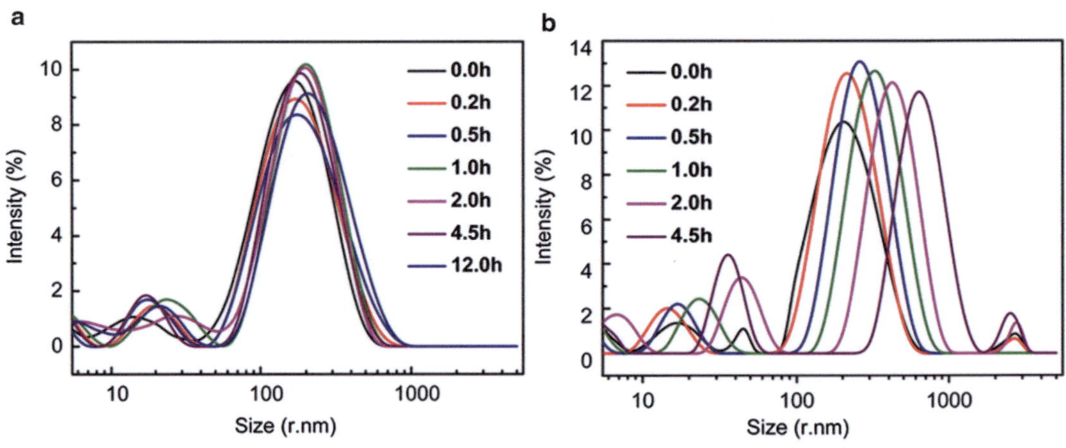

Fig. 2 Time-dependent changes in the size in diameter of the mPEG-SS-PLL45/DNA complexes during the 12 h storage (**a**) and upon exposure to 10 mM GSH (**b**) as determined by DLS (reproduced from ref. [18])

visualize nucleic acid bands at $\lambda = 254$ nm. In the presence of 10 mM GSH, we could observe the migration of negatively charged gene toward the cathode from polyplexes in different degrees.

3.3 Cytotoxicity Assay

1. Seed cells (*see* **Note 10**) (5×10^3 cells/well with 100 μL of full DMEM medium) in 96-well plate and cultivate for 24 h.
2. Dissolve mPEG-SS-PLL cationers in fresh pure DMEM (gradient concentrations of ranging from 11 to 200 mg/L or higher) and filter them by 0.22 μm microfiltration membrane. Add 200 μL of catiomers per well and cultivate for another 24 h (*see* **Note 11**).
3. Replace the medium by 200 μL of fresh full DMEM medium and 20 μL of MTT (5 mg/mL) (*see* **Note 12**). Then incubate the cells for 4 h.
4. Add 150 μL of DMSO to each well after removing the medium, and place the plate on a shaker with gentle shaking for 10 min. And then measure the optical density (OD) at 570 nm by microplate reader (*see* **Note 13**).

3.4 In Vitro Transfection

1. Seed cells (5×10^4 cells/well with 1000 μL of full DMEM medium) into 24-well plate and cultivate for 24 h (*see* **Note 14**).
2. Prepare the catiomers/pDNA complexes (*see* **Note 15**) at various weight ratios in pure DMEM. Then replace the medium by 100 μL (including 0.5 μg pDNA) of complexes and 900 μL fresh pure DMEM (*see* **Note 16**) in each well. After incubating for 4 h, substitute the medium for full medium and cultivate the cells for additional 44 h.
3. The BCA protein assay for the GFP relative quantitative determination: wash cells by PBS and crack them by 200 μL of lysis buffer at 4 °C for 30 min. Then transfer 100 μL of lysates to black 96-well plate for determining the fluorescence intensity by fluorescence spectrophotometer at excitation wavelength of 488 nm and emission wavelength of 509 nm. And use the rest 100 μL (*see* **Note 17**) for analyzing the total protein content by BCA protein assay kit. The transfection efficiency was described as fluorescence intensity (A.U.)/mg of total protein (*see* **Note 18**).
4. The FCM assay for the GFP relative quantitative determination: collect cells and wash cells by 500 μL of FCM buffer. Then collect cells again, fix in FCM stationary liquid and store at 4 °C before measurement. The transfection efficiency was described as the percentage of the GFP-positive cells detected by flow cytometer.
5. For the GFP relative qualitative determination, image the transfected cells under the inverted fluorescence microscope.

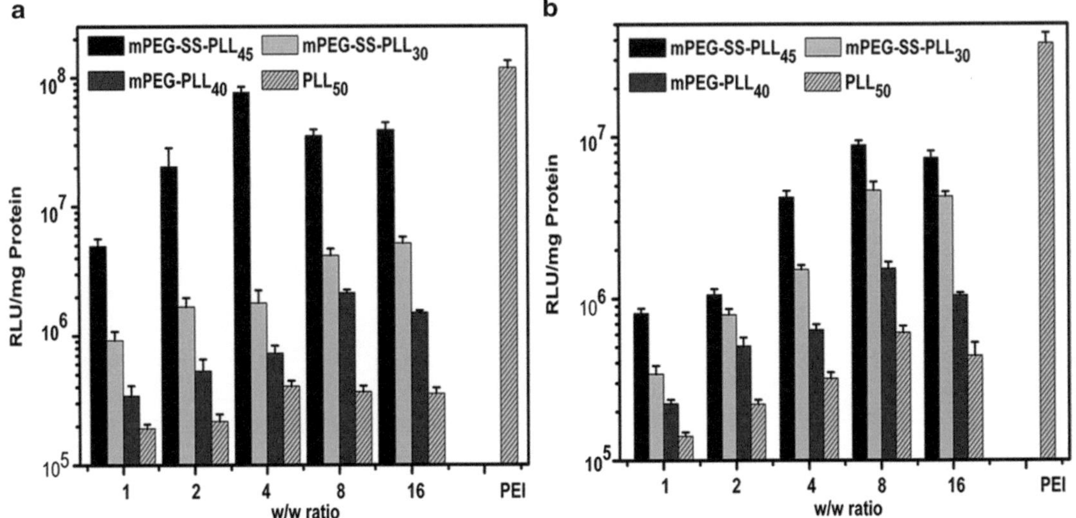

Fig. 3 Luciferase expression mediated by catiomer/DNA complexes at various w/w ratios ranging from 1 to 16 in 293T cells (**a**) and HeLa cells (**b**) (reproduced from ref. [18])

6. The quantification of transfection efficiency could also be determined by luciferase activity detection (using pGL-3). After the cell lysis, mix 20 μL of supernatant with 100 μL of luciferase substrate and detect the activity of luciferase by chemiluminescence apparatus. The transfection efficiency was described as relative light unit (RLU)/mg of total protein (measure by BCA protein assay kit) (Fig. 3).

3.5 Cellular Uptake

1. Label the pDNA by Cy3 (*see* **Note 19**): mix 10 μL of pEGFP (1.6 μg/μL), 8 μL of Label IT Tracker™, 16 μL of 10× Labeling Buffer A, and 126 μL of sterile H_2O, and incubate at 37 °C for 1 h. After centrifugation (16,414 × *g*, 30 s), continue incubation at 37 °C for another 1 h. Then add 40 μL of DEPC H_2O, 20 μL of 5 M NaCl aqueous solution and 400 μL of cold ethanol, and store at −20 °C for 2 h. Then centrifuge (49,088 × *g*, 4 °C) for 10 min to pellet the labeled DNA, and remove the supernatant carefully (do not damage the DNA pellet). Wash the pellet with 0.5 mL of ethanol (70 %), and centrifuge as above. After all ethanol removed, disperse DNA with 32 μL of sterile H_2O. At last, measure the concentration of DNA labeled by Cy3 using micro-spectrophotometer.
2. The FCM assay for quantitative determination: seed HepG2 cells (2×10^5 /well with 2000 μL of full DMEM medium) into 6-well plate, and cultivate for 24 h. Prepare the catiomers/pDNA (labeled by Cy3) complexes at each optimal transfection weight ratio in pure DMEM and treat the cells for 30 min (2 μg pDNA/well). Then replace the medium by 1.5 mL of fresh pure DMEM. After 4 h, deal with the cells and measure

the cellular uptake efficiency (referred to the FCM assay described above).

3. The CLSM assay for qualitative determination: seed and cultivate cells on the coverslip in 6-well plate for 24 h as above. But differently, after 4 h, treat the cells with PBS washing (*see* **Note 20**), 4 % paraformaldehyde fixing, DAPI dyeing, and glycerin sealing. Store at 4 °C in dark before measurement (*see* **Note 21**). Observe cells by confocal laser scanning microscopy (Cy3: at excitation wavelength = 570 nm and DAPI at emission wavelength = 340 nm) (Fig. 4).

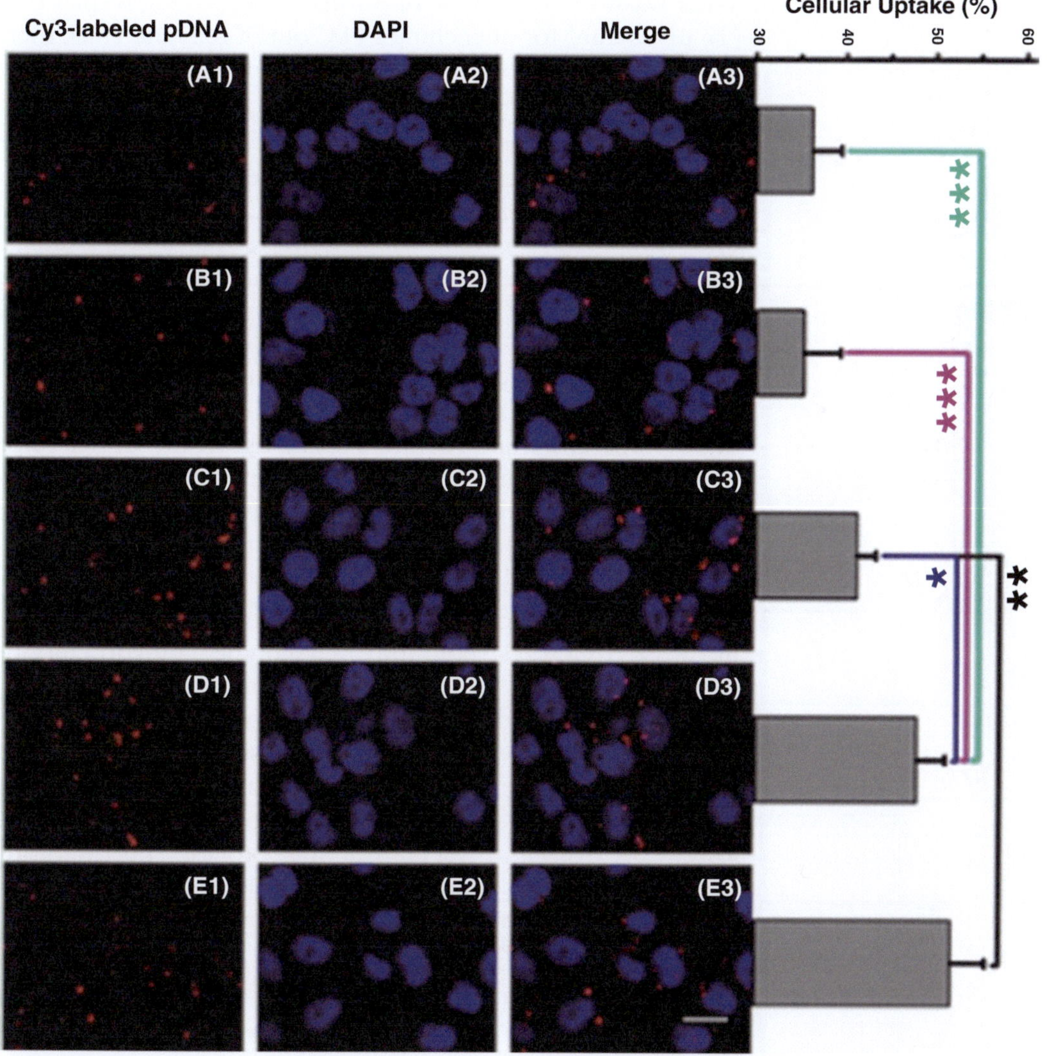

Fig. 4 Cellular uptake by CLSM and FCM in HepG2 cells. CLSM images are Cy3-labeled pDNA (*red*), DAPI dyed nucleus (*blue*). The images of mPEG-SS-Lys55/pDNA (*panel a*), mPEG-SS-Lys95/pDNA (*panel b*), mPEG-SS-Lys55-r-His20/pDNA (*panel c*), mPEG-SS-Lys95-r-His20/pDNA (*panel d*) and PEI/pDNA (*panel e*) with w/w ratios 4:1, 5:1, 3:1, 2:1, and 1.3:1 were observed (reproduced from ref. [15])

3.6 Cellular Uptake Pathway

1. 30 mg mPEG-SS-PLL, 30 mg FITC, and 400 μL of triethylamine were mixed in DMF, and stirred for 48 h in dark at room temperature. Then dialyze (MWCO 3500 Da) against water to remove the organic solvent and small molecules, then the dialysate was subjected to lyophilize to get dried FITC-labeled mPEG-SS-PLL powder.
2. Seed cells (5×10^3/well) in 96-well plate and cultivate for 24 h. Then replace the medium by 200 μL of fresh full DMEM (containing specific channel inhibitors, detailed in Materials Part) and incubate at 37 °C for 30 min. Add 5 μg FITC-labeled mPEG-SS-PLL for each well and store at 37 °C for 4 h.
3. Then wash the cells with PBS three times, 0.4 % trypan blue (15 μL, 2 min) for quenching FITC out of cells and crack them by 400 μL of lysis buffer at 4 °C for 30 min.
4. Transfer 200 μL of lysates to black 96-well plate for determining the fluorescence intensity of FITC by the Fluorospectro Photometer at excitation wavelength of 485 nm and emission wavelength of 538 nm (*see* **Note 22**).

4 Notes

1. Triphosgene is easy to decompose to toxic phosgene gas at high temperature (over 130 °C, 90 °C with moisture), so avoid high temperature and require alkaline absorbing device (e.g., 10 % aqueous sodium hydroxide). HBr possessing high toxicity tends to be oxidized when exposed to light and heat, and TFA has strong corrosion and irritation, so protection measurement and good ventilation are essential.
2. Cysteamine dihydrochloride is needed to desalt before adding in mPEG-COOH solution: Cystamine dihydrochloride (8.88 mmol) in 30 mL of deionized water was desalinated by NaOH under stirring at room temperature for 3 h. After condensation by rotary evaporation, the residue was dissolved in CH_2Cl_2 and filtrated to remove insoluble matter. Cystamine was obtained after CH_2Cl_2 removal by rotary evaporation, followed by overnight vacuum-drying process. Furthermore, ensure excess cystamine so as to avoid reaction of mPEG-COOH with the amino-groups of cystamine on both sides.
3. *N*-Carboxyanhydride (NCA) is sensitive to moisture and heat, so the synthesis and polymerization of zLL-NCA requires strictly anhydrous and oxygen-free environment. Long-term storage also should be avoided.
4. Hexane is better to be precooled before use for the recrystallization of zLL-NCA.

5. The temperature of recrystallization needs to be carefully controlled: keep the mixture at room temperature, then store at 4 °C, and at −20 °C overnight. Finally, white needle-like crystals with expected structure can be collected. If cooling too fast, the product might contain impurities.
6. During dialysis, benzyloxycarbonyl groups in $C_6H_5CH_2Br$ form were removed, with unreacted HBr/HOAc.
7. GSH maintains a millimolar concentration (3–10 mM) in the cytosol and subcellular compartments but lower concentration (~2.8 μM) in cellular exterior such as plasma [16]. So we simulated the intracellular environment by 10 mM GSH.
8. DNase (DNA enzyme) may degrade DNA, causing weak and vague strip signals, even absent.
9. For siRNA experiment, RNase-free devices were required and complex with siRNA was conducted in clean worktable. In addition, perform the experiment as soon as possible.
10. The cells should be passaged frequently and at regular intervals to maintain cells in midlog growth.
11. For the cytotoxicity assay, we used branched polyethylenimine (MW = 25000, bPEI) and blank as control. bPEI is recognized as an important gold standard in many contrast experiments.
12. Once added MTT in, the plate should be protected from light.
13. The relative cell viability was calculated by the following equation: the viability (%) = (ODsample/ODcontrol) × 100. The OD control was obtained in the absence of polymers while the OD sample was obtained in the presence of polymers.
14. The purpose was to make the cells reach 60–70 % confluence, so that cell metabolism and productivity were the most vigorous which was conducive to cellular uptake.
15. There are clues to indicate that adding reagent to DNA was shown to be tenfold more efficient than the reverse order, with regard to transfection efficiency [17].
16. To evaluate the effect of serum on transfection at a cellular level, 10 % FBS was added into the DMEM.
17. The rest lysis buffer was centrifuged (372 × *g*) for 10 min to remove large cell debris and the supernatant was separated gently.
18. The fluorescence tends to quench easily under long time of excitation.
19. The sample needs to be protected from light.
20. Try to ensure the viability and activity of cells by careful manipulation in the washing process.

21. Minimize the storage time and detect early.
22. The experiment uses FITC-labeled mPEG-SS-PLL without channel inhibitor as negative control. All operations were conducted while minimizing light exposure.

Acknowledgement

This work was financially supported by research grants from the National Natural Science Foundation of China (NSFC 21104059 and 51173136), the Fundamental Research Funds for the Central Universities (2013KJ038), and "Chen Guang" project (12CG17) founded by Shanghai Municipal Education Commission and Shanghai Education Development Foundation.

References

1. Toncheva V, Wolfert MA, Dash PR, Oupicky D, Ulbrich K, Seymour LW, Schacht EH (1998) Novel vectors for gene delivery formed by self-assembly of DNA with poly(L-lysine) grafted with hydrophilic polymers. Biochim Biophys Acta 1380:354–368
2. Rungsardthong U, Deshpande M, Bailey L, Vamvakaki M, Armes SP, Garnett MC, Stolnik S (2001) Copolymers of amine methacrylate with poly(ethylene glycol) as vectors for gene therapy. J Control Release 73:359–380
3. Mannisto M, Vanderkerken S, Toncheva V, Elomaa M, Ruponen M, Schacht E, Urtti A (2002) Structure-activity relationships of poly(L-lysines): effects of pegylation and molecular shape on physicochemical and biological properties in gene delivery. J Control Release 83:169–182
4. Langer R (1998) Drug delivery and targeting. Nature 392:5–10
5. Otsuka H, Nagasaki Y, Kataoka K (2003) PEGylated nanoparticles for biological and pharmaceutical applications. Adv Drug Deliv Rev 55:403–419
6. Riehemann K, Schneider SW, Luger TA, Godin B, Ferrari M, Fuchs H (2009) Nanomedicine—challenge and perspectives. Angew Chem Int Ed Engl 48:872–897
7. Knop K, Hoogenboom R, Fischer D, Schubert US (2010) Poly(ethylene glycol) in drug delivery: pros and cons as well as potential alternatives. Angew Chem Int Ed Engl 49:6288–6308
8. Masuda T, Akita H, Niikura K, Nishio T, Ukawa M, Enoto K, Danev R, Nagayama K, Ijiro K, Harashima H (2009) Envelope-type lipid nanoparticles incorporating a short PEG-lipid conjugate for improved control of intracellular trafficking and transgene transcription. Biomaterials 30:4806–4814
9. Nie S (2010) Understanding and overcoming major barriers in cancer nanomedicine. Nanomedicine (Lond) 5:523–528
10. Pozzi D, Colapicchioni V, Caracciolo G, Piovesana S, Capriotti AL, Palchetti S, De Grossi S, Riccioli A, Amenitsch H, Lagana A (2014) Effect of polyethyleneglycol (PEG) chain length on the bio-nano-interactions between PEGylated lipid nanoparticles and biological fluids: from nanostructure to uptake in cancer cells. Nanoscale 6:2782–2792
11. Lin S, Du F, Wang Y, Ji S, Liang D, Yu L, Li Z (2008) An acid-labile block copolymer of PDMAEMA and PEG as potential carrier for intelligent gene delivery systems. Biomacromolecules 9:109–115
12. Hatakeyama H, Ito E, Akita H, Oishi M, Nagasaki Y, Futaki S, Harashima H (2009) A pH-sensitive fusogenic peptide facilitates endosomal escape and greatly enhances the gene silencing of siRNA-containing nanoparticles in vitro and in vivo. J Control Release 139:127–132
13. Cai X, Dong C, Dong H, Wang G, Pauletti GM, Pan X, Wen H, Mehl I, Li Y, Shi D (2012) Effective gene delivery using stimulus-responsive catiomer designed with redox-sensitive disulfide and acid-labile imine linkers. Biomacromolecules 13:1024–1034
14. Wang K, Liu Y, Yi WJ, Li C, Li YY, Zhuo RX, Zhang XZ (2013) Novel shell-cross-linked micelles with detachable PEG corona for

glutathione-mediated intracellular drug delivery. Soft Matter 9:692–699

15. Zhu H, Dong C, Dong H, Ren T, Wen X, Su J, Li Y (2014) Cleavable PEGylation and hydrophobic histidylation of polylysine for siRNA delivery and tumor gene therapy. ACS Appl Mater Interfaces 6:10393–10407
16. Son S, Namgung R, Kim J, Singha K, Kim WJ (2012) Bioreducible polymers for gene silencing and delivery. Acc Chem Res 45:1100–1112
17. van Gaal EV, van Eijk R, Oosting RS, Kok RJ, Hennink WE, Crommelin DJ, Mastrobattista E (2011) How to screen non-viral gene delivery systems in vitro? J Control Release 154:218–232
18. Cai XJ, Dong HQ, Xia WJ, Wen HY, Li XQ, Yu JH, Li YY, Shi DL (2011) Glutathione-mediated shedding of PEG layers based on disulfide-linked catiomers for DNA delivery. J Mater Chem 21:14639–14645

Chapter 6

Preparation of a Cyclic RGD: Modified Liposomal SiRNA Formulation for Use in Active Targeting to Tumor and Tumor Endothelial Cells

Yu Sakurai, Tomoya Hada, and Hideyoshi Harashima

Abstract

The delivery of SiRNA is not only a challenging strategy for developing new remedies, but is also useful as an analytic tool for an in vivo phenotypic alteration by loss-of-function. Specifically, ligand-mediated SiRNA active targeting can be used to silence any gene in any organ of interest. In this chapter, we describe the preparation of an active targeting system to tumor endothelial cells (TECs) using liposomal SiRNA modified with cyclic RGD peptides. The procedure consists of essentially three steps: (1) the synthesis of a cyclic RGD peptide derivative, (2) SiRNA encapsulation into a liposomal delivery system, and (3) modification of liposomal SiRNA with a cyclic RGD derivative.

Key words Liposomal delivery system, Active targeting, Tumor, Tumor endothelial cells, SiRNA

1 Introduction

Since the discovery that small interfering RNA (SiRNA), which consists of 21-nt double stranded RNA, could induce RNA interference (RNAi) without severe immune response in mammalian cells [1], a number of studies have reported on the preparation of a SiRNA drug delivery system (DDS) using polymers, inorganic materials, and liposomes. Target organs for recently developed SiRNA DDS are mainly the liver [2]. The liver is the principle clearance organ for such macromolecules due to the fenestra in vessels, which comprises a hole with a diameter of approximately 100 nm [3]. Regarding cancer, macromolecules with a prolonged circulation time after a systemic injection passively accumulate in the tumor tissue, a process that is referred to as the enhanced permeability and retention (EPR) effect [4]. Thus, a variety of traditional DDSs exploit such "passive targeting." Actually, almost all SiRNA DDS currently under clinical study targets the liver because of the ease of targeting this organ [5].

Kato Shum and John Rossi (eds.), *SiRNA Delivery Methods: Methods and Protocols*, Methods in Molecular Biology, vol. 1364, DOI 10.1007/978-1-4939-3112-5_6, © Springer Science+Business Media New York 2016

To silence any genes in an organ, a targeting ligand needs to be introduced into an SiRNA DDS. We herein describe a method for the in vivo SiRNA active targeting of DDS to tumor endothelial cells (TECs) using a cyclic RGD peptide [6]. The procedure for displaying a specific ligand on the surface of a liposomal carrier is mainly divided into three steps: pre-modification, post-modification, and surface-reaction. In the pre-modification method, ligands are typically added to the lipid solution before liposome formulation [7]. On the other hand, ligand–lipid micelles are mixed with assembled liposomes in the post-modification method [8]. In surface-reactions, a polyethylene glycol (PEG) spacer with a reactive head group is included to lipid solution, and the ligand is then allowed to form covalent bonds by mixing the ligand and liposome solution after the liposome formulation is finished [9]. In the current section, we describe the preparation of ligand-modified liposomal DDS using the post-modification method.

2 Materials

Prepare all solutions in distilled, deionized water and use special grade chemicals.

2.1 Conjugation of cRGD Peptides with PEG Lipid

1. Cyclic RGD (–Arg Gly Asn *D*-Phe Lys-).
2. *N*-hydroxysuccinimide-PEG2,000-di-stearoyl-*sn*-glycerophosphoethanolamine (NHS-PEG-DSPE).
3. Dialysis membrane (MWCO 1000).
4. Phosphate buffered saline, pH 7.4 (PBS): Dissolve a tablet of PBS per 100 mL distilled deionized water (DDW). Pass the solvent through a microfiltration filter (0.2 μm).
5. Sinapic acid.

2.2 Encapsulation of siRNA into the MEND

1. YSK05: Synthesize YSK05, as previously reported [10].
2. Lipids: Dissolve YSK05, 1-palmitoyl-2-oleoyl-sn-glycerophosphoethanol amine (POPE) and cholesterol in 90 % tertiary butanol (DDW, vol/vol) at 20 mM. And dissolve 1,2-dimyristoyl-sn-glycerol, methoxypolyethylene glycol (molecular weight of PEG; 2000, PEG-DMG) in 90 % tertiary butanol at 2 mM.
3. SiRNA: Dissolve SiRNA in DDW at 2.0 mg/mL.
4. Citrate buffer: Weigh 2.68 g of citric acid and 1.76 g of trisodium citrate dehydrate. Add 900 mL of DDW to the powder, adjust the pH of the solution to 4.0 with 1 M aqueous sodium hydroxide. Add water to a volume of 1000 mL, and then pass the solvent through s microfiltration filter (0.2 μm).

5. Ultrafiltration filter (Amicon Ultra-15, MWCO 100,000).
6. PBS (−): Dulbecco's Phosphate-Buffered Saline for ultrafiltration.

2.3 Incorporation of cRGD-PEG Conjugates into a Lipid Membrane

1. Ultrafiltration filter (Amicon Ultra-15, MWCO 100,000).
2. 200 proof ethanol.
3. PBS, pH 5.5: Adjust pH of PBS (−) with 2 M hydrogen chloride.
4. HEPES buffer. Dissolve 119.2 mg of 2-[4-(2-hydroxyethyl) piperazin-1-yl]ethanesulfonic acid in 50 mL of DDW. Adjust pH of the solution to 7.4.
5. Qunat-iT RiboGreen RNA assay kit.
6. Dissolve 50 mg of dextran sulfate in 50 mL of DDW.
7. Dilute triton X-100 to 10 % (weight/vol) with DDW.

3 Methods

3.1 Synthesis of cRGD Peptides-PEG Lipid Derivatives

1. Mix 3.3 mg of the cyclic RGD (5.5 mmol) peptide with 15.0 mg of NHS-PEG-DSPE (5.0 mmol) in 1.0 mL of PBS (−), and then incubate the mixture with gentle mixing over night at 37 °C (*see* **Note 1**).
2. Put the mixture into dialysis membrane. Place the membrane in 500 mL of PBS (−) for 2 h, and then replace the solution with 500 mL of refresh PBS (−) for 2 h (*see* **Note 2**). Next, place the membrane in 500 mL of DDW for 2 h. Replace the solution with 500 mL of DDW and allow it to stand overnight.
3. Put the dialysate into a 50 mL tube, and allow it to congeal in a −80 °C refrigerator. Lyophilize the dialysate over night at normal temperature.
4. Dissolve the resulting powder in 200 proof ethanol at a concentration of 5 mM.
5. Add 1.0 μL of trifluoroacetic acid to 1.0 mL of DDW (0.1 % TFA water), and then add 0.1 % water and acetonitrile at a ratio of 1:1. Dissolve the sinapic acid in a 0.1 % water/acetonitrile mixture to give a concentration of 10 mg/mL.
6. Add 2.0 μL of a 5 mM RGD-PEG solution to 50 μL of the sinapic acid solution. Analyze the mixture by matrix assisted laser desorption ionization time-of-flight mass spectrometry (MALDI TOF-MS) (Fig. 1).

3.2 Preparation of Bare MEND

1. Add 100 μL of DDW, 20 μL of citrate buffer and 80 μL of 2.0 mg/mL SiRNA to a 1.7 mL tube.
2. Add 90 μL of 20 mM YSK05, 22.5 μL of 20 mM POPE, 22.5 μL of 20 mM cholesterol and 45 μL of 2 mM PEG-DMG to a 5.0 mL tube.

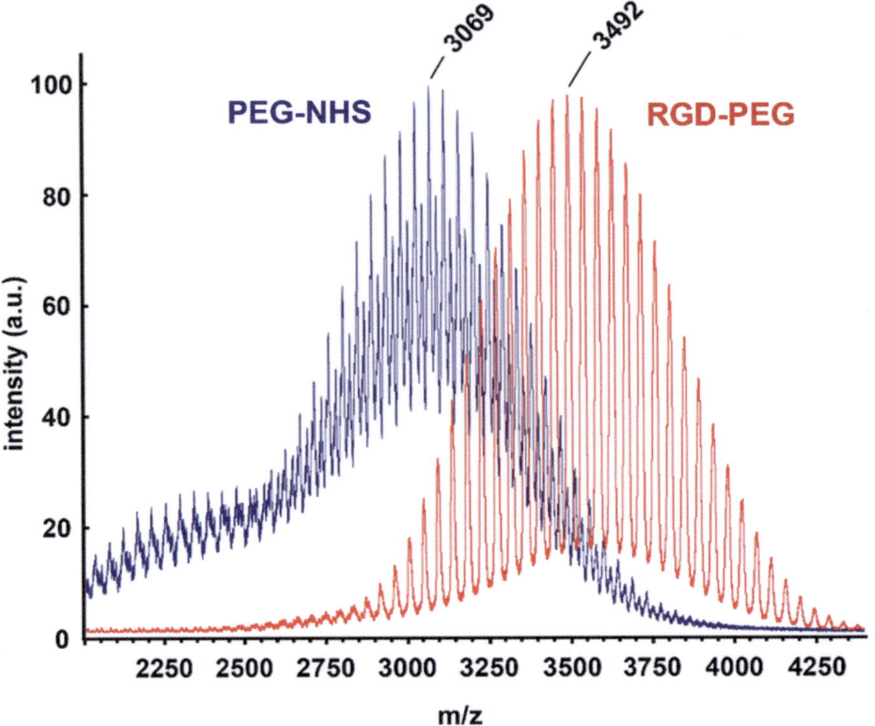

Fig. 1 Typical spectrum of PEG-NHS and RGD-PEG. The RGD peptide (molecular weight 603) was covalently bonded to the PEG-NHS (molecular weight 3069), accompanied by the desorption of NHS group

3. Add the SiRNA mixture to the lipid solution with vigorous mixing using a vortex mixer. Then, mix the SiRNA/lipid mixture with 2.0 mL of citrate buffer in a 15 mL tube with vigorous mixing using a vortex mixer (*see* **Note 3**).
4. Immediately add 4.0 mL of PBS to the diluted SiRNA/lipid mixture with vigorous mixing using a vortex mixer. Put the solution into the upper tube of an Amicon Ultra-15, and centrifuge the sample at room temperature for 20 min at 1000 × *g* (*see* **Note 4**).
5. Discard the flow-through in the lower tube, and then add 14.0 mL of DDW to the residue in the upper tube (*see* **Note 5**). Then, centrifuge the solution at room temperature for 25 min at 1000 × *g*.
6. Recover the residual MEND solution in the upper tube, and rinse the upper tube with DDW, particularly the membrane, at least 2 times. Adjust the volume of the MEND to 1.0 mL (bare MEND).

3.3 Modification of the MEND with cRGD-PEG lipid derivatives

1. Mix 1.0 mL of the bare MEND solution with 770 mL of PBS, pH 5.5 and 30 μL of RGD-PEG. To the mixture, add 200 μL of 200 proof ethanol with gentle mixing using a vortex mixer.

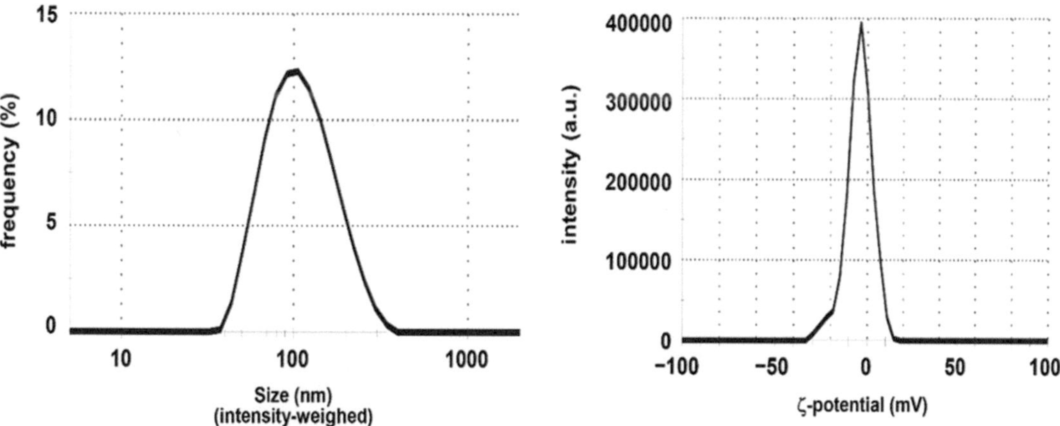

Fig. 2 The typical characteristics of the RGD-MEND. The diameter of the RGD-MEND is around 100 nm, with a homogenous particle size distribution. The ζ-potential is usually slightly negative, probably due to the leakage of siRNA from the interior of the liposome

2. Incubate the mixture at 60 °C for 30 min at 1000 rpm using a shaking incubator (*see* **Notes 6** and 7).
3. Put the mixture into the upper chamber of an Amicon Ultra-15, and centrifuge the sample at room temperature for 20 min at 1000 × *g*. Discard the flow-through in the lower tube, and then add 14.0 mL of PBS (−) to the residue in the upper tube. Centrifuge it at room temperature for 25 min at 1000 × *g*.
4. Recover the residual MEND solution in the upper tube, and rinse the upper tube with PBS (−), the membrane, at least 2 times. Adjust the volume of the MEND to 1.0 mL (RGD-MEND).
5. Determine the particle size distribution and ζ-potential of the RGD-MEND using a Zetasizer Nano ZS ZEN3600 instrument. Dilute the RGD-MEND 100 fold with PBS (−) for measurement of the particle size distribution; dilute the RGD-MEND 20-fold with HEPES buffer for measurement of the ζ-potential. Typically, RGD-MEND is around 100 nm in z-average and has a slightly negative charge (Fig. 2).
6. To quantify the ratio between SiRNA recovery and encapsulation efficiency, prepare a series of standard solutions of SiRNA (0, 400, 800, 1200, 1600, 2000 ng/mL) to prepare a standard curve and dilute the RGD-MEND solution to an appropriate concentration (<2000 ng/mL, considered as 100 % recovery and encapsulation) with HEPES buffer.
7. Mix the SiRNA standard curve solution and the diluted RGD-MEND with the assay reagent as shown below in Table 1, and then add all samples to a 96-well black microplate.
8. Measure the fluorescence at an excitation/emission 500/525 nm (Band width 5 nm) at room temperature using a Microplate reader.

Table 1
Preparation of assay reagents and an siRNA solution for siRNA quantification

	Triton X-100 (−)	Triton X-100 (+)
siRNA standard or RGD-MEND solution	100 μL	100 μL
1.0 mg/mL dextran sulfate	4 μL	4 μL
10 % triton X-100	–	2 μL
Qunat-iT RiboGreen	0.25 μL	0.25 μL
HEPES buffer	95.75 μL	93.75 μL

9. Calculate the SiRNA concentration of the RGD-MEND solution using the standard curve and estimate the ratio of recovery to encapsulation efficiency using the formulae below.

$$\text{siRNA recovery ratio}\,(\%) = \frac{\text{siRNA conc.}\,(\text{Triton}\,(+)) \times \text{recovered volume}}{\text{initial siRNA}\,(\mu\text{g})}$$

$$\text{siRNA encapsulation efficiency}\,(\%) = \frac{\left[\text{siRNA conc.}\,(\text{Triton}(+)) - \text{siRNA conc.}\,(\text{Triton}(-))\right]}{\text{siRNA conc.}\,(\text{Triton}(+))} \times 100$$

4 Notes

1. NHS-PEG-DSPE cannot be removed from the reaction solution by dialysis unlike a peptide, because NHS-PEG-DSPE forms micelles in water. Make sure that an excess amount of peptide is added to the reaction mixture.
2. It is better to first perform a dialysis against PBS (−). This is because the sudden change in ionic strength can lead to the nonspecific binding of the peptide to the membrane.
3. The RGD-PEG solution should be stored at −80 °C. RGD-PEG may undergo degradation when stored as a higher temperature, such as 4 or −20 °C.
4. Vigorous mixing is important. If the agitation of the lipid and SiRNA is not robust, aggregates may form and precipitate.
5. At the second ultrafiltration step, a solution without buffering capacity should be used. In the next RGD-PEG modification, it is important to maintain the pH at 5.5. Otherwise, SiRNA may leak from the interior of the RGD-MEND.

6. Centrifugation time should be minimized, since centrifugation may affect the characteristics of the MEND.
7. The efficiency of RGD-PEG incorporation is strongly dependent on the incubation time, temperature and ethanol concentration.

Acknowledgement

This work was supported in part by Grant-in-Aid for Young Scientists (Start-up) from Japan Society for the Promotion of Science (JSPS) (Grant number 25893001) and a Grant-in-Aid for Research on Medical Device Development from the Ministry of Health, Labour and Welfare of Japan (MHLW), grants from the Special Education and Research Expenses of the Ministry of Education, Culture, Sports, Science and Technology of Japan and Japan Science and Technology Agency, Adaptable and Seamless Technology Transfer Program: JST A-STEP through target-driven R & D, Contract No. AS251Z01439Q. We thank Dr. Milton S. Feather for his helpful advice in writing the English manuscript.

References

1. Elbashir SM, Harborth J, Lendeckel W, Yalcin A, Weber K, Tuschl T (2001) Duplexes of 21-nucleotide RNAs mediate RNA interference in cultured mammalian cells. Nature 411:494–498
2. Kanasty R, Dorkin JR, Vegas A, Anderson D (2013) Delivery materials for siRNA therapeutics. Nat Mater 12:967–977
3. Jacobs F, Wisse E, De Geest B (2010) The role of liver sinusoidal cells in hepatocyte-directed gene transfer. Am J Pathol 176: 14–21
4. Maeda H, Nakamura H, Fang J (2013) The EPR effect for macromolecular drug delivery to solid tumors: improvement of tumor uptake, lowering of systemic toxicity, and distinct tumor imaging in vivo. Adv Drug Deliv Rev 65:71–79
5. Yin H, Kanasty RL, Eltoukhy AA, Vegas AJ, Dorkin JR, Anderson DG (2014) Non-viral vectors for gene-based therapy. Nat Rev Genet 15:541–555
6. Sakurai Y, Hatakeyama H, Sato Y, Hyodo M, Akita H, Ohga N, Hida K, Harashima H (2014) RNAi-mediated gene knockdown and anti-angiogenic therapy of RCCs using a cyclic RGD-modified liposomal-siRNA system. J Control Release 173:110–118
7. Akinc A, Querbes W, De SM, Qin J, Frank-Kamenetsky M, Jayaprakash KN, Jayaraman M, Rajeev KG, Cantley WL, Dorkin JR, Butler JS, Qin LL, Racie T, Sprague A, Fava E, Zeigerer A, Hope MJ, Zerial M, Sah DWY, Fitzgerald K, Tracy MA, Manoharan M, Koteliansky V, de Fougerolles A, Maier MA (2010) Targeted delivery of RNAi therapeutics with endogenous and exogenous ligand-based mechanisms. Mol Ther 18:1357–1364
8. Wang M, Lowik DWPM, Miller AD, Thanou M (2009) Targeting the urokinase plasminogen activator receptor with synthetic self-assembly nanoparticles. Bioconjug Chem 20:32–40
9. Gray BP, Li SZ, Brown KC (2013) From phage display to nanoparticle delivery: functionalizing liposomes with multivalent peptides improves targeting to a cancer biomarker. Bioconjug Chem 24:85–96
10. Sato Y, Hatakeyama H, Sakurai Y, Hyodo M, Akita H, Harashima H (2012) A pH-sensitive cationic lipid facilitates the delivery of liposomal siRNA and gene silencing activity in vitro and in vivo. J Control Release 163:267–276

Chapter 7

A Multifunctional Envelope-Type Nano Device Containing a pH-Sensitive Cationic Lipid for Efficient Delivery of Short Interfering RNA to Hepatocytes In Vivo

Yusuke Sato, Hideyoshi Harashima, and Michinori Kohara

Abstract

Various types of nanoparticles have been developed with the intent of efficiently delivering short interfering RNA (siRNA) to hepatocytes to date. To achieve efficient SiRNA delivery, various aspects of the delivery processes and physical properties need to be considered. We recently developed an original lipid nanoparticle, a multifunctional envelope-type nano device (MEND) containing YSK05, a pH-sensitive cationic lipid (YSK05-MEND). The YSK05-MEND with SiRNA in its formulation showed hepatocyte-specific uptake and robust gene silencing in hepatocytes after intravenous administration. Here, we describe the procedure used in the preparation and characterization method of the YSK05-MEND.

Key words Multifunctional envelope-type nano device, pH-sensitive cationic lipid, YSK05, Delivery, Hepatocytes, Short interfering RNA, Gene silencing, RNA interference

1 Introduction

Since the discovery of small interfering RNA (SiRNA) [1], many researchers have attempted to develop SiRNA delivery carriers in order to realize RNA interference (RNAi) medicines. Given the fact that liver genetic disorders are responsible for over 4000 human diseases [2], many types of SiRNA carriers for hepatocytes have been developed. However, only several carriers have succeeded in near complete gene silencing using a single dose [3–7]. Lipid nanoparticles with pH-sensitive cationic properties have been well characterized and appear to be the most potent for use in conjunction with hepatocytes. The lipid nanoparticles are electrostatically near neutral in the blood stream and are taken up by hepatocytes, converted to a cationic form in endosomes, an acidic vesicular compartment, and then fused with the endosomal membrane, resulting in successful cytosolic SiRNA delivery. To achieve these processes, the fusogenic activity of the pH-sensitive cationic

Kato Shum and John Rossi (eds.), *SiRNA Delivery Methods: Methods and Protocols*, Methods in Molecular Biology, vol. 1364, DOI 10.1007/978-1-4939-3112-5_7, © Springer Science+Business Media New York 2016

lipid should be high [7]. The acid dissociation constant (pKa) of the nanoparticle should be carefully adjusted to 6.2–6.5 [8]. The particle size should be controlled less than 100 nm [9]. And the stability of the particle in the blood stream should be guaranteed.

We recently synthesized a fusogenic and pH-sensitive cationic lipid, YSK05 [7, 10–12]. The pKa of YSK05 is approximately 6.5, which is within the optimal range for delivering SiRNA to hepatocytes. We also optimized the procedure for the preparation of the YSK05-MEND, in which we refer to as the *tert*-BuOH injection method. YSK05-MENDs can be prepared on a small scale (160 μg SiRNA/batch) with high reproducibility. Typically, the mean diameter (number-weighted) of the YSK05-MEND can be controlled from 50 to 60 nm, as evidenced by dynamic light scattering analysis, which is the optimal size range for hepatocyte-targeting.

2 Materials

Prepare all solutions using ultrapure water (by purifying deionized water to attain a sensitivity of 18 MΩ cm at 25 °C) and analytical reagents.

2.1 Formation of the MEND

1. 90 % *tert*-BuOH: Mix *tert*-BuOH and water at a 9:1 volume ratio (*see* **Note 1**). Store at room temperature.
2. 10 mM YSK05: Weigh 62.8 mg of YSK05 in a 15 mL conical tube and dissolve it in 10 mL of 90 % *tert*-BuOH. Store at −20 °C (Fig. 1).

YSK05

Cholesterol

$_{m}$PEG$_{2k}$-DMG

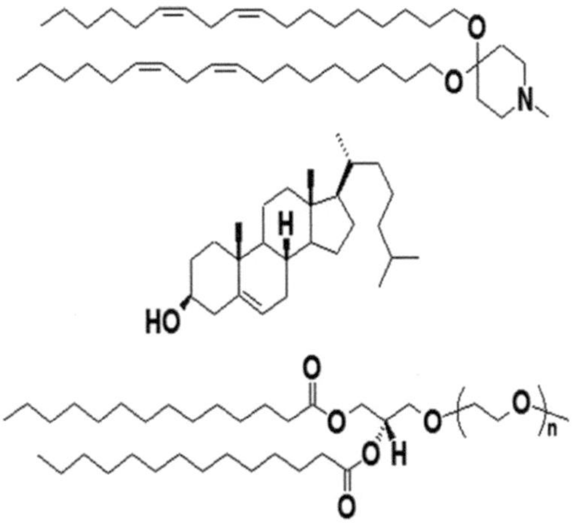

siFVII

Sense strand: 5'-**GGA ucA ucu cAA Guc uuA cT*T**-3'
Anti-sense strand: 5'-**GuA AGA cuu GAG AuG Auc cT*T**-3'
(capital letter: natural, small letter: 2'-F, *: phosphorothioate linkage)

Fig. 1 Structures of lipids and a sequence of siRNA

3. 10 mM Cholesterol: Weigh 38.7 mg of cholesterol in a 15 mL conical tube and dissolve it in 10 mL of 90 % *tert*-BuOH (*see* **Note 2**). Store at −20 °C (Fig. 1).
4. 2 mM 1,2-dimyristoyl-sn-glycerol methoxypolyethyleneglycol 2000 ($_{m}PEG_{2k}$-DMG): Weigh 50.0 mg of $_{m}PEG_{2k}$-DMG in a 15 mL conical tube and dissolve it in 10 mL of 90 % *tert-BuOH* (*see* **Note 3**). Store at −20 °C (Fig. 1).
5. 0.8 mg/mL SiRNA: Dissolve lyophilized SiRNA (HPLC glade) in water to become 0.8 mg/mL. Store at −20 °C (*see* **Note 4**) (Fig. 1).
6. 20 mM citrate buffer: Add about 90 mL of water to a glass beaker. Weigh 384.2 mg of citric acid and transfer it to the glass beaker. Mix and adjust the pH to 4.0 with sodium hydroxide. Make up to 100 mL with water. Pass through a 0.2 μm syringe filter. Store at 4 °C.
7. PBS (−): Dissolve 5 PBS (−) tablets in 500 mL water. Pass through a 0.2 μm syringe filter. Store at 4 °C.
8. 25G needle.
9. 1 mL syringe.
10. Amicon Ultra-15 centrifugal filter unit (MWCO 100 kDa).

2.2 Characterization of the MEND

1. 10 mM HEPES buffer: Add about 90 mL water to a glass beaker. Weigh 238.3 mg of HEPES and transfer it to the glass beaker. Mix and adjust the pH to 7.4 with NaOH. Make up to 100 mL with water. Pass through a 0.2 μm syringe filter. Store at 4 °C.

2.3 RiboGreen Assay

1. 1 mg/mL sodium dextran sulfate: Dissolve 10 mg of sodium dextran sulfate in 10 mL of 10 mM HEPES buffer. Store at 4 °C.
2. 10 % Triton X-100: Dissolve 1 g of Triton X-100 in 9 mL of water. Store at room temperature.
3. RiboGreen. Store at 4 °C.
4. 96-well plate (black).

3 Methods

Carry out all procedures at room temperature unless otherwise specified.

3.1 MEND Formulation

1. Mix 210 μL of 10 mM YSK05, 90 μL of 10 mM cholesterol, 45 μL of 2 mM $_{m}PEG_{2k}$-DMG and 55 μL of 90 % *tert*-BuOH in a 5 mL tube.

2. Add 200 μL of 0.8 mg/mL SiRNA dropwise to the mixed lipid solution under vigorous mixing using a vortex mixer. Continue the vigorous mixing for 5 s after adding the solutions (*see* **Note 5**).
3. Collect the SiRNA–lipid mixed solution using a 1 mL syringe with a 25G needle.
4. Inject the SiRNA–lipid mixed solution into 2 mL of 20 mM citrate buffer in a 15 mL conical tube under vigorous mixing for 3–5 s (*see* **Note 6**). Maintain the vigorous mixing for 5 s after the injection is finished to create a completely homogeneous solution (*see* **Note 7**).
5. Add 3.5 mL of PBS (−) to the MEND solution under vigorous mixing (*see* **Note 8**). Maintain the vigorous mixing for 5 s to create a completely homogenous.
6. Transfer the diluted MEND solution to an Amicon ultra-15 centrifugal filter unit in which 7 mL of PBS (−) has been added. Wash a 15 mL conical tube with 3.5 mL of PBS (−) and mix with the MEND solution in the Amicon.
7. Centrifuge the Amicon at 1000 × *g* for 21 min (*see* **Note 9**).
8. Discard the filtered solution and dilute the concentrated MEND solution with 10 mL of PBS (−).
9. Centrifuge the Amicon at 1000 × *g* for 23 min (*see* **Note 9**).
10. Collect the concentrated MEND solution. Rinse the filters of the Amicon with an appropriate volume (300–500 μL) of PBS (−) by pipetting and combine with the MEND solution (*see* **Note 10**).
11. Make up MEND solution to 1 mL with PBS (−). Store at 4 °C.

3.2 Characterization of the MENDs

1. Add 1 μL of a MEND solution to 49 μL of PBS (−) in a 1.7 mL tube.
2. Add 50 μL of the diluted MEND solution to a cuvette (Marvern Instruments, UK). Place the cuvette on Zetasizer Nano ZS ZEN3600 (Marvern Instruments, UK) and start the process to measure particle size by dynamic light scattering (Fig. 2, Table 1).
3. Add 2 μL of the MEND solution to 748 μL of 10 mM HEPES buffer in a 1.7 mL tube.
4. Put the 750 μL of the diluted MEND solution into a disposable cuvette (Marvern Instruments, UK) (*see* **Note 11**). Place the cuvette in a Zetasizer Nano ZS ZEN3600 (Marvern Instruments, UK) and start the process for measuring the ζ-potential by electrophoretic light scattering (Fig. 2, Table 1).

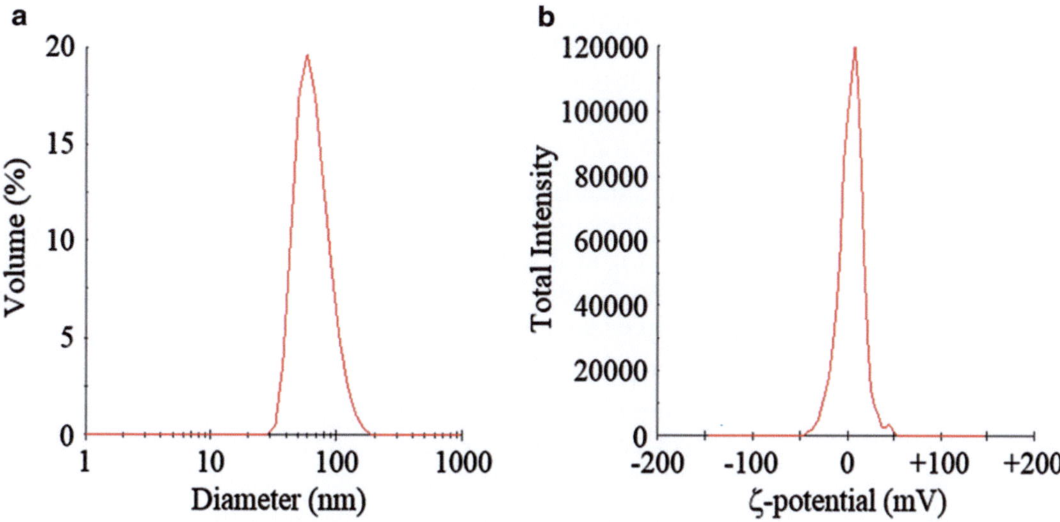

Fig. 2 Typical diameter (Volume-weighted) (**a**) and ζ-potential (**b**) distribution if the YSK05-MEND

Table 1
The summery of the typical characteristics of the YSK05-MEND

Polydispersity index	0.081
Intensity-weighted mean diameter	83.2 nm
Volume-weighted mean diameter	67.1 nm
Number-weighted mean diameter	55.0 nm
ζ-potential	+3.32 mV
siRNA encapsulation ratio	98.7 %
Total siRNA concentration	135.6 μg/mL

3.3 RiboGreen Assay

1. Preparation of standard curve samples. Add 1.00 μL of a 0.8 mg/mL solution of SiRNA to 799 μL of 10 mM HEPES buffer to give a solution containing 1000 ng SiRNA/mL in a 1. 7 mL tube. Then, add 300 μL of the 1000 ng/mL SiRNA solution and HEPES buffer to make a 500 ng/mL SiRNA solution. Following the above procedure, make up the following solutions: 300 μL of 250, 125, and 62.5 ng/mL SiRNA solution in a 1.7 mL tube. Add 300 μL of HEPES buffer to a 1.7 mL tube for use as a blank sample (0 ng/mL SiRNA solution).
2. Add 5.00 μL of the MEND solution to 995 μL of 10 mM HEPES buffer in a 1.7 mL tube. (The theoretical SiRNA concentration should be 800 ng/mL, assuming that the SiRNA recovery rate is 100 %.)
3. Prepare two kinds of RiboGreen solution. For the RiboGreen solution (Triton+), mix 40 μL of a 1 mg/mL solution of

sodium dextran sulfate, 20 μL of 10 % Triton X-100, 2.5 μL of RiboGreen and 937.5 μL of HEPES buffer in a 1.7 mL tube. For the RiboGreen solution (Triton−), mix 40 μL of a 1 mg/mL solution of sodium dextran sulfate, 2.5 μL of RiboGreen and 957.5 μL of HEPES buffer in a 1.7 mL tube.

4. Add the standard curve samples and measurement samples to a 96-well plate at a volume of 100 μL per well (For each sample, add it in 2 wells.). Then, add 100 μL of the RiboGreen solution (Triton+) to 1 well of each sample and 100 μL of the RiboGreen solution (Triton−) to another well of each sample.
5. Shake the plate for about 20 s using a shaker at 800 rpm.
6. Measure fluorescent intensity derived from RiboGreen using a fluorescent microplate leader, Varioskan Flash. (excitation wave length: 500 nm, emission wave length: 525 nm)
7. Calculation of the total SiRNA concentration and SiRNA encapsulation ratio. Convert the unit of the measurement sample from fluorescent intensity to μg SiRNA/mL using the SiRNA standard curve. The values in the Triton+ and Triton− sample indicate the total SiRNA concentration and the total unencapsulated SiRNA concentration, respectively. Calculate the SiRNA encapsulation ratio using a following formula,

$$\text{siRNA encapsulation ratio}\,(\%) = 100 \times \left(1 - \frac{\text{Unencapsulated siRNA concentration}}{\text{Total siRNA concentration}}\right)$$

(Table 1).

4 Notes

1. Pure *tert*-BuOH can form a solid at room temperature (melting point: 25–26 °C). To circumvent this, warm up pure *tert-BuOH* to 40 °C before use.
2. Cholesterol tends to form aggregates in 90 % *tert*-BuOH. Sonicate the aggregate using a bath-type sonicator.
3. The average molecular weight of $_{m}$PEG$_{2k}$-DMG can vary with each lot. Therefore, calculate the appropriate weight needed in each lot. The value described in the text (25.0 mg) is accurate when the average molecular weight of $_{m}$PEG$_{2k}$-DMG is 2500.
4. Wear a mask and gloves when handling SiRNA. Dispense 200 μL of an 0.8 mg/mL SiRNA solution to each 1.7 mL tube in order to avoid freeze/thaw cycles.
5. Confirm that the resulting solution is optically clear and not milky. There are two cases for the solution to become milky.

The first is contamination by salt in the SiRNA solution. The presence of salt causes the precipitation of SiRNA in an alcoholic solution analogous to the ethanol precipitation procedure. In this case, add a small volume (10–50 μL) of DDW to the SiRNA–lipid mixed solution under vigorous mixing until the solution becomes optically clear. The second cause is due to unexpected SiRNA–lipid interactions. In this case, add a small volume (10–100 μL) of 90 % *tert*-BuOH to the SiRNA–lipid mixed solution under vigorous mixing until is becomes optically clear.

6. Put the edge of the needle on the inside wall of a 15 mL tube (bottom than half of the citrate buffer) in order to promote mixing efficiency. Avoid allowing the syringe to touch the inside wall of the 15 mL tube in order assure that mixing occurs.
7. As the viscosity of *tert*-BuOH is high, additional mixing for several seconds is needed for the solution to reach homogeneity. Heterogeneous mixing causes the YSK05-MEND to develop a heterogeneous size distribution.
8. To prevent conditions of both a high alcohol concentration and a neutral pH, which causes the YSK05-MEND to become unstable, rapidly add PBS (−) to the MEND solution under vigorous mixing.
9. Confirm that the volume of the MEND solution is 500 μL or less. When the volume exceeds 500 μL, centrifuge again for an appropriate period of time at 1000×*g*.
10. A portion of the MEND particles adsorb weakly to filters of the Amicon. Recover them by rinsing.
11. Confirm that no air babbles are present in the cuvette.

Acknowledgement

This work was supported in parts by grants from the Special Education and Research Expenses of the Ministry of Education, Culture, Sports, Science and Technology (MEXT) of Japan, and a Grants-in-Aid for Research Activity Start-up from the Japan Society for Promotion of Science (JSPS). The authors also wish to thank Dr. Milton S. Feather for his helpful advice in writing the English manuscript.

References

1. Elbashir SM, Harborth J, Lendeckel W, Yalcin A, Weber K, Tuschl T (2001) Duplexes of 21-nucleotide RNAs mediate RNA interference in cultured mammalian cells. Nature 411:494–498
2. McClellan J, King MC (2010) Genetic heterogeneity in human disease. Cell 141:210–217
3. Adami RC, Seth S, Harvie P, Johns R, Fam R, Fosnaugh K, Zhu TY, Farber K, McCutcheon M, Goodman TT, Liu Y, Chen Y, Kwang E,

Templin MV, Severson G, Brown T, Vaish N, Chen F, Charmley P, Polisky B, Houston ME (2011) An amino acid-based amphoteric liposomal delivery system for systemic administration of siRNA. Mol Ther 19:1141–1151

4. Semple SC, Akinc A, Chen JX, Sandhu AP, Mui BL, Cho CK, Sah DWY, Stebbing D, Crosley EJ, Yaworski E, Hafez IM, Dorkin JR, Qin J, Lam K, Rajeev KG, Wong KF, Jeffs LB, Nechev L, Eisenhardt ML, Jayaraman M, Kazem M, Maier MA, Srinivasulu M, Weinstein MJ, Chen QM, Alvarez R, Barros SA, De S, Klimuk SK, Borland T, Kosovrasti V, Cantley WL, Tam YK, Manoharan M, Ciufolini MA, Tracy MA, de Fougerolles A, MacLachlan I, Cullis PR, Madden TD, Hope MJ (2010) Rational design of cationic lipids for siRNA delivery. Nat Biotechnol 28:172–176
5. Love KT, Mahon KP, Levins CG, Whitehead KA, Querbes W, Dorkin JR, Qin J, Cantley W, Qin LL, Racie T, Frank-Kamenetsky M, Yip KN, Alvarez R, Sah DWY, de Fougerolles A, Fitzgerald K, Koteliansky V, Akinc A, Langer R, Anderson DG (2010) Lipid-like materials for low-dose, in vivo gene silencing. Proc Natl Acad Sci U S A 107:1864–1869
6. Wooddell CI, Rozema DB, Hossbach M, John M, Hamilton HL, Chu QL, Hegge JO, Klein JJ, Wakefield DH, Oropeza CE, Deckert J, Roehl I, Jahn-Hofmann K, Hadwiger P, Vornlocher HP, McLachlan A, Lewis DL (2013) Hepatocyte-targeted RNAi therapeutics for the treatment of chronic hepatitis B virus infection. Mol Ther 21:973–985
7. Sato Y, Hatakeyama H, Sakurai Y, Hyodo M, Akita H, Harashima H (2012) A pH-sensitive cationic lipid facilitates the delivery of liposomal siRNA and gene silencing activity in vitro and in vivo. J Control Release 163:267–276
8. Jayaraman M, Ansell SM, Mui BL, Tam YK, Chen JX, Du XY, Butler D, Eltepu L, Matsuda S, Narayanannair JK, Rajeev KG, Hafez IM, Akinc A, Maier MA, Tracy MA, Cullis PR, Madden TD, Manoharan M, Hope MJ (2012) Maximizing the potency of siRNA lipid nanoparticles for hepatic gene silencing in vivo. Angew Chem Int Ed Engl 51:8529–8533
9. Jacobs F, Wisse E, De Geest B (2010) The role of liver sinusoidal cells in hepatocyte-directed gene transfer. Am J Pathol 176:14–21
10. Watanabe T, Hatakeyama H, Matsuda-Yasui C, Sato Y, Sudoh M, Takagi A, Hirata Y, Ohtsuki T, Arai M, Inoue K, Harashima H, Kohara M (2014) In vivo therapeutic potential of Dicer-hunting siRNAs targeting infectious hepatitis C virus. Sci Rep 4
11. Hatakeyama H, Murata M, Sato Y, Takahashi M, Minakawa N, Matsuda A, Harashima H (2014) The systemic administration of an anti-miRNA oligonucleotide encapsulated pH-sensitive liposome results in reduced level of hepatic microRNA-122 in mice. J Control Release 173:43–50
12. Hayashi Y, Suemitsu E, Kajimoto K, Sato Y, Akhter A, Sakurai Y, Hatakeyama H, Hyodo M, Kaji N, Baba Y, Harashima H (2014) Hepatic monoacylglycerol O-acyltransferase 1 as a promising therapeutic target for steatosis, obesity, and type 2 diabetes. Mol Ther Nucleic Acids 3:e154

Chapter 8

Bioreducible Poly(Beta-Amino Ester)s for Intracellular Delivery of SiRNA

Kristen L. Kozielski and Jordan J. Green

Abstract

RNA interference (RNAi) is a powerful tool to target and knockdown gene expression in a sequence-specific manner. RNAi can be achieved by the intracellular introduction of SiRNA; however, intracellular SiRNA delivery remains a challenging obstacle. Herein we describe the use of bioreducible nanoparticles formed using poly(beta-amino ester)s (PBAEs) for safe and efficient SiRNA delivery. Methods for polymer synthesis, nanoparticle formation, and SiRNA delivery using these particles are described. A template protocol for nanoparticle screening is presented and can be used to quickly optimize SiRNA delivery for novel applications.

Key words Polymer, Nanoparticle, SiRNA, RNA interference, Gene delivery, Transfection

1 Introduction

Sequence-specific gene knockdown via RNA interference (RNAi) has the potential to treat or cure many diseases caused by overexpression or dysregulation of a given gene [1]. RNAi can be induced by intracellular delivery of small interfering RNA (SiRNA) complementary to the targeted gene's mRNA. However, SiRNA delivery remains a challenge [2, 3]. Viral delivery methods, although effective, raise several safety concerns due to their inherent immunogenicity and tumorigenicity [4]. Nonviral methods, such as lipid-based, inorganic, and polymeric nanoparticles, typically can avoid such issues but are often less effective at SiRNA delivery [5].

Poly(beta-amino ester)s (PBAEs), a class of cationic polymer, are well-established as safe and effective DNA delivery vectors [6–9]. PBAEs are promising potential candidates for SiRNA delivery, as they can bind negatively charged nucleic acids into nanoparticles, promote cellular uptake, and achieve endosomal escape due to their positive charge and high buffering capacities. However, SiRNA with 21–23 bp is much shorter than the average plasmid used for DNA delivery (typically 4–10 kpb), and therefore has

Kato Shum and John Rossi (eds.), *SiRNA Delivery Methods: Methods and Protocols*, Methods in Molecular Biology, vol. 1364, DOI 10.1007/978-1-4939-3112-5_8,

much less multivalency to electrostatically bind to cationic polymers. This is one reason why cationic polymers may be unable to stably form nanoparticles with SiRNA [10, 11]. Additionally, SiRNA must be released at its site of action in the cytoplasm in order for effective RNAi to take place [12].

Bioreducible PBAEs have previously been found to overcome these challenges and to successfully deliver SiRNA in human glioblastoma cells [13, 14]. Methods to deliver SiRNA to new cell types are fundamental to the development of safer and more efficient SiRNA delivery vehicles. Investigating the SiRNA delivery capabilities of these polymers in untested cell types is important to discovering their potential for the treatment of various diseases. This chapter discusses a method to synthesize bioreducible PBAEs and assess their ability to safely and effectively deliver SiRNA to a new cell type.

2 Materials

2.1 Bioreducible Monomer

1. Bis(2-hydroxyethyl) disulfide.
2. Triethylamine (TEA).
3. Acryloyl chloride.
4. Tetrahydrofuran (THF), anhydrous.
5. Dichloromethane (DCM).
6. Sodium carbonate (Na_2CO_3).
7. Sodium sulfate (Na_2SO_4).
8. Coarse porosity filter paper.

2.2 Polymer Synthesis

1. 4-amino-1-butanol (S4).
2. 2-(3-(aminopropyl)amino)methanol (E6).
3. 1-(3-aminopropyl)-4-methylpiperazine (E7).
4. Teflon-lined screw cap glass vials, 5 mL.
5. Teflon-coated magnetic micro stir bars.
6. Incubator/oven capable of reaching 95 °C.

2.3 Cell Culture

1. Cell line positive for green fluorescent protein (GFP) or other fluorescent protein (*see* **Note 1**). Example cells in experiments described below will be human glioblastoma cells isolated from intraoperative samples [15], and transfected to constitutively express GFP [9].
2. Cell culture medium: DMEM-F12 containing 10 % fetal bovine serum (FBS) and 1 % antibiotic-antimycotic.
3. Trypsin.

4. Phosphate buffered saline (PBS).
5. Hemocytometer.

2.4 siRNA Delivery Experiment

1. SiRNA delivery polymers from synthesis in Subheading 2.2.
2. Dimethyl sulfoxide (DMSO), >99.7 %, anhydrous.
3. 3 M Sodium acetate (NaAc), pH 5.2, sterile.
4. AllStars Hs Cell Death SiRNA.
5. *Silencer*® eGFP SiRNA.
6. *Silencer*® Negative Control No. 1 SiRNA.
7. Clear 96-well plate with lid, flat bottom, tissue culture treated, sterile.
8. Clear 96-well plate with lid, round bottom, sterile.
9. Fluorescence plate reader.
10. Fluorescence microscope.
11. Flow cytometer, preferably with a HyperCyt® Autosampler.

3 Methods

3.1 Bioreducible Monomer Synthesis

Researchers should be familiar with basic organic chemistry laboratory techniques such as separation of organic and aqueous phases, drying organic solutions, and completing a reaction in anhydrous conditions. All work should be conducted in a chemical fume hood, and appropriate safety precautions should be followed for each chemical used.

1. Fill 1000 mL three-neck round bottom flask with 450 mL THF, and fill a 100 mL graduated cylinder with 50 mL THF. Clamp round bottom flask to ring stand.
2. Dissolve 12.2 mL hydroxyethyl disulfide and 37.5 mL TEA in THF in round bottom flask. Add a stir bar and begin to magnetically stir.
3. Cover all outlets of round bottom flask with rubber septa and flush with nitrogen for 5 min by attaching the nitrogen tank to the left-most outlet with a needle and inserting an outlet needle to the right-most outlet. Leave nitrogen flowing and cover round bottom flask with aluminum foil.
4. Dissolve 24.4 mL acryloyl chloride in 50 mL THF in graduated cylinder.
5. Clamp glass syringe (without plunger) with needle onto the ring stand and lower so that needle goes through the top rubber septum on the round bottom flask.
6. Fill syringe with acryloyl chloride/THF solution to allow solution to add to round bottom flask dropwise. Continue adding

until all acryloyl chloride is added. A pale yellow precipitate will form during the reaction; this is TEA HCl.

7. Remove glass syringe from top septum and allow nitrogen to flow for an additional 10 min.
8. Turn nitrogen off and remove needle. Leave an outlet needle in one septum attached to a balloon. This will keep air from entering the reaction but will allow for any slight pressure changes during the reaction.
9. Allow reaction to continue while stirring overnight (at least 12 h).
10. Use a glass funnel and filter paper to remove the TEA HCl, and wash the precipitate with an additional 100 mL THF.
11. Move solution to a 1000 mL round bottom flask. Remove THF via rotary evaporation. You will be left with a yellow, viscous liquid.
12. Dissolve liquid in 200 mL DCM and move to separatory funnel.
13. Wash solution with 200 mL of 0.2 M Na_2CO_3, mix in separatory funnel, move organic phase (the lower layer) to a round bottom flask. The aqueous phase (upper layer) may be discarded.
14. Repeat Na_2CO_3 washes four additional times, followed by three washes in deionized water.
15. Weigh a clean, dry 500 mL round bottom flask and record its weight.
16. Following the final separation, use Na_2SO_4 to dry the contents of the organic phase. Use a glass funnel and filter paper to remove the Na_2SO_4.
17. Move the organic phase into the clean round bottom flask, and remove the DCM using rotary evaporation. Weigh the flask to determine the product yield.
18. The final product, disulfanediylbis(ethane-2,1-diyl) (BR6), will be a yellow, viscous fluid and should yield approximately 10 g. The authors recommend using H^1-NMR to confirm the purity of BR6, in particular checking for any remaining solvent and acrylic acid.

3.2 Polymer Synthesis

1. Weigh 800 mg of monomer DMSO (or other diacrylate monomer) synthesized in Subheading 3.1, into a 5 mL screw cap glass vial.
2. Add 1059 mg DMSO to BR6 and vortex to dissolve.
3. Add amine monomer to acrylate monomer at an acrylate/amine stoichiometric ratio of 1.05:1; this will yield acrylate-terminated base polymers. To make polymers BR6-S4-E6

(R646) or BR6-S4-E7 (R647), add 259 mg of monomer S4 to 800 mg BR6.

4. Add a teflon-coated magnetic stir bar to the vial.
5. Polymerize while magnetically stirring in an oven at 90 °C for 24 h.
6. To end-cap the polymers, dilute base polymers to 166.7 mg/mL in DMSO, dilute end-caps to 0.5 M, and combine 480 μL polymer with 320 μL end-cap. This will yield a final polymer concentration of 100 mg/mL. To make polymer R647, for example, dilute base polymer BR6-S4 to 166.7 mg/mL with 480 μL total volume, and dilute 25.2 μL E7 in 320 μL DMSO, then combine.
7. Shake at room temperature for 1 h.
8. Store polymers at −20 °C in the presence of a desiccant until ready to use.

3.3 siRNA Delivery Experiment

Researchers should be familiar with basic sterile cell culture techniques such as growing, passing, and plating cells. All work should be completed in a laminar flow biosafety cabinet using sterilized reagents and equipment. The following protocol is for testing SiRNA formulations in a 96-well plate. Each nanoparticle formulation is tested using both the targeting SiRNA and a scrambled control RNA (scRNA), each in quadruplicate. The following example experiment will test two polymers, R646 and R647, each at two different concentrations (300 and 150 μg/mL), and each with an SiRNA dose of 90 and 45 nM (*see* **Notes 1–3**).

1. Twenty-four hours prior to transfection, seed cells into a clear tissue culture 96-well plate at 15,000 cells/well in a volume of 100 μL per well. Leave Rows E–H, column 1 free of cells and filled only with 100 μL of media.
2. Dilute 3 M sodium acetate to 25 mM by adding 4.2 mL of 3 M sodium acetate into 495.8 mL sterile deionized water.
3. Thaw SiRNA and scRNA stock solutions (each stored at 50 μM). Dilute each using 7.1 μL stock RNA in 322.9 μL sodium acetate to make a 1080 nM RNA dilution. (This will give you a final concentration of 90 nM once the RNA is diluted by half during nanoparticle formation, and then by six once the particles are added to cells.) To make the 540 nM dilution, dilute 110 μL of 1080 nM RNA in 110 μL sodium acetate. Aliquot 50 μL of each RNA dilution into a round-bottom 96-well Master Plate following Fig. 1.
4. Thaw polymers R646 and R647 made in Subheading 3.2 (each stored at 100 mg/mL). Dilute each using 11.9 μL stock polymer in 318.1 μL sodium acetate to make a 3600 μg/mL

Master Plate

	1	2	3	4	5	6	7	8	9	10	11	12
A		R646 300 μg/mL	R646 150 μg/mL	R647 300 μg/mL	R647 150 μg/mL							
B												
C		siRNA 90 nM	siRNA 90 nM	siRNA 90 nM	siRNA 90 nM	siRNA 45 nM	siRNA 45 nM	siRNA 45 nM	siRNA 45 nM			
D		scRNA 90 nM	scRNA 90 nM	scRNA 90 nM	scRNA 90 nM	scRNA 45 nM	scRNA 45 nM	scRNA 45 nM	scRNA 45 nM			
E												
F												
G												
H												

Cell Plate

	1	2	3	4	5	6	7	8	9	10	11	12
A	untreated	R646 300 μg/mL	R646 150 μg/mL	R647 300 μg/mL	R647 150 μg/mL	R646 300 μg/mL	R646 150 μg/mL	R647 300 μg/mL	R647 150 μg/mL	untreated	untreated	untreated
B	untreated	R646 300 μg/mL	R646 150 μg/mL	R647 300 μg/mL	R647 150 μg/mL	R646 300 μg/mL	R646 150 μg/mL	R647 300 μg/mL	R647 150 μg/mL	untreated	untreated	untreated
C	untreated	R646 300 μg/mL	R646 150 μg/mL	R647 300 μg/mL	R647 150 μg/mL	R646 300 μg/mL	R646 150 μg/mL	R647 300 μg/mL	R647 150 μg/mL	untreated	untreated	untreated
D	untreated	R646 300 μg/mL	R646 150 μg/mL	R647 300 μg/mL	R647 150 μg/mL	R646 300 μg/mL	R646 150 μg/mL	R647 300 μg/mL	R647 150 μg/mL	untreated	untreated	untreated
E	No cells	R646 300 μg/mL	R646 150 μg/mL	R647 300 μg/mL	R647 150 μg/mL	R646 300 μg/mL	R646 150 μg/mL	R647 300 μg/mL	R647 150 μg/mL	untreated	untreated	untreated
F	No cells	R646 300 μg/mL	R646 150 μg/mL	R647 300 μg/mL	R647 150 μg/mL	R646 300 μg/mL	R646 150 μg/mL	R647 300 μg/mL	R647 150 μg/mL	untreated	untreated	untreated
G	No cells	R646 300 μg/mL	R646 150 μg/mL	R647 300 μg/mL	R647 150 μg/mL	R646 300 μg/mL	R646 150 μg/mL	R647 300 μg/mL	R647 150 μg/mL	untreated	untreated	untreated
H	No cells	R646 300 μg/mL	R646 150 μg/mL	R647 300 μg/mL	R647 150 μg/mL	R646 300 μg/mL	R646 150 μg/mL	R647 300 μg/mL	R647 150 μg/mL	untreated	untreated	untreated

Fig. 1 High throughput siRNA delivery experiment plate layout

dilution (which will yield a final polymer concentration of 300 μg/mL). Dilute 110 μL of the 3600 μg/mL dilution in 110 μL NaAc to make a 1800 μg/mL dilution. Aliquot 105 μL of each polymer to the Master Plate following Fig. 1.

5. Using the multichannel pipette, move 50 μL of polymer from Row A of the Master Plate to the RNA in Row C; pipette vigorously several times. Repeat with RNA in Row D. Allow nanoparticles to form for 10 min.
6. Prior to adding nanoparticles, use a fluorescence plate reader to measure the pre-transfection fluorescence of the cells in each well. Remove the media and add 100 μL of serum-free media. Following Fig. 1 and using a multichannel pipette, add 20 μL of nanoparticles in Row C of the Master Plate to Rows A–D of the Cell Plate. Add 20 μL of nanoparticles in Row D of the Master Plate to Rows E–H of the Cell Plate. Gently shake the Cell Plate on a flat surface to promote mixing. Place cells in a cell culture incubator for the duration of the incubation.
7. Following 1–4 h incubation, remove the nanoparticle-containing media using a multichannel pipette. The incubation time should be optimized for any new cell type used. Replenish cells with 100 μL of fresh, previously warmed media, and replace cells in the incubator.

3.4 Measurement of Gene Knockdown

The following protocol requires cells that are constitutively GFP positive and a flow cytometer equipped with a Hypercyt® Autosampler or other autosampler. If one is not available, the researcher may follow the protocol as written, but will be required to prepare the samples in the traditional manner for flow cytometry, in which individual samples are moved into tubes. The researcher should be familiar with flow cytometry and analysis of flow cytometric data.

1. Using the fluorescence plate reader, the GFP knockdown should be tracked each day following transfection until the peak day can be determined. Percent knockdown (%KD) can be determined by first averaging and then subtracting the GFP fluorescence readings of wells containing media only from all samples. Using these new values, the average of SiRNA-treated and scRNA-treated wells should be calculated, and %KD can be found by subtracting the SiRNA-treated values from scRNA-treated values, normalizing to scRNA-treated values, then multiplying by 100.
2. Repeat the SiRNA knockdown experiment, and on the previously determined peak day of knockdown, use a fluorescence microscope to visualize GFP expression in each group.
3. Prepare the cells for flow cytometry by removing the media from the wells in a 96-well plate, adding 100 μL PBS per well, removing the PBS, and adding 30 μL trypsin. Wait until the cells have released from the plate, add 170 μL of Flow Buffer (2 % FBS in PBS), and move all 200 μL in each well to the corresponding wells of a round-bottom 96-well plate. Centrifuge the round-bottom plate to spin cells down, then remove 170 μL of Flow Buffer from each well, and resuspend cells by pipetting to increase the cell concentration. Use the flow cytometer and autosampler to measure the FL1 GFP fluorescence levels of the cells in each sample.
4. Cells should be gated for as during normal flow cytometric analysis. FL1 geometric mean fluorescence should be calculated for each well (Fig. 2). Using these values, calculate %KD as with the data acquired from the plate reader, by normalizing SiRNA-treated values to scRNA-treated values.

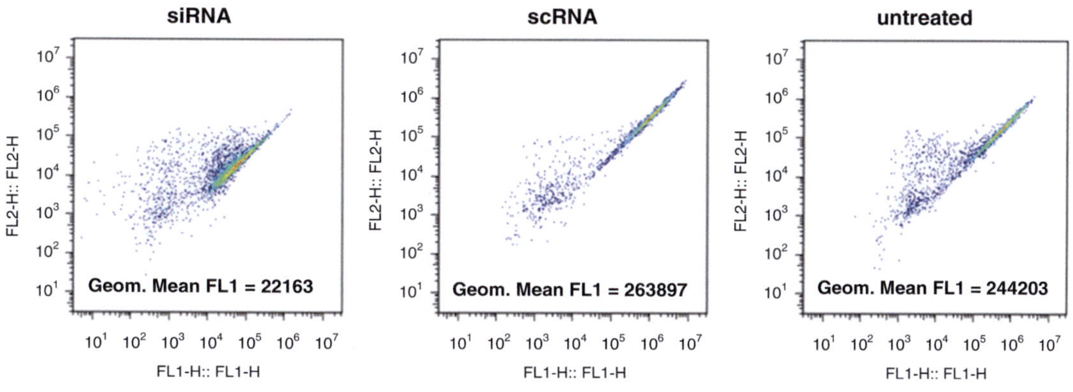

Fig. 2 Example flow cytometry results of GFP+ glioblastoma cells treated with polymer R647 at 180 μg/mL with 20 nM of either GFP siRNA or scRNA

4 Notes

1. The above protocol assumes that the cells used are or can be made to express GFP or another fluorescent protein. If this is difficult or impossible, several other options can be utilized. One option is to instead use a death positive control SiRNA such as AllStars Death Control SiRNA available from Qiagen. If this SiRNA is used, SiRNA delivery efficacy can be determined by comparing the toxicity of SiRNA-treated cells to scRNA-treated cells, either by cell counting or a proliferation assay.
2. If SiRNA delivery efficacy is lower than desired, increasing polymer or SiRNA concentration or increasing incubation time may increase efficacy.
3. If toxicity is higher than desired, lowering SiRNA or polymer concentration, lowering incubation time, or increasing the serum concentration in the media can lower toxicity.

Acknowledgements

This work was supported in part by the NIH (1R01EB016721). K.L.K. also thanks the NIH Cancer Nanotechnology Training Center (R25CA153952) at the JHU Institute for Nanobio technology and the NIH NCI Ruth L. Kirschstein NRSA Predoctoral Fellowship (F31CA196163) for fellowship support.

References

1. Fire A, Xu SQ, Montgomery MK, Kostas SA, Driver SE, Mello CC (1998) Potent and specific genetic interference by double-stranded RNA in Caenorhabditis elegans. Nature 391:806–811
2. Wu W, Sun M, Zou GM, Chen J (2007) MicroRNA and cancer: current status and prospective. Int J Cancer 120:953–960
3. Yadav S, van Vlerken LE, Little SR, Amiji MM (2009) Evaluations of combination MDR-1 gene silencing and paclitaxel administration in biodegradable polymeric nanoparticle formulations to overcome multidrug resistance in cancer cells. Cancer Chemother Pharmacol 63:711–722
4. Thomas CE, Ehrhardt A, Kay MA (2003) Progress and problems with the use of viral vectors for gene therapy. Nat Rev Genet 4:346–358
5. Pack DW, Hoffman AS, Pun S, Stayton PS (2005) Design and development of polymers for gene delivery. Nat Rev Drug Discov 4:581–593
6. Green JJ, Langer R, Anderson DG (2008) A combinatorial polymer library approach yields insight into nonviral gene delivery. Acc Chem Res 41:749–759
7. Lynn DM, Langer R (2000) Degradable poly (beta-amino esters): synthesis, characterization, and self-assembly with plasmid DNA. J Am Chem Soc 122:10761–10768
8. Sunshine JC, Akanda MI, Li D, Kozielski KL, Green JJ (2011) Effects of base polymer hydrophobicity and end-group modification on polymeric gene delivery. Biomacromolecules 12:3592–3600
9. Tzeng SY, Guerrero-Cazares H, Martinez EE, Sunshine JC, Quiñones-Hinojosa A, Green JJ (2011) Non-viral gene delivery nanoparticles based on Poly (beta-amino esters) for treatment of glioblastoma. Biomaterials 32:5402–5410
10. Hagerman PJ (1997) Flexibility of RNA. Annu Rev Biophys Biomol Struct 26:139–156

11. Kebbekus P, Draper DE, Hagerman P (1995) Persistence length of RNA. Biochemistry 34:4354–4357
12. Kawasaki H, Taira K (2003) Short hairpin type of dsRNAs that are controlled by tRNAVal promoter significantly induce RNAi-mediated gene silencing in the cytoplasm of human cells. Nucleic Acids Res 31:700–707
13. Kozielski KL, Tzeng SY, Green JJ (2013) A bioreducible linear poly(beta-amino ester) for siRNA delivery. Chem Commun 49:5319–5321
14. Kozielski KL, Tzeng SY, de Mendoza BAH, Green JJ (2014) Bioreducible cationic polymer-based nanoparticles for efficient and environmentally triggered cytoplasmic siRNA delivery to primary human brain cancer cells. ACS Nano 8:3232–3241
15. Guerrero-Cázares H, Chaichana K, Quiñones-Hinojosa A (2009) Neurosphere culture and human organotypic model to evaluate brain tumor stem cells. In: Yu JS (ed) Cancer stem cells. Humana Press, NY, pp 73–83

Chapter 9

Preparation of Polyion Complex Micelles Using Block Copolymers for SiRNA Delivery

Hyun Jin Kim, Meng Zheng, Kanjiro Miyata, and Kazunori Kataoka

Abstract

Polyion complex (PIC) micelles can be prepared through the spontaneous assembly of cationic block copolymers with oppositely charged short interfering RNAs (SiRNAs). Their core–shell architectures offer a delivery platform for vulnerable SiRNA, improving their biological activities for medicinal applications such as tumor-targeted therapy. Here, we report a protocol for the preparation of SiRNA-loaded PIC micelles using a poly(ethylene glycol)-*block*-poly(aspartamide) derivative, providing the physicochemical criteria for well-defined micellar formulation. In addition, we describe protocols for a stability assay for SiRNA-loaded PIC micelles in the presence of serum using fluorescence correlation spectroscopy and a luciferase assay for cultured cancer cells stably expressing luciferase, thus providing the biological criteria for further medicinal applications.

Key words RNAi, SiRNA delivery, Polyion complex micelle, Block copolymer, Fluorescence correlation spectroscopy

1 Introduction

RNA interference (RNAi), which is a post-transcriptional gene regulation mechanism triggered by small interfering RNA (SiRNA), has attracted much attention as a potential therapeutic technique [1]. However, the therapeutic efficacy of SiRNA is limited because of its rapid nuclease degradation, renal filtration during circulation in the bloodstream, and inefficient cellular uptake [2–4]. To overcome these biological hurdles, block copolymers comprised of neutral and hydrophilic poly(ethylene glycol) (PEG) and polycations have been highlighted as promising delivery vehicles. In aqueous milieu, cationic block copolymers electrostatically bind to oppositely charged SiRNA to form polyion complex (PIC) micelles (Fig. 1). The charge-neutralization between cationic segments and SiRNA sequesters them into PIC cores, and hydrophilic PEG chains align toward the aqueous phase, generating a unique core–shell structure [5]. This core–shell structure protects SiRNA

Kato Shum and John Rossi (eds.), *SiRNA Delivery Methods: Methods and Protocols*, Methods in Molecular Biology, vol. 1364, DOI 10.1007/978-1-4939-3112-5_9,

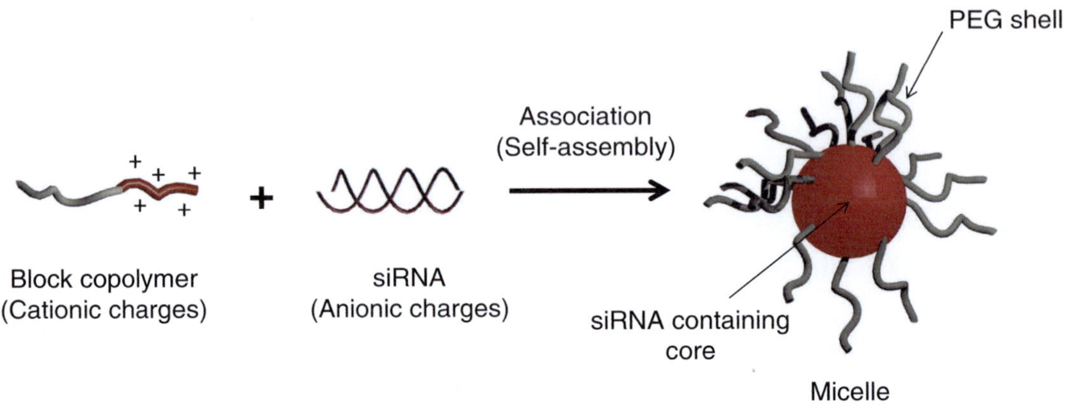

Fig. 1 Schematic illustration of PICm (siRNA) formation

payloads from enzymatic degradations and reduces unfavorable interactions with biomacromolecules, allowing passive tumor targeting through enhanced permeability and retention (EPR) effect [6, 7]; systemically administered PIC micelles can pass across the loosely organized vascular wall surrounding tumor tissues and preferably accumulate in the tumor tissue. Furthermore, PIC micelles can be highly functionalized by introducing (1) core-stabilizing moieties (e.g., cholesterol and stearoyl moieties), (2) targeting ligands (e.g., cyclic RGD peptide), and (3) bioresponsive (or acidic pH-, redox potential-, and ATP concentration-responsive) moieties (e.g., hydrazon, disulfide, and phenylboronic acid moieties) for enhanced biological activities [8–14].

The above functionalities of PIC micelles rely on the controlled self-assembly of block copolymers to have a regulated size of several tens of nanometers and a narrow size distribution. In this regard, the micelle formation can be characterized by static and dynamic light scattering (SLS and DLS) in terms of scattered light intensity (SLI), hydrodynamic size, and size distribution (or polydispersity index (PDI)) [15, 16]. SLI is dominantly influenced by changes in an apparent molecular weight of macromolecules and their self-assemblies and thus can be used as an indicator for self-assembly [17]. In particular, the micelle formation in an oppositely charged ionomer pair contributes to a drastic increase in SLI, compared with separately prepared non-associated ionomers. On the other hand, the preformed micelle structure needs to be maintained in extracellular conditions to retain its structural advantages; sufficiently stabilized PIC micelles in extracellular conditions permit significant cellular internalization of SiRNA payloads [10], which is directed toward more efficient gene silencing in target cells [8]. In this regard, fluorescence correlation spectroscopy (FCS) is useful for a stability assay of PIC micelles because FCS can determine

Table 1
Physicochemical criteria for well-defined PICm(siRNA) formulations

Parameter	Sufficient condition
Hydrodynamic (cumulant) diameter	Less than 100 nm
PDI	Less than 0.1
SLI	Significantly higher than that from free ionomers
Encapsulation of siRNA	No detectable free siRNA

diffusion coefficients (*D*s) of fluorescent dye-labeled macromolecules even under highly diluted and serum conditions.

In this chapter, we introduce standard protocols for the preparation and characterization of SiRNA-loaded PIC micelles (PICm(SiRNA)) using a representative cationic block copolymer, PEG-*block*-poly[*N'''*-(*N''*-{*N'*-[*N*-(2-aminoethyl)-2-aminoethyl]-2-aminoethyl}-2-aminoethyl)aspartamide]-cholesteryl (PEG-PAsp(TEP)-Chol) [18, 19]. The important parameters for micelle formation, including SLI, hydrodynamic size, and PDI, as well as the encapsulation of SiRNA during micelle formation, were determined as a function of a residual molar ratio of amines in the cationic copolymer to phosphates in SiRNA (hereafter, termed N/P ratio) (Table 1). The *D* values of PICm(Cy3-labeled SiRNA) in the presence of serum were then time-dependently monitored by FCS to examine their serum resistance. A luciferase assay was further described to determine the gene silencing efficiency of the PICm(SiRNA) formulation. These protocols are important to characterize SiRNA delivery vehicles and are thus widely applicable for the self-assembly of SiRNA prepared with newly synthesized cationic block copolymers (*see* **Note 1**).

2 Materials

2.1 Preparation of Cationic Block Copolymer and siRNA

1. Cationic block copolymer.
2. Firefly luciferase SiRNA (sense: 5′-CUU ACG CUG AGU ACU UCG AdTdT-3′; antisense: 5′-UCG AAG UAC UCA GCG UAA GdTdT-3′) (siLuc).
3. Control (scramble) SiRNA (sense: 5′-UUC UCC GAA CGU GUC ACG UdTdT-3′; antisense: 5′-ACG UGA CAC GUU CGG AGA AdTdT-3′) (siScr).
4. 10 mM Hepes buffer, pH 7.3.
5. 0.01 M HCl (2 L ultrapure water + 2 mL of 5 M hydrogen chloride, prechilled).

6. Dialysis membrane (molecular weight cutoff: 3.5 kDa).
7. Dialysis tubing closures.
8. 2 L glass beaker.
9. 100 mL round-bottom flask.
10. Liquid N_2.
11. Freeze dryer.
12. Stir bar.

2.2 Size and Scattered Light Intensity (SLI) Measurement

1. Zetasizer (Malvern Instruments).
2. Low-volume quartz cuvette (ZEN2112, Malvern Instruments).
3. Acetone.
4. Dryer.
5. 0.1 M sodium hydroxide solution.

2.3 Gel Retardation Assay

1. 20 % Novex polyacrylamide TBE gel.
2. SYBR Green II RNA Gel Stain.
3. 50 % glycerol: Dilute glycerol (0.5 mL) with 10 mM Hepes buffer, pH 7.3 (0.5 mL) and store at 4 °C.
4. XCell Surelock Mini-Cell electrophoresis module.
5. 1× TBE buffer: Add 100 mL 10× TBE buffer to 900 mL ultra-pure water.
6. Molecular Imager FX.
7. Opaque plastic tray.

2.4 Stability Assay

1. A confocal laser-scanning microscope (LSM 510, Carl Zeiss) equipped with a ConfoCor 3 module and a Zeiss C-Apochromat 40× water objective should be used. A He-Ne laser (543 nm) is used for the excitation of Cy3-labeled SiRNA and the emission is detected through a 560–615 nm band pass filter. Autocorrelation curves of each sample are collected from ten measurements during a sampling time of 10 s and are fitted with a Zeiss ConfoCor 3 software package to calculate the *D* of Cy3-labeled SiRNA.
2. Rhodamine 6G.
3. 8-well Lab-Tek chambered borosilicate cover glass.
4. Cy3-labeled firefly luciferase SiRNA (*see* **Note 2**).
5. 10 % fetal bovine serum (FBS) in PBS (1 mL of FBS mixed with 9 mL of PBS).

2.5 Transfection to HeLa-Luc Cells

1. Human cervical cancer cells stably expressing luciferase (HeLa-Luc).
2. 75-cm^2 cell culture flask.

3. Humidified incubator, maintained at 37 °C and 5 % CO_2.
4. High-glucose DMEM.
5. FBS.
6. Penicillin/streptomycin solution (P/S).
7. 1× Trypsin-EDTA solution.
8. Cell counter.
9. 15 mL centrifuge tubes.
10. Centrifuge equipped with swinging bucket rotor.
11. 96-well culture plate with flat-bottom.
12. Cell lysis buffer.
13. Luciferase assay system.
14. Luminescence microplate reader.
15. Opaque 96-well plate with flat-bottom.

3 Methods

Carry out all procedures at room temperature with autoclaved tips and 1.5 mL tubes, but it is not necessary work on a clean bench. Block copolymer solutions should be mixed with SiRNA solutions at various N/P ratios to select the PICm (SiRNA) that fulfills all physicochemical criteria. If such an N/P ratio is not found, you may need to change the degree of polymerization (DP) in the polycation and/or PEG segment. In addition, you may need to redesign the chemical structure of the polycationic segment. We found that not all cationic block copolymers could form well-defined micelles that fulfill the criteria.

3.1 Micelle Formation

3.1.1 Preparation of Cationic Block Copolymer

1. Synthesize a block copolymer, PEG-PAsp(TEP)-Chol, or prepare other cationic polymers [18]. Here, PEG-PAsp(TEP)-Chol with PEG (MW = 12,000) and PAsp(TEP) (DP = 57) is used as an example (*see* **Note 3**).
2. Dissolve the block copolymer in ultrapure water (10–30 mL) and load the polymer solution into a dialysis tube (*see* **Note 4**).
3. Dialyze the solution against 2 L of 0.01 M hydrogen chloride at 4 °C for 4 h in a glass beaker.
4. Discard old 0.01 M HCl and fill with fresh 2 L of 0.01 M hydrogen chloride.
5. Dialyze at 4 °C for another 4 h.
6. Discard old 0.01 M HCl and fill with fresh 2 L of 0.01 M hydrogen chloride.
7. Dialyze at 4 °C overnight.
8. Discard old 0.01 M hydrogen chloride and fill with 2 L of ultrapure water.

9. Dialyze at 4 °C for 2 h to remove excess hydrogen chloride.
10. Discard old water and fill with fresh 2 L of ultrapure water.
11. Dialyze at 4 °C for 2 h (*see* **Note 5**).
12. Carefully open the dialysis tube and collect the polymer solution in a 100 mL round-bottom flask.
13. Freeze the polymer solution in liquid N_2.
14. Lyophilize the sample in a freeze dryer.
15. Collect and store polymer powder at –20 °C.

3.1.2 Calculation of N/P Ratio

1. Total amine molar concentration in 1 mg/mL polymer solution (*see* **Note 6**): Obtain the MW of the cationic block copolymer. The MW of the counter-ions against the ionized functional groups in the cationic block copolymer should be also included. Count the total number of amino groups that can be protonated (*see* **Note 7**). Calculate the amine molar concentration (mM) in 1 mg/mL polymer solution (Fig. 2, Table 2).

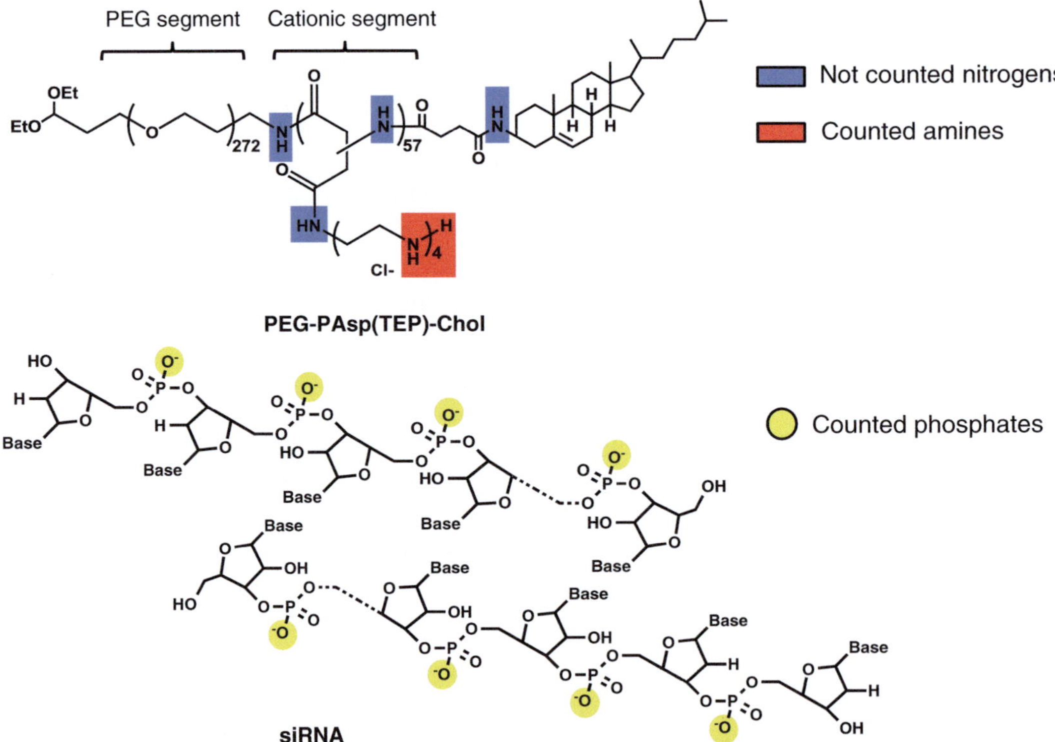

Fig. 2 Chemical structures of PEG-PAsp(TEP)-Chol and siRNA (21mer/21mer). Amines and phosphates that should be used for calculating N/P ratios are indicated by *red* and *yellow* colors, respectively

Table 2
Example of the calculation of amine and phosphate concentrations

MW of PEG-PAsp(TEP)-Chol including Cl^- as counter-ions	MW of PEG + MW of Asp(TEP) unit × DP of PAsp(TEP) + MW of Cl^- × DP of PAsp(TEP) × Number of Cl^- in Asp(TEP) unit + MW of Cholesteryl moiety 12,097 + 290 × 57 + 35.5 × 57 × 4 + 413 = 36,721 g/mol
Amine concentration in 1 mg/mL	$\frac{1\text{mg} / \text{mL}}{\text{MW of PEG} \quad \text{PAsp}(\text{TEP}) \quad \text{Chol}}$ $\times$ Number of amines in Asp(TEP) unit $\times$ DP of PAps(TEP) $\frac{1\text{mg} / \text{mL}}{36{,}721\text{g} / \text{mol}} \times 4 \times 57 = 6.22 \times 10^{3}\,\text{M} = 6.21\text{mM}$
Phosphate concentration in 20 μM siRNA	siRNA concentration × number of phosphates in siRNA 20 μM × 40 = 800 μM

Table 3
Volumes of polymer, siRNA, and buffer solutions required for N/P ratios of 3, 4, and 5

N/P ratio	Volume of 1 mg/mL Polymer (μL)	Volume of 20 μM siRNA (μL)	10 mM Hepes buffer pH 7.3 (μL)
3	1.9	5	13.1
4	2.6	5	12.4
5	3.2	5	11.8

2. Total phosphate molar concentration in 20 μM SiRNA solution. Some manufacturers provide 21mer/21mer SiRNA without the phosphate group in the 3′-end. In this case, the total number of phosphate groups in one SiRNA is 40. If not, it is 42. Calculate the phosphate molar concentration (μM) in 20 μM SiRNA solution.
3. Prepare solutions according to desired N/P ratio (Table 3).
4. Calculate the desired volume of 1 mg/mL polymer solution and SiRNA solution to obtain a 5 μM PICm(SiRNA) solution, as described below:

$$\begin{aligned}&\text{Volume of 1 mg / mL polymer solution } (\mu\text{L})\\&\quad = \text{N / P ratio} \times \text{phosphate concentration } (\mu\text{M})\\&\qquad \times \text{volume of siRNA solution } (\mu\text{L}) \times 10^{-3} \text{ / amine concentration } (\text{mM}).\end{aligned}$$

3.1.3 PICm(siRNA) Formation

1. Dissolve 1 mg cationic block copolymer in 1 mL of 10 mM Hepes buffer, pH 7.3 (*see* **Note 8**).
2. Prepare SiRNA solution (20 μM) in 10 mM Hepes buffer, pH 7.3.
3. Mix the polymer solution sequentially with 10 mM Hepes buffer, pH 7.3 and the SiRNA solution to obtain a PICm(SiRNA) solution (5 μM SiRNA concentration) (Table 3).
4. Place the mixture at room temperature for 1 h. Micelles are then ready to use. Micelle solution can be stored at 4 °C overnight to use the next day. Do not freeze micelle solution (*see* **Note 9**).

3.2 Size and SLI Measurement

1. Load 20 μL of ultrapure water into a low-volume quartz cuvette.
2. Measure the derived count rate and confirm that it is less than 100 kcps with no attenuator mode (fully opened slit) (*see* **Note 10**).
3. Remove ultrapure water and gently wash inside with acetone. Completely dry the cuvette with gentle heating.
4. Prepare PICm(SiRNA) (5 μM SiRNA concentration) at various N/P ratios according to Subheading 3.1.
5. Load one sample (20 μL) into a low-volume quartz cuvette. Measure size/PDI in dynamic light scattering (DLS) mode and SLI in static light scattering (SLS) mode.
6. Remove the sample and wash the quartz cuvette with ultrapure water twice. Gently wash inside with acetone and completely dry the cuvette with gentle heating.
7. Load the next sample (20 μL) into the quartz cuvette. Measure size/PDI in DLS mode and SLI in SLS mode.
8. Analyze the size, PDI, and SLI (*see* **Notes 11** and **12**) (Fig. 3).

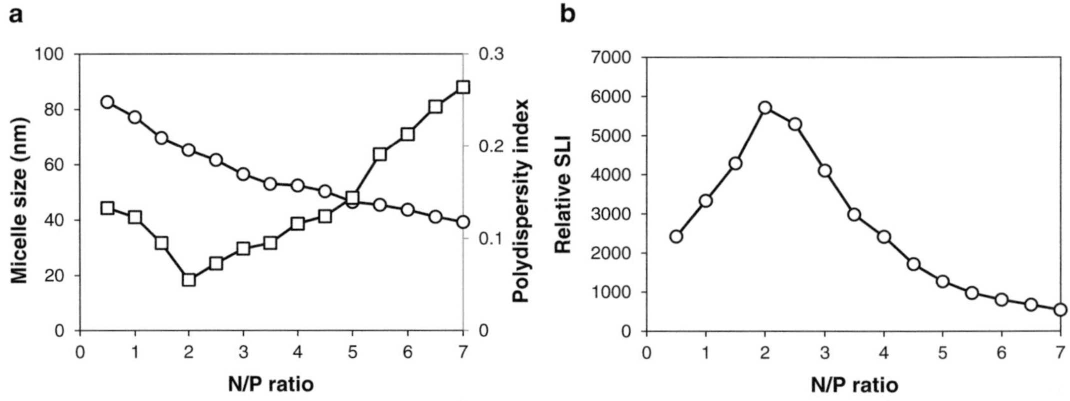

Fig. 3 (**a**) Size (*open circle*) and PDI (*open square*) of PEG-PAsp(TEP)-Chol PICm (siRNA) measured by DLS. (**b**) Relative SLI of PEG-PAsp(TEP)-Chol PICm (siRNA) measured by SLS

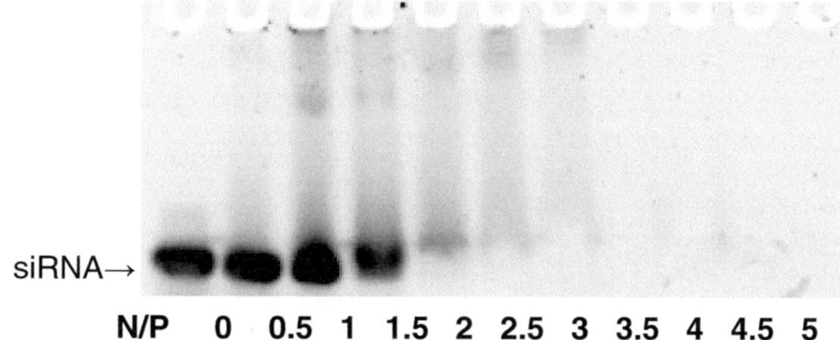

Fig. 4 Gel retardation assay of PEG-PAsp (TEP)-Chol PICm (siRNA) with increasing N/P ratios. The disappearance of free siRNA band is observed at N/P = 2

3.3 Gel Retardation Assay

1. Prepare PICm(SiRNA) (5 μM SiRNA concentration) at various N/P ratios according to Subheading 3.1.
2. Mix 1.5 μL of 5 μM PICm(SiRNA), 1 μL of 50 % glycerol, and 7.5 μL of 10 mM Hepes buffer, pH 7.3.
3. Assemble electrophoresis modules in 1× TBE buffer (*see* **Note 13**).
4. Load samples (10 μL) into 20 % TBE polyacrylamide gel.
5. Electrophorese the gel at 100 V for 60 min.
6. Dilute SYBR Green II RNA Gel stain (3 μL) in ultrapure water (30 mL) in an opaque plastic tray (*see* **Note 14**).
7. Stain the gel with SYBR Green II RNA Gel Stain solution for 10 min and analyze the stained gel using an imaging system.
8. Analyze the association between the polymer and SiRNA (*see* **Note 15**) (Fig. 4).
9. Choose N/P ratios satisfying all physicochemical criteria of size, PDI, SLI, and encapsulation efficiency.

3.4 Stability Assay by FCS

1. Prepare 100 nM Rhodamine 6G in 10 mM Hepes buffer, pH 7.3.
2. Place 100 μL of Rhodamine 6G solution into an 8-well Lab-Tek chambered borosilicate cover glass.
3. Place the chambered cover glass on the microscope stage and adjust the position where the He-Ne laser (543 nm) beams at the sample solution in a well.
4. Appropriately adjust laser intensity and Rhodamine 6G concentration by monitoring count rate and counts per molecules, according to the manufacturer's instructions.
5. Measure the diffusion time of Rhodamine 6G ($\tau_{\text{D, Rhodamine 6G}}$), as well as the structure parameter.
6. Prepare PICm (SiRNA) (5 μM SiRNA concentration) containing Cy3-labeled SiRNA according to Subheading 3.1.

7. For the measurement of initial D of PIC micelles without incubation, dilute the PICm (Cy3-labeled SiRNA) up to 100 nM SiRNA with 10 mM Hepes buffer, pH 7.3.
8. Place 100 μL of sample solution into an 8-well Lab-Tek chambered borosilicate cover glass.
9. Place the chambered cover glass on the microscope stage and adjust the position where the He-Ne laser (543 nm) beams at the sample solution in a well.
10. Appropriately adjust laser intensity by monitoring count rate and counts per molecules.
11. Measure $\tau_{D,\ \mathrm{micelle}}$ based on the predetermined structure parameter.
12. For the serum stability assay, dilute the PICm (Cy3-labeled SiRNA) up to 100 nM SiRNA with 10 % FBS in PBS and incubate at 37 °C.
13. Place 100 μL of sample solutions into an 8-well Lab-Tek chambered borosilicate cover glass for a designated incubation time.
14. Place the chambered cover glass on the microscope stage and adjust the position where the He-Ne laser (543 nm) beams at the sample solution in a well.
15. Measure $\tau_{D,\ \mathrm{micelle}}$ based on the predetermined structure parameter.
16. Convert the obtained τ_D values to D using the following equation:

$$D_{\mathrm{micelle}} \times \tau_{\mathrm{D,\,micelle}} = D_{\mathrm{Rhodamine\,6G}} \times \tau_{\mathrm{D,\,Rhodamine\,6G}}$$

where $D_{\mathrm{Rhodamine\ 6G}}$ is 414 $\mu m^2/s$ according to the literature [20] and $\tau_{D,\ \mathrm{micelle}}$ and $\tau_{D,\ \mathrm{Rhodamine\ 6G}}$ are obtained from the above experiments. The representative data of the stability assay can be referred to in the literature [10, 18].

3.5 Transfection to HeLa-Luc Cells

1. On day zero, Culture cells in a 75 cm^2 cell culture flask with DMEM containing 10 % FBS and 1 % P/S (hereafter referred to growth DMEM) until cells are healthy and approximately 80 % confluency.
2. Aspirate growth DMEM and gently wash cells once with 5 mL of room temperature PBS.
3. Add 5 mL of 1× Trypsin-EDTA solution to the flask and incubate at 37 °C for 5 min.
4. Resuspend the cells with 5 mL growth DMEM and transfer them into a 15 mL conical centrifuge tube.
5. Centrifuge at 500 × g for 5 min at 4 °C, aspirate supernatant, and resuspend cells in 3–5 mL growth DMEM.

6. Count cells and calculate the volume of growth DMEM required for 3×10^3 cells/well in 96-well culture plates (*see* **Note 16**).
7. Plate cells in 96-well plates and return to the incubator.
8. On day 1, Separately prepare PICm (siLuc) and PICm (siScr).
9. Dilute 5 μM PICm (SiRNA) solution to the proper transfection concentration. For 200 nM SiRNA in a 96-well plate, mix 16 μL of 5 μM PICm (SiRNA) with 24 μL of 10 mM Hepes buffer, pH 7.3 in order to prepare 10 μL of 2 μM PICm (SiRNA) per well ($n=4$).
10. Carefully aspirate growth DMEM from the 96-well plate and add fresh pre-warmed growth DMEM (90 μL/well).
11. Add PICm (SiRNA) solution (10 μL) to each well and gently tap the 96-well culture plate to disperse micelle solution (*see* **Note 17**).
12. Return the plate to the incubator.
13. On day 3, carefully aspirate growth DMEM.
14. Gently wash cells with 100 μL/well of room temperature PBS and aspirate the PBS.
15. Add cell lysis buffer (50 μL/well) and incubate for 1 h at room temperature.
16. Place 20 μL/well of cell lysates in an opaque flat bottom 96-well plate and add luciferase assay kit solution (30–50 μL/well).
17. Measure luminescence intensity with a luminescence microplate reader.
18. Analyze and normalize the luminescence intensity to determine the gene silencing effect (*see* **Note 18**).

4 Notes

1. To apply these procedures to other cationic block copolymers or homopolymers, start by calculating the MW of the polymers and counting the number of protonatable amines at a neutral pH. Carefully consider experimental data based on micelle size, PDI, SLI, and encapsulation of SiRNA. The readers also may refer to **Note 8** to apply these procedures to other functional polymers.
2. Cy5-labeled SiRNA (a 633 nm He-Ne laser), Alexa 647-labeled SiRNA (a 633 nm He-Ne laser), and Alexa 546-labeled SiRNA (a 543 nm He-Ne laser) can also be used with appropriate emission filters. The *D* of Atto655-maleimide is applied to calculate the *D* of PICm (SiRNA) containing Cy5-labeled SiRNA or Alexa 647-labeled SiRNA [20].

3. Protonation of newly synthesized cationic polymers with HCl is important because complete encapsulation of SiRNA can be obtained at the calculated stoichiometric amount of the polymers and SiRNA. Dialysis of the polymer against acetic acid produces protonated amine, bearing acetate (CH_3COO^-) as a counter-ion. However, the resulting polymer needs less/more amount of polymers to obtain neutral N/P ratio than calculated. It is assumed that (1) the amount of acetate is not fully counted in the MW calculation of the polymer, and (2) acetate may prohibit self-assembly between the polymer and SiRNA.
4. Use at least two times the amount of dialysis tubing because the volume of polymer solution will increase during dialysis from osmosis. For example, use at least 60 mL of dialysis tubing for 30 mL polymer solution. When the polymer is dissolved in organic solvents (DMSO, ethanol, or methanol), at least three times extra volume of dialysis tubing is recommended.
5. Remove excess hydrogen chloride completely. When large amounts of polymers are dialyzed, it is recommended that the polymer solution should be dialyzed for longer periods or additional exchanges with fresh ultrapure water should be performed.
6. Polymer solutions with higher concentrations may be necessary to prepare PICm (SiRNA) with a concentration of more than 5 μM SiRNA, if required.
7. Protonation degree of amino groups in polycation can be determined by a potentiometric titration [19, 21].
8. Polymers can be dissolved in a solution of organic solvent and 10 mM Hepes buffer (pH 7.3) (1:1 v/v ratio). Common organic solvents are methanol, ethanol, dimethylsulfoxide, *N*,*N*-dimethylformamide, and *N*-methylpyrrolidone. Organic solvents can be removed by dialysis after micelle formation is complete.
9. When PICm(SiRNA) needs to be further stabilized by disulfide crosslinking, check the physicochemical characteristics (size, PDI, and SLI) at this point and go to the next stabilizing procedure. For example, the micelle solution may need to be transferred to a dialysis tube and dialyzed against the proper buffer for days to obtain disulfide cross-linked micelles [8].
10. When the derived count rate of ultrapure water is over 100 kcps or the glass window in a low-volume quartz cuvette is dirty, fill cuvette with 0.1 M sodium hydroxide solution and incubate it overnight. Wear gloves and carefully pipet the solution 10–20 times. Remove the 0.1 M sodium hydroxide solution and wash the cuvette with excess ultrapure water. Check SLI with no attenuator mode.

11. Find N/P ratios where the sizes of PICm(SiRNA) are below 100 nm "AND" PDI values are less than 0.1, simultaneously. Confirm that the SLI has a peak around a neutral N/P. PEG-PAsp(TEP)-Chol has neutral N/P = 2 because two of four primary/secondary amines in PAsp(TEP) side chains are protonated at pH = 7.3 [19, 21]. Choose an N/P ratio of PICm(SiRNA) higher than the neutral N/P ratio. For example, the suitable N/P ratios of PEG-PAsp(TEP)-Chol PICm(SiRNA) are 2–3.5 from Fig. 3. PICm (SiRNA) prepared at an extremely high N/P ratio may have a "free polymer effect"; free polymers (not bound to micelles) are likely to affect cultured cells. Homo cationic polymers without PEG segments may form aggregates of micrometer sizes at neutral N/P ratios.
12. It is recommended to compare SLI of the same concentration of free polymer in the absence of SiRNA with that of micelles at a designated N/P ratio. The samples can be prepared by adding the same volume of 10 mM Hepes buffer instead of SiRNA in Table 3. This can provide evidence that the polymers associate with SiRNA to assemble into higher molecular weight structures.
13. Remove air bubbles in wells of 20 % TBE polyacrylamide gels using a gel-loading tip on a pipet before loading samples.
14. It is recommended to prepare fresh 1× SYBR Green II each day it is used.
15. Confirm that the free SiRNA band disappears at the N/P ratios selected in DLS/SLS experiments. Generally, the free SiRNA band disappears at neutral or higher N/P ratios. Smear bands at higher molecular weight positions are observed as higher molecular weight structures between the polymers and SiRNAs.
16. For real-time PCR or flow cytometry analysis, plate the cells at 1×10^5 cells/well in 6-well culture plates. Add fresh prewarmed growth DMEM (1.9 mL/well) and micelles (100 μL/well).
17. We prefer sequentially adding growth DMEM and micelles to the cells because the micelles begin to degrade as soon as they are added to growth DMEM. Avoid preparing a combined growth DMEM and micelle solution, then adding it to the cells.
18. Luciferase gene silencing of PICm (siLuc) should be compared with that of PICm (siScr). Luminescence intensities in cells treated with both PICm (siLuc) and PICm (siScr) are further normalized with those in cells treated with growth DMEM, which confirms there are no adverse side effects (or off-target effects) derived from the delivery carrier.

Acknowledgements

This work was financially supported by the Funding Program for World-Leading Innovative R&D in Science and Technology (FIRST) (JSPS), Grants-in-Aid for Scientific Research of MEXT (JSPS KAKENHI Grant Numbers 25000006 and 25282141), and the Center of Innovation (COI) Program (JST). K. Miyata would like to express his thanks to Mochida Memorial Foundation for Medical and Pharmaceutical Research.

References

1. Fire A, Xu SQ, Montgomery MK et al (1998) Potent and specific genetic interference by double-stranded RNA in *Caenorhabditis elegans*. Nature 391:806–811
2. Yin H, Kanasty RL, Eltoukhy AA et al (2014) Non-viral vectors for gene-based therapy. Nat Rev Genet 15:541–555
3. Williford JM, Wu J, Ren Y et al (2014) Recent advances in nanoparticle-mediated siRNA delivery. Annu Rev Biomed Eng 16:347–370
4. Resnier P, Montier T, Mathieu V et al (2013) A review of the current status of siRNA nanomedicines in the treatment of cancer. Biomaterials 34:6429–6443
5. Kataoka K, Harada A, Nagasaki Y (2001) Block copolymer micelles for drug delivery: design, characterization and biological significance. Adv Drug Deliv Rev 47:113–131
6. Cabral H, Matsumoto Y, Mizuno K et al (2011) Accumulation of sub-100 nm polymeric micelles in poorly permeable tumours depends on size. Nat Nanotechnol 6:815–823
7. Matsumura Y, Maeda H (1986) A new concept for macromolecular therapeutics in cancer chemotherapy: mechanism of tumoritropic accumulation of proteins and the antitumor agent smancs. Cancer Res 46:6387–6392
8. Christie RJ, Matsumoto Y, Miyata K et al (2012) Targeted polymeric micelles for siRNA treatment of experimental cancer by intravenous injection. ACS Nano 6:5174–5189
9. Pittella F, Cabral H, Maeda Y et al (2014) Systemic siRNA delivery to a spontaneous pancreatic tumor model in transgenic mice by PEGylated calcium phosphate hybrid micelles. J Control Release 178:18–24
10. Kim HJ, Miyata K, Nomoto T et al (2014) siRNA delivery from triblock copolymer micelles with spatially ordered compartments of PEG shell, siRNA-loaded intermediate layer, and hydrophobic core. Biomaterials 35:4548–4556
11. Kim HJ, Ishii A, Miyata K et al (2010) Introduction of stearoyl moieties into a biocompatible cationic polyaspartamide derivative, PAsp(DET), with endosomal escaping function for enhanced siRNA-mediated gene knockdown. J Control Release 145:141–148
12. Naito M, Ishii T, Matsumoto A et al (2012) A phenylboronate-functionalized polyion complex micelle for ATP-triggered release of siRNA. Angew Chem Int Ed 51:1–6
13. Suma T, Miyata K, Anraku Y et al (2012) Smart multilayered assembly for biocompatible siRNA delivery featuring dissolvable silica, endosome-disrupting polycation, and detachable PEG. ACS Nano 6:6693–6705
14. Takemoto H, Miyata K, Hattori S et al (2013) Acidic pH-responsive siRNA conjugate for reversible carrier stability and accelerated endosomal escape with reduced IFNα-associated immune response. Angew Chem Int Ed 52:6218–6221
15. Harada A, Kataoka K (1995) Formation of polyion complex micelles in an aqueous milieu from a pair of oppositely charged block copolymers with poly(ethylene glycol) segments. Macromolecules 28:5294–5299
16. Kataoka K, Togawa H, Harada A et al (1996) Spontaneous formation of polyion complex micelles with narrow distribution from antisense oligonucleotide and cationic block copolymer in physiological saline. Macromolecules 29:8556–8557
17. Harada A, Kataoka K (2008) Selection between block- and homo-polyelectrolytes through polyion complex formation in aqueous medium. Soft Matter 4:162–167
18. Kim HJ, Ishii T, Zheng M et al (2014) Multifunctional polyion complex micelle featuring enhanced stability, targetability, and endosome escapability for systemic siRNA delivery to subcutaneous model of lung cancer. Drug Deliv Transl Res 4:50–60

19. Suma T, Miyata K, Ishii T et al (2012) Enhanced stability and gene silencing ability of siRNA-loaded polyion complexes formulated from polyaspartamide derivatives with a repetitive array of amino groups in the side chain. Biomaterials 33:2770–2779
20. Műller CB, Loman A, Pacheco V et al (2008) Precise measurement of diffusion by multi-color dual-focus fluorescence correlation spectroscopy. EPL 83:46001
21. Uchida H, Miyata K, Oba T et al (2011) Odd-even effect of repeating aminoethylene units in the side chain of N-substituted polyaspartamides on gene transfection profiles. J Am Chem Soc 133:15524–15532

Chapter 10

Delivery of Small Interfering RNAs to Cells via Exosomes

Jessica Wahlgren, Luisa Statello, Gabriel Skogberg, Esbjörn Telemo, and Hadi Valadi

Abstract

Exosomes are small membrane bound vesicles between 30 and 100 nm in diameter of endocytic origin that are secreted into the extracellular environment by many different cell types. Exosomes play a role in intercellular communication by transferring proteins, lipids, and RNAs to recipient cells.

Exosomes from human cells could be used as vectors to provide cells with therapeutic RNAs. Here we describe how exogenous small interfering RNAs may successfully be introduced into various kinds of human exosomes using electroporation and subsequently delivered to recipient cells. Methods used to confirm the presence of siRNA inside exosomes and cells are presented, such as flow cytometry, confocal microscopy, and Northern blot.

Key words Exosomes, Extracellular vesicles, Electroporation, siRNA delivery, Plasma exosomes, Exosome delivery, Exosome uptake, Post-transcriptional gene silencing

1 Introduction

Exosomes are nano-sized vesicles, 30–100 nm in size produced by various cell types including dendritic cells [1], B cells [2], T cells [3, 4], mast cells [5], epithelial cells [6], and various tumor cells [7]. These vesicles have been found in different tissues and body fluids such as peripheral blood [8], thymus [9], urine [10], malignant effusions [11], and bronchoalveolar lavage fluid [12]. It has been shown that exosomes are involved in signal transduction, including antigen presentation to T-cells [13], tolerance development [6], and transfer of RNA [14].

In 2007, Valadi et al. discovered that exosomes contain a substantial amount of RNA [14]. They concluded that exosomal RNA mainly consists of mRNA and microRNA but contains a very low (or no) amount of rRNA. Moreover, Valadi et al. demonstrated that exosomes were capable of transporting different RNA species in a selective way to a number of target cells. These findings opened up a new research field of how cells communicate [14].

Kato Shum and John Rossi (eds.), *SiRNA Delivery Methods: Methods and Protocols*, Methods in Molecular Biology, vol. 1364, DOI 10.1007/978-1-4939-3112-5_10,

We have succeeded to transfect exogenous small interfering RNA (siRNA) into various human cell types, e.g., monocytes and lymphocytes. Moreover, we have shown that exosomes are capable of delivering siRNA to recipient cells causing selective gene silencing [15]. Since exosomes are capable of shuttling RNAs between cells, we hypothesized that these vesicles could be used as vectors in gene therapy to provide cells with therapeutic nucleic acids e.g., therapeutic siRNA. However, to use exosomes as vectors, the siRNA must be introduced into these vesicles. Thus, in this study, the aim was to develop a method for introducing foreign nucleic acids i.e., siRNA into the exosomes. Subsequently, these vesicles containing the exogenous siRNA would be used as vectors to transport the foreign nucleic acids to the target cells. In order for the exosomes to act as vectors in gene therapy, the therapeutic genetic material needs to be introduced into the exosomes. Since siRNA inhibits translation of mRNA and is of a similar size as microRNA, the most common RNA found in exosomes, it was a suitable synthetic nucleic acid for this application. Electroporation and chemical transfection were used to insert double stranded siRNAs into exosomes, which were isolated from human immortalized cell lines and plasma. Specifically, plasma exosomes from healthy blood donors, a lung cancer cell line, and HeLa cells were used. The identification of exosomes was confirmed by a size determination device measuring particle's size between 1 and 1000 nm in diameter and by flow cytometry detecting CD9, CD63, and CD8 proteins, which are known to be present in many exosome types. Taken together, these experiments showed that the isolated vesicles were exosomes and they were used in subsequent transfection and electroporation experiments.

First, we used chemical transfection where the siRNA was embedded in lipid micelles and incubated with the exosomes. The results produced false positives, since it was not possible to determine whether the exosomes or the micelles delivered exogenous siRNA to the target cells. Next, we examined electroporation for introducing siRNA into the exosomes. Isolated exosomes were mixed with siRNA molecules and exposed to an electric field pulse. To find the optimal settings for electroporation, concentrations of siRNA and exosomes, voltage and capacitance values were varied. Flow cytometry and confocal microscopy confirmed the presence of siRNA inside the exosomes using fluorescently tagged siRNA. Furthermore, siRNA designed to inhibit MAPK-1 enabled antibody detection of a decrease of MAPK-1 protein with Western blot, and sequence specific detection with Northern blot, showing that the heterologous siRNA resided inside the exosomes and that the siRNA was functional. Hence, we could conclude that electroporation was a successful method for introducing siRNAs into exosomes.

The next step was to investigate (a) if the transfected exosomes could deliver the siRNA to the target cells and (b) if the siRNA was functional in the recipient cells by causing a selective gene silencing. Initially, a fluorescently labeled siRNA was introduced into human plasma exosomes using electroporation. Subsequently, these exosomes were used as vectors to deliver the heterologous siRNAs to target cells. Monocytes and lymphocytes isolated from peripheral blood mononuclear cells (PBMC) were used as recipient cells. Taken together flow cytometry, confocal microscopy, and Northern blot showed that plasma exosomes seemed to be capable of delivering siRNA to recipient cells.

Since the delivery of heterologous siRNA to cells was successful, the ability of siRNA to cause post-transcriptional gene silencing in cells was investigated using exosomes as vectors. Herein, the functionality of MAPK-1 siRNAs delivered to cells via exosomes was investigated using Western blot. Subsequent measurements showed that the exosome-delivered siRNA caused selective gene silencing in recipient cells. Hence, we showed that the MAPK-1 protein was downregulated in recipient cells co-cultured with electroporated exosomes. Furthermore, we investigated the kinetics of exosome-delivery of siRNA to monocytes and lymphocytes. The results showed that high concentrations of electroporated exosomes could more effectively downregulate the expression of MAPK-1 in both lymphocytes and more prominently in monocytes, which may be due to the phagocytic capabilities of monocytes.

Our results showed for the first time that exosomes from human tissue can be used as vectors to deliver exogenous nucleic acids, i.e., siRNA to cells from human blood and this may have vast potentials in clinical applications, e.g., gene therapy. This notion is further supported by a recent paper where they show targeted delivery of siRNA using exosomes in a mouse model [16]. The method of gene therapy is still lacking an efficient vector with minimal immunogenic concerns. Here, we show that human exosomes that carry heterologous siRNAs could potentially be used as autologous vectors for delivery of therapeutic genetic material to various cells and tissues in the body. Thus, autologous exosomes may be safe non-immunogenic vectors in gene therapy.

2 Materials

2.1 Isolation of Exosomes

1. Cell line: HTB-177 lung cancer cell line.
2. Cell culture medium: RPMI1640 for HTB-177, supplemented with 2 mM L-glutamine, 100 μg/ml streptomycin/streptavidin, and 10 % exosome depleted FBS. 175 cm^2 cell culture flasks.
3. Buffy coat (blood treated with anticoagulants) from healthy donors.

4. 50 ml tubes, centrifuge with rotor for 50 ml tubes, 50 ml syringe needle, 2.1 × 80 mm, scissors and 0.2 μm filters.
5. Ultracentrifuge tubes (25 × 89 mm).
6. Dulbecco's PBS.
7. Ultracentrifuge, e.g., Beckman OptimaLE-80K or Optima™ L-100 and Rotor 70 Ti.

2.2 Characterization of Exosomes Using Flow Cytometry

1. 4 μm sized aldehyde/sulfate latex beads (4 % w/v).
2. Dilution of beads: Take 5 μl latex beads from stock (40 μg/μl) and dilute to 250 μl with PBS to prepare a working solution. Take 100 μl from working solution and dilute with 900 μl PBS (final bead concentration 0.2 μg/μl). Take 50 μl of the dilution/sample (5 μg beads/sample).
3. Antibody coated aldehyde/sulfate latex beads: Add 25 μl latex beads from stock to a 1.5 ml tube. Add 100 μl MES buffer. Centrifuge at 4000 × *g* for 15 min. Dissolve the pellet in 100 μl MES buffer. Centrifuge at 4000 × *g* for 20 min. Dissolve the pellet thoroughly in 100 μl MES buffer. Mix 12.5 μg antibody with 25 μl MES buffer. Add the beads to the antibody solution. Incubate overnight with gentle agitation at room temperature. Centrifuge at 4000 × *g* for 20 min. Remove supernatant, dissolve the pellet in 100 μl PBS and centrifuge as before. Repeat this step twice. Dissolve the beads in 100 μl PBS storage buffer. Store the beads at 4 °C.
4. MES buffer pH 6.0 (250 ml): 1.33 g MES (2-[*N*-Morpholino] ethanesulfonic acid), 50 ml dH_2O. Dissolve by stirring at 50 °C. Add 31 ml NaOH. Add 169 ml dH_2O when the mixture has reached room temperature.
5. PBS storage buffer pH 7.2 (1 L): 1 g glycine, 1 g sodium azide and dissolve in 1 L Dulbecco's PBS.
6. 1 M Glycine (10 ml): Dissolve 0.7507 g glycine (H_2NCH_2COOH; formula weight = 75. 07 g/mole) in Dulbecco's PBS.
7. Antibodies against exosomal proteins: e.g., CD9, CD63, and CD81.

2.3 Characterization of Exosomes Using Size Determination

1. A Brunel microscope equipped with a 20 times objective and a Marlin camera. A NanoSight LM10 and a computer with nanoparticle tracking analysis (NTA) software.
2. 1 ml disposable syringes.
3. Exosomes isolated as in Subheading 3.1.
4. Dulbecco's PBS.

2.4 Electroporation of Exosomes

1. UV-irradiated cuvettes for electroporation.
2. Electroporation device (Gene pulser II system).

3. Exosomes isolated as in Subheading 3.1.
4. Small interfering RNA (siRNA) against sequence of interest for mRNA silencing. We used siRNA against mitogen-activated protein (MAPK). For detection with flow cytometry and confocal microscopy we used fluorescently tagged siRNA and for Western blot we used siRNA without a fluorescent tag.
5. CytoMix transfection buffer: 120 mM KCl, 0.15 mM $CaCl_2$, 10 mM K_2HPO_4/KH_2PO_4, pH 7.6, 25 mM HEPES, pH 7.6, 2 mM EGTA, pH 7.6, and 5 mM $MgCl_2$ (pH adjusted with potassium hydroxide) [17].

2.5 Preparation of Peripheral Blood Mononuclear Cells from Buffy Coat

1. Buffy coat, i.e., blood treated with anti-coagulants for enabling density centrifugation, from healthy donors.
2. Lymphoprep (Axis-Shield Poc AS) or other density gradient buffers able to separate peripheral blood mononuclear cells (PBMC).

2.6 Isolation of Monocytes and Lymphocytes

1. MACS buffer: Dulbecco's PBS without Ca^{2+} and Mg^{2+}, 0.5 % FBS, and 2 mM EDTA.
2. 10× Ammonium chloride buffer pH 7.3 for lysing red blood cells (1 L): 82.6 g ammonium chloride (NH_4Cl), 10 g potassium hydrogen carbonate ($KHCO_3$), and 0.336 g EDTA Na-salt (Titriplex III). Dissolve in dH_2O.
3. Cell separation equipments including magnetic beads coated with antibodies against CD14, separation columns and separation magnet, e.g., CD14 microbeads, LS column and MidiMACS magnet (Miltenyi Biotec).
4. Cell counting equipment, e.g., Sysmex K21-N cell counter.

2.7 Exosome Delivery Assay

1. Cell culture plates of suitable size for your application.
2. Phorbol-myristate-acetate (PMA).
3. Exosomes electroporated with siRNA of interest. Non-electroporated exosomes for use as negative control sample.
4. Incubator at 37 °C with 5 % CO_2 atmosphere.

2.8 RNA Isolation from Cells and Exosomes

1. TRIzol reagent.
2. Chloroform.
3. Glycogen.
4. Isopropanol.
5. 75 % ethanol kept at 4 °C.
6. Vacuum device for aspiration.
7. RNase-free water.

2.9 Northern Blot for Detection of siRNA in Exosomes and Cells

1. Filter papers for blotting.
2. Nylon hybridization transfer membrane (Hybond-N+).
3. Slot blot device (e.g., Bio-Dot SF, Bio-Rad).
4. Digoxigenin labeled RNA-probe designed to be complementary to the siRNA used.
5. Denhardt's solution 100× (100 ml): 2 g Ficoll 400, 2 g polyvinylpyrrolidone, 2 g bovine serum albumin (BSA) fraction V, add dH_2O up to 100 ml. Filter and store at –20 °C in 10 ml tubes.
6. 10× NBC buffer (1 L): 0.5 M Boric acid, 10 mM Na_3-citrate, 50 mM NaOH. Add 0.1 % diethylpyrocarbonate (DEPC). Incubate overnight and autoclave the day after.
7. 20× SSC (Sodium chloride/sodium citrate) 1 L: 3 M NaCl (175 g/L), 0.3 M Na_3-citrate (dihydrate) (88 g/L) in 800 ml dH_2O. Adjust pH to 7.0 with 1 M HCl and adjust volume to 1 L with dH_2O finally autoclave to sterilize.
8. Phosphate buffer: 0.5 M sodium phosphate, pH 6.5.
9. Pre-hybridization buffer (20 ml): Freshly made each time: 5 ml 20× SSC, 0.4 ml phosphate buffer, 2 ml 100× Denhardt's solution, 0.6 ml denatured salmon sperm DNA (3 mg/ml) (*see* **Note 1**), 8 ml dH_2O and 4 ml 10 % sodium dodecyl sulfate (SDS). Add SDS just before use.
10. Hybridization solution (20 ml): Freshly made each time: 5 ml 20× SSC, 0.4 ml phosphate buffer, 2 ml 100× Denhardt's solution, 0.6 ml denatured salmon sperm DNA (3 mg/ml), 4 ml dH_2O, 4 ml polyethylene glycol (PEG) 4000, 4 ml SDS (10 %). Add SDS just before use.
11. Hybridization temperature: A good hybridization is achieved with a temperature of about 5 °C below melting temperature (Tm) for the oligonucleotide used. Tm is calculated as following: $Tm = 4(G+C) + 2(A+T)$.
12. Detection Kit for Dig-marked probes (Roche Applied Science).
13. Maleic acid buffer: 0.1 M Maleic acid, 0.15 M NaCl. Adjust pH to 7.5 with NaOH.
14. Washing buffer: Add 0.3 % (v/v) Tween 20 to Maleic acid buffer.
15. Detection buffer: 0.1 M Tris–HCl, 0.1 M NaCl, pH 9.5.
16. Blocking reagent: Dissolve Blocking reagent (bottle 5 in the detection kit for Dig-marked probes in **step 12**) in Maleic acid buffer to a final concentration of 10 % (w/v). Use shaking and heating either on a heating block or in a microwave oven. Autoclave the solution.

17. Blocking solution: Prepare a 1× working solution by diluting blocking reagent 10× in Maleic acid buffer.
18. Antibody solution: Centrifuge the antibody for 5 min at 9300 ×*g* in the original vial prior to each use. Pipette the necessary amount carefully from the top of the supernatant. Dilute anti-digoxigenin conjugated to alkaline phosphatase 1:5000 (150 mU/ml) in blocking solution.
19. Color substrate solution: Add 200 μl of the nitro blue tetrazolium/5-Bromo-4-chloro-3-indolyl phosphate (NBT/BCIP) stock solution (vial 4 in the detection kit for Dig-marked probes in **step 12**) to 10 ml of Detection buffer.

2.10 Protein Isolation from Cells and Exosomes

1. TRIzol reagent.
2. Milder protein extraction buffer, e.g., ProteoJet lysis buffer from Fermentas.
3. Ethanol 95–100 %.
4. Table top centrifuge.
5. 0.3 M guanidine hydrochloride.
6. Urea solution: 4 M urea, 0.5 % SDS, 1 M Tris–HCl pH 4.8, pipette and vortex.
7. Water-bath sonicator.
8. Extraction buffer: 50 mM 0.1 M Tris–HCl, pH 7.5, 0.1 % SDS.

2.11 Western Blot for Detection of siRNA Gene Silencing in Exosomes and Cells

1. Slot/dot blot device, e.g., Bio-Dot from Bio-Rad.
2. Filter papers and polyvinylidene difluoride (PVDF) membranes.
3. Tris-buffered saline buffer (TBS-T): 50 mM Tris–HCl pH 8, 150 mM NaCl, and 0.05 % Tween 20.
4. 10× transfer buffer stock: 30.3 g Tris base, 144.1 g glycine, H_2O 1 L. 1× transfer buffer: 100 ml 10× stock, 500 ml H_2O, 200 ml methanol or 100 % ethanol, H_2O 1 L. Store at 4 °C

2.12 Confocal Microscopy of Cells and Exosomes

1. Exosomes electroporated with siRNA tagged with a fluorochrome.
2. Diluted latex beads from Subheading 3.2, **step 2**.
3. Membrane staining reagent, e.g., CellMask Deep Red Plasma Membrane stain from Invitrogen.
4. Nucleus staining reagent, e.g., Hoechst 33342 or DAPI.
5. Mounting medium for fluorescence, e.g., Prolong Gold.
6. Confocal laser scanning microscopy, e.g., LSM 700 from Carl Zeiss.

2.13 Flow Cytometry for Detection of siRNA in Exosomes and Cells

1. Dulbecco's PBS with 1 % FBS.
2. Beriglobin or other immunoglobulin for blocking.

3. Fluorescently tagged antibodies to identify the cells of choice.
4. Viability marker for flow cytometry.
5. Flow cytometer.

3 Methods

3.1 Isolation of Exosomes

There are several methods for isolating exosomes. However, there are still no isolation methods that can guarantee total purity of the yielded vesicles. Exosomes are usually isolated from body fluids and cell culture supernatant by the use of differential centrifugation. The isolation of vesicles of endosomal origin we refer to as exosomes has been performed with differential centrifugation in three steps both from cells grown to confluence in culture medium and from plasma diluted 1:1 (*see* **Note 2**).

1. For exosome isolation from cell lines, grow cells in culture medium suitable for the cell type of interest supplemented with 10 % fetal bovine serum (FBS) and 100 mg/ml streptomycin/ penicillin. The FBS should be depleted from exosomes, to avoid the presence of any exosomes in the medium other than those produced by the cells (*see* **Note 3**). Transfer (pour to save time and plastics) the supernatant from cells cultured in 175 cm^2 bottles yielding 40 ml supernatant to 50 ml tubes (Fig. 1, step 1a).
2. For exosome isolation from plasma, add one buffy coat bag of about 40 ml to a 50 ml tube. Centrifuge at 900 × *g* at room temperature for 20 min with no breaks, i.e., zero deceleration. Remove the plasma (the upper phase) and avoid touching the cell phase (lower white/red). Transfer the plasma to a 50 ml tube and isolate exosomes as for cells from **step 8** below diluting the plasma 1:1 with PBS in the ultracentrifuge tubes (Fig. 1, step 1b).
3. Centrifuge at 300–400 × *g* for 10 min at 4 °C to collect viable cells. If you do not need viable cell pellet or isolated exosomes from plasma, centrifuge at 3000 × *g* for 15 min to remove cells (Fig. 1, step 2).
4. Transfer the supernatant to ultracentrifuge tubes with a syringe and needle. Fill up the tube with PBS if not enough supernatant to fill up the last tube (*see* Fig. 1, step 3 and **Note 4**).
5. Centrifuge at 16,500 × *g* for 30 min at 4 °C to remove cell compartments and organelles (*see* Fig. 1, step 4 and **Note 5**).
6. Harvest the supernatant (avoid the pellet) with syringe and needle. Remove the needle and attach it to a 0.2 μm filter. Put the needle/filter assembly back on the syringe and transfer the supernatant to new ultracentrifuge tubes. Fill all the way up

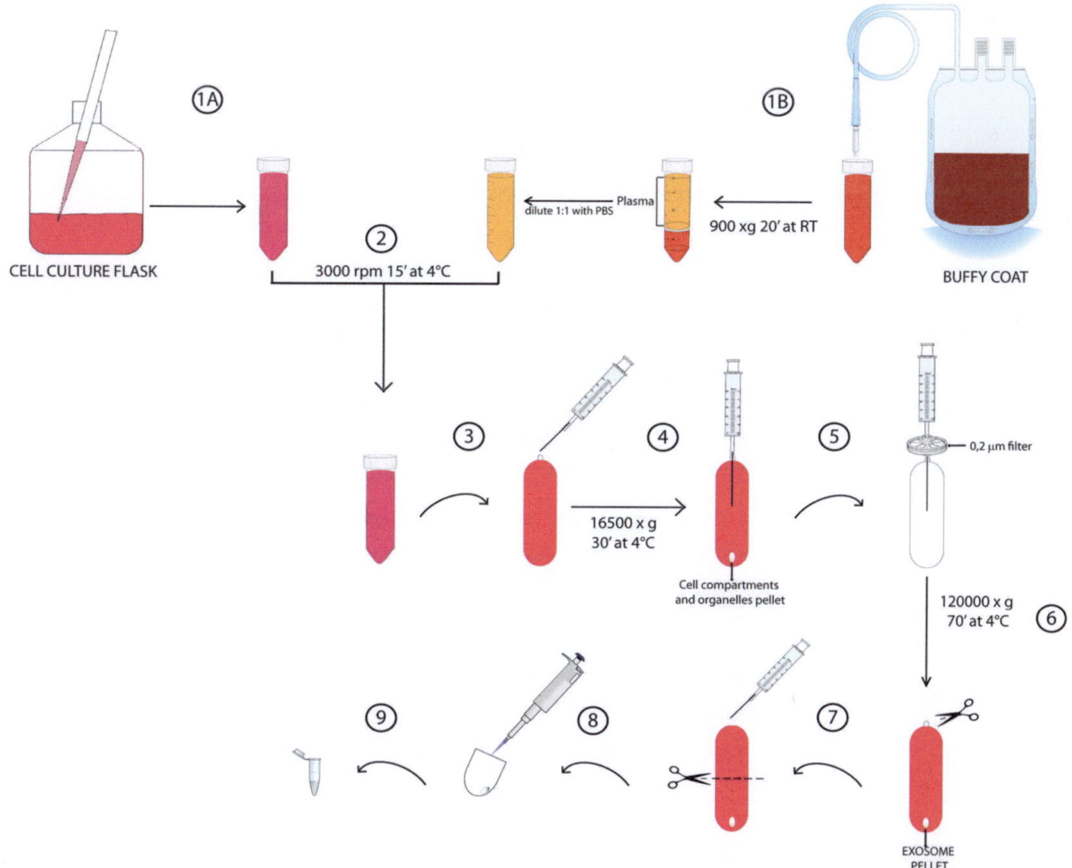

Fig. 1 Exosome isolation workflow. Transfer cell culture medium (step 1a) or plasma after centrifugation of buffy coat (step 1b) to a 50 ml tube. Pellet the cell debris at 3000 rpm, 15′ at 4 °C and transfer to an ultracentrifuge tube, centrifuge at 16,500 × *g*, for 30 min at 4 °C and collect the supernatant being careful not to detach the pellet containing cell organelles and cell compartments. Transfer the supernatant in a new ultracentrifuge tube using a 0.2 μm filter and pellet the exosomes at 120,000 × *g*, 70′ at 4 °C; cut the sealed top of the tube and remove part of the supernatant with a syringe, then cut the tube in half and pour the rest of the supernatant. Let air-dry the exosome pellet and resuspend with PBS; store the resuspended exosome sample at −80 °C or use directly for the application of interest

and avoid bubble formation, fill up with PBS if necessary (*see* Fig. 1, step 5 and **Note 6**).

7. Melt and fuse the top of the tubes with a tube topper/heating block and metal cap on the tube tip. Be careful not to break off the tip of the tube. Put the tubes on ice.
8. Pellet the exosomes by ultracentrifugation at 120,000 × *g* for 70 min at 4 °C. Keep the tubes on ice (*see* Fig. 1, step 6 and **Note 7**).
9. Cut the sealed top of the ultracentrifuge tubes. Remove three thirds of the supernatant with a syringe and needle (*see* Fig. 1, step 7).

10. Cut the tubes in half with a pair of scissors and pour off the rest of the supernatant from the bottom part of the tube (*see* Fig. 1, step 8).
11. Turn the bottom part of the tubes upside down on a tissue and let them air-dry for a couple of minutes.
12. Dissolve the exosome pellet in PBS (*see* Fig. 1, step 9 and **Note 8**).
13. Store the exosomes in a −80 °C freezer if not used directly.

3.2 Characterization of Exosomes Using Flow Cytometry

Exosomes are too small to be reliably analyzed directly by flow cytometry. A standard approach is to bind the exosomes to latex beads of a size within the detection range of a flow cytometer (usually starting from 0.5 μm). The exosomes can then be labeled with fluorescent antibodies and analyzed with flow cytometry.

1. Prepare exosome samples and use an excess of exosomes, if possible, to achieve an effective coating of the beads (*see* **Note 9**).
2. Add 5 μg beads/sample from diluted beads.
3. Incubate at 37 °C for 30 min and then overnight at 4 °C or 3 h at room temperature on a rotator or shaker.
4. Centrifuge at 4000 × *g* for 10 min (check the pellet and if it is not visible you might need to centrifuge again). Resuspend the pellet in 100 μl PBS.
5. Alternatively, exosomes may be adhered to beads coated with antibodies against exosome marker protein as in Subheading 2.2, **item 1**.
6. Add 20 μl of 1 M glycine (to get a final concentration of 100 mM) and incubate for 30 min at room temperature.
7. Fill up and wash in PBS with 0.1–1 % FBS staining buffer as in **step 11** and resuspend in 80 μl of PBS/FBS staining buffer.
8. Add 20 μl antibody or isotype of BDs pre-diluted antibody to each sample (*see* **Note 10**).
9. Incubate for 30–60 min at 4 °C.
10. Fill up and wash with PBS/FBS buffer.
11. Resuspend in 200 μl PBS/FBS. Analyze with flow cytometry (*see* **Note 11**).

3.3 Characterization of Exosomes Using Size Determination

1. Dilute exosome sample in PBS to reach a proximal concentration of 2–10 × 10^8 exosomes/ml (*see* **Note 12**).
2. Inject sample into NanoSight LM10 instrument using the syringe, in order to fill the instrument chamber a minimum required volume of 0.3 ml. Locate zero position with its thumbprint pattern next to the laser line.
3. Focus the sample, choose settings (minimum expected particle size, detection threshold, blur, extract background and minimum track length) and start the measurement [18].

3.4 Electroporation of Exosomes

1. Prepare electroporation cuvettes (we use 0.4 cm cuvettes) by radiating them with UV-light for 1 min and put them on ice (dirty cuvettes are put in 70 % ethanol for later rinsing in dH_2O).
2. Mark sample tubes.
3. When isolating exosomes for electroporation, it is better to resuspend the exosome pellet in CytoMix buffer. After resuspending the exosome pellet, read the exosome concentration and dilute in CytoMix buffer to get a 0.25–1 μg/μl concentration.
4. Add 90 μl exosomes per sample and 10 μl of siRNA with a final concentration of 2 μmole/ml. Keep the sample tubes on ice at all times (*see* Fig. 2).
5. Transfer the samples to ice cold electroporation cuvettes.
6. Electroporate the samples. Use Bio-Rad electroporation device (Gene pulser II system) and set the electroporation device to 150 V and 100 μF (*see* Fig. 2).
7. Transfer electroporated suspensions to new tubes and proceed to downstream applications, e.g., adding them to cells or binding them to latex beads (*see* Fig. 2).

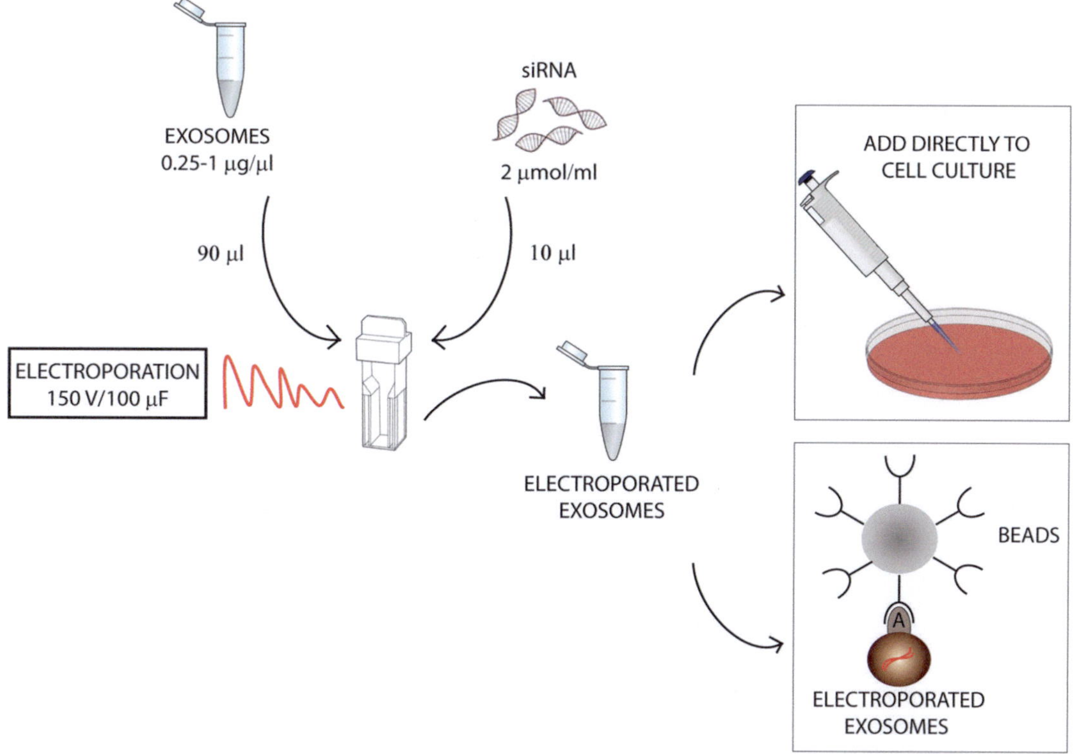

Fig. 2 Electroporation of exosomes. 90 μl of exosomes (0.25–1 μg/μl) and 10 μl of siRNA (2 μmol/ml) are transferred to a 0.4 cm electroporation cuvette. Samples are electroporated at 150 V and 100 μF. The electroporated sample can be directly added to cell culture or bound to beads for downstream applications

3.5 Preparation of Peripheral Blood Mononuclear Cells from Buffy Coat

1. Dilute 25 ml buffy coat with 25 ml Dulbecco's PBS.
2. Add 15 ml Lymphoprep to four 50 ml tubes.
3. Carefully pipette the buffy coat on top of Lymphoprep reagent.
4. Centrifuge for 20 min at 800 × *g*, at room temperature with no deceleration.
5. Recover peripheral blood mononuclear cells (PBMC) from the plasma/Lymphoprep interface and transfer the cells to a new tube.
6. Wash the PBMC three times with isolation buffer by centrifugation at 400 × *g* for 8 min at 2–8 °C.
7. Resuspend the PBMC in 20 ml isolation buffer and count the cells.

3.6 Isolation of Monocytes and Lymphocytes

1. Wash PBMC in MACS buffer, count cells and transfer 200×10^6 cells to a new tube.
2. Centrifuge at 300 × *g* for 10 min at 4 °C. Remove the supernatant and resuspend the pellet in 10 ml 1× ammonium chloride in a 50 ml tube.
3. Incubate for 7 min at room temperature (*see* **Note 13**).
4. Add MACS buffer up to 50 ml and centrifuge at 300 × *g* for 10 min at 4 °C.
5. Count the cells again and transfer 100×10^6 cells to a new tube. Centrifuge as in **step 4**.
6. Remove the supernatant and resuspend the cells in 800 μl MACS buffer in a 10 ml tube.
7. Add 200 μl $CD14^+$ beads and incubate for 15 min at 4 °C.
8. Wash the mixture in 10 ml MACS buffer at 300 × *g* for 10 min at 4 °C. Remove the supernatant completely.
9. Resuspend the pellet in 500 μl MACS buffer and put on ice.
10. Place the MidiMACS magnet in the holding plate.
11. Rinse the LS-column with 3 ml MACS buffer by putting a 10 ml tube underneath and press down the MACS buffer through the column.
12. Apply the cell suspension onto the column. Collect unlabeled $CD14^-$ cells (i.e., defined as lymphocytes) into a 10 ml tube. Just leave the suspension to go through drop by drop, do not press down.
13. Wash the column three times by adding 3 ml MACS buffer and leave it to pass through drop by drop. Only add new buffer when the column reservoir is empty. Put the tube on ice.
14. Remove the column from the separator and place it on a new 10 ml tube called $CD14^+$ (monocytes). Add 5 ml MACS buffer

onto the column. Immediately flush out the magnetically labeled cells by firmly pushing the plunger into the column.

15. Count both $CD14^-$ and $CD14^+$ cells (*see* **Note 14**).

3.7 Exosome Delivery Assay

1. Seed monocytes and lymphocytes separately in 6-well or 24-well plates (*see* **Note 15**).
2. Stimulate the cells with Phorbol-myristate-acetate (PMA) at a final concentration of 1 μg/ml medium (*see* **Note 16**).
3. Add electroporated plasma exosome samples to corresponding wells. Pulse the cells with non-electroporated exosomes from plasma as a negative control.
4. Incubate the cells for 36 h at 37 °C with 5 % CO_2 (*see* **Note 17**).

3.8 RNA Isolation from Cells and Exosomes

For RNA isolation from cells, the manufacturer's protocol was used for the TRIzol reagent. However, the protocol was slightly modified for exosome isolation as described below.

1. This protocol is set for 1 ml TRIzol reagent used (*see* **Note 18**).
2. Lyse the exosome pellet directly with 1 ml TRIzol and keep at room temperature for 10–15 min to allow a complete lysis (*see* **Note 19**).
3. Add 200 μl of chloroform, shake vigorously and keep at room temperature for 2–3 min, then centrifuge at 12,000 × *g* at 4 °C for 15 min (*see* **Note 20**).
4. Save the aqueous phase to a new tube, add 1–2 μl of glycogen and vortex. Add 500 μl isopropanol, vortex and keep at room temperature for 10 min. Precipitate the samples overnight at −80 °C.
5. Centrifuge the samples at 12,000 × *g* at 4 °C for 10 min; discard the supernatant and add 1 ml ethanol (75 % ice cold). Incubate for at least 20 min at −20 °C.
6. Centrifuge at 8000–12,000 × *g* for 10 min at 4 °C. Pour off the supernatant and aspirate the leftover drops with a vacuum pump being careful not to aspirate the pellet, which should be white in color because of the glycogen (*see* **Note 21**).
7. Air-dry the pellet for 5–10 min and resuspend in RNase free water (10 μl starting amount if you do not see any pellet). Store at −80 °C.

3.9 Northern Slot Blot for Detection of siRNA in Exosomes and Cells

1. Clean all the parts of the Bio-Dot device. Do not use ethanol.
2. Cut out a membrane and five filter papers to fit in the Bio-Dot device.
3. Pre-wet the membrane in 5× SSC buffer. Place the membrane on top of the five filters and place the whole stack under the sealing gasket.

4. For loading of the RNA into the Bio-Dot device use 20 μg RNA in 5 μl. Mix with 2 μl 10× NBC, 3 μl formaldehyde, and 10 μl formamide. Incubate for 5 min at 65 °C to denature the RNA. Put the sample tubes on ice.
5. Load the denatured RNA into each well in the Bio-Dot device. Add 500 μl of 2× SSC buffer into each well. Attach the Bio-Dot device to vacuum and turn on the vacuum after each application of sample/buffer.
6. Remove the membrane and fix/crosslink the RNA to the membrane by placing it on a UV-table and run (120,000 μJ/cm^2) during 1 min (do not exceed 1 min).
7. To perform pre-hybridization, rinse a hybridization container/flask with DEPC-H_2O. Insert the RNA membrane into the container with the RNA facing the inside. Pour 20 ml of the pre-hybridization solution into the container.
8. Incubate at hybridization temperature: for the Dig probe it will be 65 °C; calculate the temperature for the probe (it should be around 55–70 °C). Incubate for 2–4 h.
9. To continue with the hybridization pour off the pre-hybridization solution, add the desired amount of probe into the hybridization solution (we use a final concentration of 2 nmole/L probe in 20 ml hybridization solution). Hybridize overnight at hybridization temperature.
10. Wash the membranes in 1× SSC with 0.1 % SDS twice at room temperature for 10 min and once at hybridization temperature for 10 min. Let dry between filter papers.
11. To continue with the immunodetection of the Dig-probe after the hybridization, rinse the membrane one time in washing buffer. Incubate for 30 min in 100 ml blocking solution. Incubate for 30 min in 20 ml Antibody solution. Wash twice for 15 min with 100 ml of washing buffer. Equilibrate 2–5 min in 20 ml Detection buffer.
12. Incubate the membrane in 10 ml freshly prepared Color substrate solution in an appropriate container in the dark. Do not shake during color development (*see* **Note 22**).
13. Stop the reaction when desired spot or band intensities are achieved by using 50 ml of sterile H_2O or TE-buffer. The results can be documented by photocopying the wet filter or by photography (*see* **Note 23**).

3.10 Protein Isolation from Cells and Exosomes

1. From the exosome samples: proteins can be isolated from the same sample after removing the remaining aqueous phase from RNA isolation with TRIzol reagent (*see* **Note 24**). For isolation from TRIzol samples add 300 μl of ethanol. Mix by inversion several times. Incubate for 5 min at room temperature.

2. Centrifuge at 2000 × *g* for 5 min at 4 °C to pellet DNA which for exosomes will be very small or not visible. Transfer the supernatant containing the proteins into a 2 ml tube by pipetting (*see* **Note 25**).
3. Add 1.3 ml of isopropanol mix by inverting the tube.
4. Incubate 10 min at room temperature allowing a thick precipitate to form.
5. Centrifuge at 12,000 × *g* for 10 min at 4 °C to pellet the proteins. Remove the supernatant by pouring or pipetting to make that the pellet is not moving.
6. Add 1 ml of 0.3 M guanidine hydrochloride in 95 % ethanol to the protein pellet. Incubate for 20 min at room temperature (*see* **Note 26**). Centrifuge at 7600 × *g* for 5 min at 4 °C and remove the supernatant. Repeat this step twice (three washes in total).
7. Add 2 ml ethanol 100 % to protein pellet and vortex. Incubate for 20 min at room temperature. Centrifuge at 7600 × *g* for 5 min at 4 °C and remove the ethanol.
8. Let the pellet air-dry for 5–10 min. Resuspend the pellet by adding 100–200 μl urea solution. Put the samples in water-bath sonicator to dissolve the pellet.
9. Centrifuge at 10,000 × *g* for 10 min at 4 °C to sediment insoluble material. Transfer the supernatant containing the proteins to a new tube.
10. Direct extraction from the exosome isolate is performed by diluting the exosome pellet in 50–75 μl of extraction buffer. Heat the mixture 90–95 °C for 5 min. Vortex the mixture thoroughly and centrifuge at 10,000 × *g* for 2–4 min. Add more extraction buffer if the lysate is too thick, centrifuge again. Transfer the supernatant containing the proteins to a new tube.
11. The protein concentration can be measured with any established method available in your laboratory. We use the Bio-Rad Protein Assay Kit or the Qubit protein assay from Invitrogen.

3.11 Western Slot Blot for Detection of siRNA Gene Silencing in Exosomes and Cells

To detect exosome specific proteins in the exosome isolate and to detect the posttranscriptional effect of the siRNA electroporated into exosomes and taken up by recipient cells, we used Western slot blot. Slot blot is a faster method than standard Western blot and is applicable when separation of proteins according to size is not needed or when the antibody is verified to be highly specific for the target protein. Load the same amount of extracted proteins (lysate) from each sample to be able to compare quantities of the expressed protein.

1. Clean all the parts of the Bio-Dot device with dH_2O. Do not use ethanol.
2. Cut five filter papers and one PVDF membrane to fit in the Bio-Dot device.
3. Prepare sample by adding protein extraction buffer to make all samples in the same volume.
4. Prepare the membrane by incubating in 5 % ethanol for 5 s, 5 min in dH_2O and 10–20 min in cold transfer buffer on a shaker. Pre-wet the filter papers in cold transfer buffer. Put five filter papers in the Bio-Dot device using tweezers. Put the membrane on top of the filters.
5. Put the lid on by tighten the screws cross wise. Fasten the vacuum tubing and make sure there is no leakage. Apply vacuum. Load 500 μl transfer buffer in unused wells wait until all buffer have passed through to control vacuum function.
6. Load the samples according to pre-decided loading scheme. Let all liquid pass through before turning off the vacuum.
7. Fixate the membrane on UV-board for 1 min.
8. For immunostaining, block the membrane in 5 % skimmed milk/TBST for 1 h at room temperature on a shaker or overnight at 4 °C.
9. Wash the membrane for 10 min in TBST three times, at room temperature on a shaker.
10. Mark the membrane with pencil and cut the lower right corner for orientation.
11. Add primary antibody in 5 % skimmed milk/TBST (*see* **Note 27**). Incubate for 1 h at room temperature on a shaker. Make sure the membrane is covered. Wash as in **step 8**.
12. Add secondary antibody to 5 % skimmed milk/TBST (*see* **Notes 27** and **28**). Wash as in **step 8**.
13. For detection of protein bands use any method available in your laboratory (*see* **Note 29**).

3.12 Confocal Microscopy of Cells and Exosomes

1. Preparing samples for detection of siRNA inside the exosomes after electroporation: Dilute beads as in **step 2** in Subheading 3.2. Add 40 μl diluted beads/sample. Incubate for 30 min at 37 °C. Incubate at 4 °C overnight or for 2 h at room temperature. Wash twice in 200 μl PBS and centrifuge at 4000 × *g* for 10 min at 4 °C. Resuspend in 200 μl PBS. Analyze with confocal microscopy in the channel used for the fluorochrome of interest. We used siRNA tagged with Alexa Fluor-488.
2. Wash harvested monocytes and lymphocytes with PBS to analyze cells from exosome delivery.

3. Immobilize the cells on a microscope slide by using Cytospin.
4. Add cell membrane staining reagent, e.g., CellMask Deep Red Plasma Membrane stain from Invitrogen at a final concentration of 2 μg/ml in PBS. Incubate for 3 min at 37 °C. Wash twice in PBS.
5. Add nucleus staining reagent, e.g., Hoechst 33342 at a concentration of 2 μg/ml.
6. Mount the slides in Prolong Gold.
7. Analyze samples with a confocal laser scanning microscopy, e.g., LSM 700 from Carl Zeiss.

3.13 Flow Cytometry for Detection of siRNA in Exosomes and Cells

1. For flow cytometry of electroporated exosomes: add 5 μg beads to each sample as in Subheading 3.2, **item 2** and incubate at 37 °C for 30 min and then at 4 °C overnight or for 3 h at room temperature on a rotator or shaker. Analyze on a flow cytometer in the channel where the fluorochrome attached to the siRNA of choice can be detected (we used Alexa-flour 488 tagged siRNA but others are available).
2. For flow cytometry of cells pulsed with exosomes electroporated with a fluorescent siRNA, the cells need to be harvested.
3. Lymphocytes are non-adherent and are collected by pipetting to a new tube.
4. Monocytes are adherent and need to be incubated with trypsin to detach. Add trypsin directly to the well and put in 37 °C for 5 min at the time until they loosen from the plate. Resuspend and transfer to new tube.
5. Wash cells twice in PBS with 1 % FBS by resuspending and centrifuging at 300 ×*g* for 5 min at 4 °C.
6. Fc-block for 15 min at room temperature with Beriglobin at 1 mg/ml (*see* **Note 30**).
7. Stain cells with cell type specific surface antibodies (*see* **Note 31**).
8. Incubate for 20–30 min at 4 °C protected from light. Wash as in **step 5**.
9. Stain with viability marker of choice (e.g., Live/dead fixable violet dead cell stain kit from Invitrogen).
10. Wash as in **step 5** and analyze on a flow cytometer.

4 Notes

1. Denature DNA/RNA by heating in 100 °C during 5 min and then quickly cool the sample on ice and keep it on ice.
2. To get a high yield of exosomes we culture cells in 175 cm^2 flasks.

3. To deplete the FBS from exosomes thaw 500 ml bottle of FBS at room temperature or at 37 °C. Heat-inactivate the FBS in water bath at 56 °C for 1 h. Transfer the FBS with syringe and needle to polyallomer tubes. Fill all the way up and avoid bubbles (do not fill up with PBS here since you do not want the FBS to be diluted). Melt and fuse the top of the tubes with a tube topper or heating block and metal cap, be sure not to break off the top. Put on ice. Centrifuge at 120,000 × *g* for 90 min at 4 °C. Put on ice. Cut the top of the tubes. Sterile filter the supernatant with a syringe and needle and 0.2 μm filters to 50 ml tubes. Fill up to 50 ml in each tube. Put in −20 °C freezer for storage.
4. Only fill up to the top "line" visible in the upper part of the polyallomer tube, in this way there is no need to seal the tube (*see* Fig. 1.).
5. Do not forget to cap the empty spaces in the rotor as well and make sure the O-ring is in place on the lid. Circle the pellet with marker pen and put on ice. The pellet can be invisible or transparent when isolating exosomes from plasma. For cultured cells the exosome pellet usually has some color remaining from the culture medium.
6. In this way vesicles and aggregates larger than 200 nm will be removed.
7. At this point it might be useful to circle the exosome pellet with a pen to highlight it for the resuspension step.
8. Add 200–300 μl PBS. We have used 150 μl PBS as a starting point (1× dilution) then we just added more depending on how much starting material and the size of the pellet. Take first 150 μl and dissolve the pellet of one tube, take it up and put in the next tube etc. Transfer to 1.5 ml tube and repeat the procedure one more time to get any remaining exosomes. If there are a lot of bubbles when resuspending it helps to spin down in the 1.5 ml tube to make room. Add TRIzol directly to the pellet for RNA isolation.
9. We use an exosome concentration of 0.6 μg/μl in 200 μl PBS for adhesion to beads.
10. We use the recommended volume of pre-diluted antibody from BD biosciences using 20 μl of antibody solution in 90 μl PBS, which lies in the range of 6 μg/ml to 50 μg/ml per sample. For this application titration of antibody is not crucial since it is not a quantitative measurement.
11. We use a FACSCantoII flow cytometer from BD. We routinely check the presence of three different tetraspanins when characterizing exosomes from a new cell type. We prepare one sample for each tetraspanin and use antibodies pre-conjugated to the same fluorochrome enabling the use the same isotype control for all samples (assuming the IgG is correct).

12. If using the NanoSight with no previous knowledge or experience of particle concentration in your sample, it is advisable to start with a low volume and dilute further with PBS if needed after achieving a concentration from the NanoSight. Just draw your sample out of the chamber with the syringe and dilute further and finally inject the diluted sample into the chamber for reanalysis.
13. Do not incubate longer time since it will cause lysis of the other cells. Use a timer.
14. Expected yield of monocytes is 10–30 % of total PBMC but it varies between donors.
15. We seed at a density of 3×10^6 cells/well in 3 ml in 6-well plates and of 0.5×10^6 cells/well in 0.5 ml in 24 well plates.
16. Leave the cells to settle for some hours. In the meantime prepare and electroporate the exosomes as described in Subheadings 4.1 and 4.3.
17. We could also see delivery and effect at 24, 48 and 72 h. 36 h was used to detect post-transcriptional gene silencing and 24 h was used to detect delivery of siRNA to cells.
18. The amount of TRIzol is decided based on the size of the exosome pellet. Generally 1 ml TRIzol is good unless you do not really see the exosome pellet then 500 μl is sufficient or even 250 μl. If decreasing the amount of TRIzol you have to scale down all the other reagents in the following steps.
19. Electroporated exosomes should be isolated by binding to latex beads as in Subheading 3.2 but incubated for 1 h at room temperature before washing twice in PBS. TRIzol should be added directly to bead/exosome pellet. For adherent cells add TRIzol directly to culture wells after removing the supernatant.
20. At this point it can happen that after centrifugation the aqueous phase is not visible, and instead there is a thick white fatty phase indicating that more TRIzol needs to be added. An exosome pellet is mainly rich in membranes and this can cause the release of a high amount of fat, in this case add more TRIzol, shake and vortex and repeat **step 2**. After this step, the phenolic phase may be used for exosomal protein extraction.
21. Be careful at this step since the pellet is generally small and it detaches easily, you might need to recentrifuge. Alternatively, aspirate by capillarity instead of using the vacuum pump, this way you will be sure the pellet will not be aspirated accidentally.
22. The color precipitate starts to form within a few minutes and the reaction is usually complete after 16 h. The membrane can be exposed to light for short time periods to monitor color development.

23. If you want to re-probe the membrane it should never be allowed to dry, store it in a plastic bag. If you want to maintain the color, store the membrane in TE-buffer and do not allow it to dry. If you do not want to re-probe the membrane dry it at 15–25 °C for storage. Color fades upon drying. To revitalize color soak the membrane in TE-buffer.
24. Protein isolation from cells can be performed with a milder extraction reagent to keep the proteins in their native state. It is important to check the datasheet of the antibody and make sure that it can detect denatured proteins if not it is essential to use a mild extraction buffer. In our application we used ProteoJet Lysis buffer.
25. This phenol-ethanol containing supernatant can be stored for several months at −80 °C.
26. Samples can be stored like this for 1 month at 4 °C or 1 year at −20 °C.
27. The optimal percentage of milk in the antibody solution might differ based on the specificity and sensitivity of the primary and secondary antibody used, testing of the optimal percentage is recommended.
28. Do not use skimmed milk when using antibodies for infrared detection. Incubate for 1 h at room temperature on a shaker.
29. We use chemiluminescence detection with ECL plus Kit from Amersham or infrared imaging depending on the conjugate of the secondary detection antibody.
30. One important way to minimize nonspecific binding is by the use of a so-called blocking reagent. A blocking reagent contains a high concentration of immunoglobulin that will bind to the Fc-receptors on cells like monocytes and some T cells, thereby blocking the nonspecific binding of the staining antibody reagents to these receptors. The blocking reagent should be immunoglobulin from the species whose cells you are staining. Alternatively, one can purchase specific antibodies directed against the Fc receptors of the cells in question.
31. We used antibodies against CD19, CD3 (to identify the lymphocytes) and CD14 (to identify the monocytes).

References

1. Thery C, Regnault A, Garin J, Wolfers J, Zitvogel L, Ricciardi-Castagnoli P, Raposo G, Amigorena S (1999) Molecular characterization of dendritic cell-derived exosomes. Selective accumulation of the heat shock protein hsc73. J Cell Biol 147:599–610
2. Raposo G, Nijman HW, Stoorvogel W, Liejendekker R, Harding CV, Melief CJ, Geuze HJ (1996) B lymphocytes secrete antigen-presenting vesicles. J Exp Med 183:1161–1172
3. Blanchard N, Lankar D, Faure F, Regnault A, Dumont C, Raposo G, Hivroz C (2002) TCR activation of human T cells induces the production of exosomes bearing the TCR/CD3/zeta complex. J Immunol 168:3235–3241

4. Wahlgren J, Karlson Tde L, Glader P, Telemo E, Valadi H (2012) Activated human T cells secrete exosomes that participate in IL-2 mediated immune response signaling. PLoS One 7, e49723
5. Skokos D, Le Panse S, Villa I, Rousselle JC, Peronet R, David B, Namane A, Mecheri S (2001) Mast cell-dependent B and T lymphocyte activation is mediated by the secretion of immunologically active exosomes. J Immunol 166:868–876
6. Karlsson M, Lundin S, Dahlgren U, Kahu H, Pettersson I, Telemo E (2001) "Tolerosomes" are produced by intestinal epithelial cells. Eur J Immunol 31:2892–2900
7. Mears R, Craven RA, Hanrahan S, Totty N, Upton C, Young SL, Patel P, Selby PJ, Banks RE (2004) Proteomic analysis of melanoma-derived exosomes by two-dimensional polyacrylamide gel electrophoresis and mass spectrometry. Proteomics 4:4019–4031
8. Caby MP, Lankar D, Vincendeau-Scherrer C, Raposo G, Bonnerot C (2005) Exosomal-like vesicles are present in human blood plasma. Int Immunol 17:879–887
9. Skogberg G, Gudmundsdottir J, van der Post S, Sandstrom K, Bruhn S, Benson M, Mincheva-Nilsson L, Baranov V, Telemo E, Ekwall O (2013) Characterization of human thymic exosomes. PLoS One 8, e67554
10. Pisitkun T, Shen RF, Knepper MA (2004) Identification and proteomic profiling of exosomes in human urine. Proc Natl Acad Sci U S A 101:13368–13373
11. Andre F, Schartz NE, Movassagh M, Flament C, Pautier P, Morice P, Pomel C, Lhomme C, Escudier B, Le Chevalier T et al (2002) Malignant effusions and immunogenic tumour-derived exosomes. Lancet 360:295–305
12. Admyre C, Grunewald J, Thyberg J, Gripenback S, Tornling G, Eklund A, Scheynius A, Gabrielsson S (2003) Exosomes with major histocompatibility complex class II and co-stimulatory molecules are present in human BAL fluid. Eur Respir J 22: 578–583
13. Sprent J (2005) Direct stimulation of naive T cells by antigen-presenting cell vesicles. Blood Cells Mol Dis 35:17–20
14. Valadi H, Ekstrom K, Bossios A, Sjostrand M, Lee JJ, Lotvall JO (2007) Exosome-mediated transfer of mRNAs and microRNAs is a novel mechanism of genetic exchange between cells. Nat Cell Biol 9:654–659
15. Wahlgren J, Karlson TD, Brisslert M, Vaziri Sani F, Telemo E, Sunnerhagen P, Valadi H (2012) Plasma exosomes can deliver exogenous short interfering RNA to monocytes and lymphocytes. Nucleic Acids Res 40, e130
16. Alvarez-Erviti L, Seow Y, Yin H, Betts C, Lakhal S, Wood MJ (2011) Delivery of siRNA to the mouse brain by systemic injection of targeted exosomes. Nat Biotechnol 29: 341–345
17. van den Hoff MJ, Moorman AF, Lamers WH (1992) Electroporation in 'intracellular' buffer increases cell survival. Nucleic Acids Res 20:2902
18. Gardiner C, Ferreira YJ, Dragovic RA, Redman CW, Sargent IL (2013) Extracellular vesicle sizing and enumeration by nanoparticle tracking analysis. J Extracell Vesicles 2

Chapter 11

Dendrimer Nanovectors for SiRNA Delivery

Xiaoxuan Liu and Ling Peng

Abstract

Small interfering RNA (SiRNA) delivery remains a major challenge in RNAi-based therapy. Dendrimers are emerging as appealing nonviral vectors for SiRNA delivery thanks to their well-defined architecture and their unique cooperativity and multivalency confined within a nanostructure. We have recently demonstrated that structurally flexible poly(amidoamine) (PAMAM) dendrimers are safe and effective nanovectors for SiRNA delivery in various disease models in vitro and in vivo. The present chapter showcases these dendrimers can package different SiRNA molecules into stable and nanosized particles, which protect SiRNA from degradation and promote cellular uptake of SiRNA, resulting in potent gene silencing at both mRNA and protein level in the prostate cancer cell model. Our results demonstrate this set of flexible PAMAM dendrimers are indeed safe and effective nonviral vectors for SiRNA delivery and hold great promise for further applications in functional genomics and RNAi-based therapies.

Key words Dendrimer, Nonviral vectors, RNA interference, SiRNA delivery

1 Introduction

Since the discovery of RNA interference (RNAi), small interfering RNA (SiRNA) is emerging as a novel therapeutics for treating diverse diseases ranging from viral infections to hereditary disorders and cancers [1, 2]. However, SiRNA is highly negatively charged and hydrophilic, hence not ready to cross cell membranes and enter into cells. In addition, naked SiRNA is rapidly degraded by various enzymes circulating in the bloodstream. Consequently, safe and efficient delivery, which can prevent SiRNA degradation and bring SiRNA to the site of interest, is the key issue to realize the clinical implementation of RNAi therapeutics [3, 4].

Both viral and nonviral vectors have been applied for SiRNA delivery [3, 5]. Although viral vectors are very effective, the increasing concerns over their safety and the high cost relating to their complicated manipulation substantiate the need to develop alternative nonviral vectors. Comparing to viral vectors, nonviral vectors bear the advantages of being noninfectious and eliciting

Kato Shum and John Rossi (eds.), *SiRNA Delivery Methods: Methods and Protocols*, Methods in Molecular Biology, vol. 1364, DOI 10.1007/978-1-4939-3112-5_11, © Springer Science+Business Media New York 2016

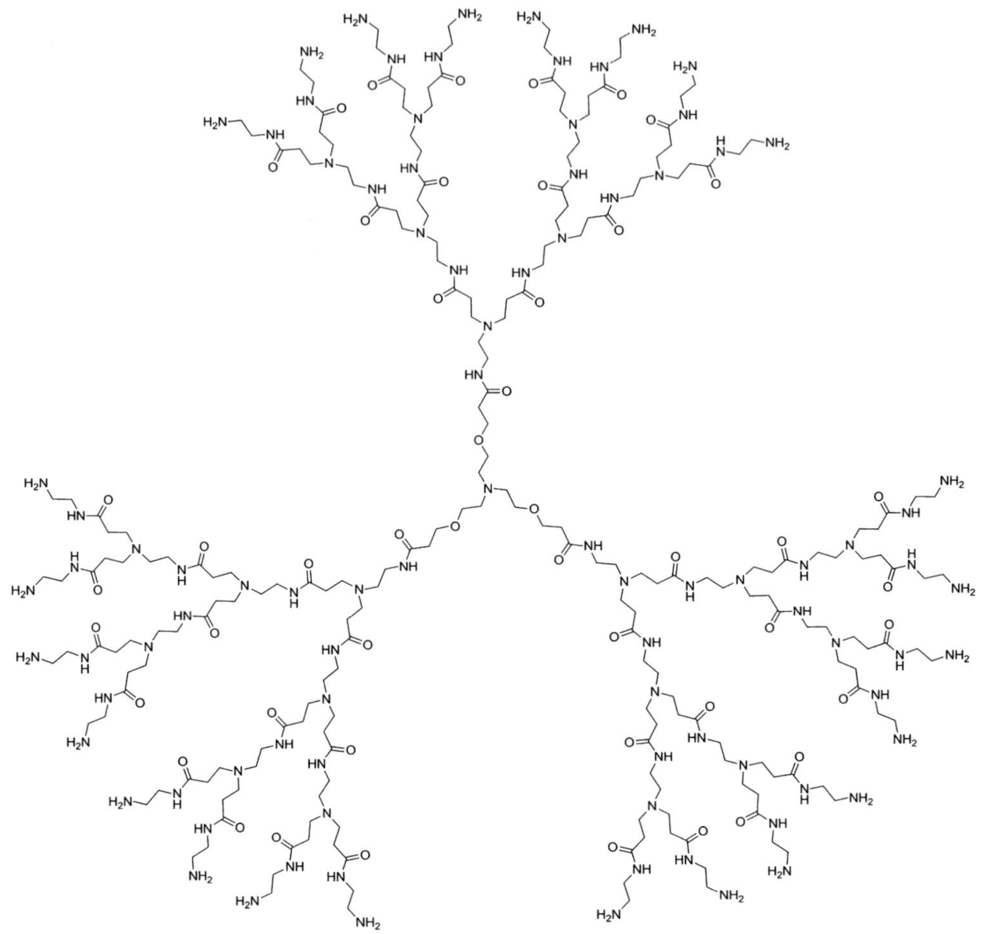

Fig. 1 Structurally flexible triethanolamine (TEA) core poly(amidoamine) (PAMAM) dendrimer of generation 3 (**G**$_3$). Adapted with permission from Mol. Pharmaceutics 2012; 9: 470–481. Copyright 2012 American Chemical Society

no or only weak immune responses alongside their convenient preparation and use.

Among various nonviral vectors developed for SiRNA delivery in the past years, dendrimers, a special type of synthetic macromolecules [6], are emerging as ideal nonviral vectors for SiRNA delivery by virtue of their unique well-defined structure and multivalent cooperativity [7]. Inspired from the DNA binding proteins, histones, which undergo conformational changes when binding to DNA to form nucleosomes, we have been particularly interested in structurally flexible poly(amidoamine) (PAMAM) dendrimers for SiRNA binding and SiRNA delivery, and hence established a series of PAMAM dendrimers bearing an extended triethanolamine (TEA) core [8] (Fig. 1). Under physiological conditions, these dendrimers have positively charged amine functionalities at the dendrimer surface, enabling condensation with negatively charged

SiRNA molecules through electrostatic interactions. It is to note that these dendrimers also harbor tertiary amines in the interior, which can be protonated in acidic endosomes and favorably promote SiRNA release via the "proton sponge" effect [9]. The so-released SiRNAs eventually join the RNAi machinery to initiate the gene silencing process. Importantly, the triethanolamine (TEA) dendrimer core featuring an extended structure, grants the dendrimers with flexible structure accompanied with less densely packed branching units and greater access for water and SiRNA, thereby favoring both SiRNA binding and release processes [10, 11]. These dendrimers have proved to be excellent vectors for the functional delivery of diverse RNAi molecules in different cell types and various disease models in vitro and in vivo [11–17].

In this chapter, we describe the dendrimer-mediated delivery of SiRNAs targeting the heat-shock protein 27 (Hsp27) in prostate cancer PC-3 cells as an model example [11, 13, 17]. In these delivery systems, Hsp27 SiRNA molecules have been packaged by the dendrimer vectors to form stable and compact nanoparticles, which can protect SiRNA from degradation and promote cellular uptake of SiRNA. The Hsp27 SiRNA delivered by dendrimers has led to effective gene silencing at both mRNA and protein level. Therefore, these TEA-core PAMAM dendrimers are indeed effective vectors for SiRNA delivery and hold great promise for applications in both functional genomics and RNAi-based therapies.

2 Materials

Triethanolamine (TEA)-core PAMAM generation 5 and 7 dendrimers are used (*see* **Note 1**). Make the stock solution with desired amount of water and store in aliquots at –20 °C (*see* **Notes 2** and **3**). Different types of SiRNA molecules and scramble SiRNAs are used as follows: Hsp27 SiRNA and scramble SiRNA (Thermo Fisher Scientific), Hsp27 dicer substrate SiRNA and scramble dicer substrate SiRNA (Integrated DNA Technologies), Hsp27 sticky SiRNA and scramble sticky SiRNA (Eurogentec). Make the stock solution with desire amount of RNase-free water and store in aliquots at –20 °C (for short-term storage) or –80 °C (for long-term storage).

2.1 Gel Shift Assays

1. Agarose. Store at room temperature.
2. 10× TBE. Store at room temperature.
3. 0.5× TBE running buffer: Dilute 10× TBE with water. Store at room temperature.
4. 10 μg/μL RNase A. Dilute to 0.01 μg/μL. Store in aliquots at –20 °C.

5. 10 mg/mL Ethidium bromide (EB) (*see* **Note 4**). Store at 4 °C.
6. Loading dye.
7. Herolab EASY CCD camera (type 429 K) (Herolab).

2.2 In Vitro Transfection

1. Different transfection plates: 4-well glass chamber slides (Labtek, Nunc), 3.5 cm dishes (Nunc), 6 cm dishes (Nunc).
2. 1×PBS.
3. Opti-MEM tranfection medium (Gibco).

2.3 Live-Cell Confocal Microscopy

1. Fluorescent labeled SiRNAs: Alexa488 SiRNA (Life technologies), Cy3 dicer substrate SiRNA (Integrated DNA Technologies), Alexa647 sticky SiRNA (Eurogentec). Make the stock solution with desire amount of RNase-free water and store in aliquots at −20 °C (for short-term storage) or −80 °C (for long-term storage) in the dark.
2. Hoechst 34580 (Invitrogen): Prepare 10 mg/mL solution in water and store in aliquots at −20 °C in the dark.
3. Zeiss LSM 510 Meta laser scanning confocal microscope equipped with inverted Zeiss Axiovert 200 M stand (Carl Zeiss GmbH).

2.4 Quantitative Real-Time (qRT)-PCR

1. TRIzol (*see* **Note 5**). Store at 4 °C.
2. AllPrep DNA/RNA/Protein Mini Kit. Store at room temperature.
3. PCR primers. Prepare 10 mM stock solution in water and store in aliquots at −20 °C.
4. ImProm-II™ Reverse Transcription kit. Store at −20 °C.
5. 2× SYBR Premix Ex Taq. Store at −20 °C in the dark.
6. RNase-free water. Store at −20 °C.
7. LightCycler 2.0 instrument.

2.5 Sodium Dodecyl Sulfate Polyacrylamide Gel

1. 40 % acrylamide/bis-acrylamide solution. This is a neurotoxin when unpolymerized. Take appropriate measures to prevent exposure. Store at 4 °C.
2. *N*,*N*,*N*,*N*′-Tetramethyl-ethylenediamine (TEMED). Store at 4 °C.
3. Ammonium persulfate (APS): Prepare 10 % solution in water and immediately freeze in single-use (150 μL) aliquots at −20 °C.
4. 1.5 M Tris–HCl, pH 8.8. Store at room temperature.
5. 1.0 M Tris–HCl, pH 6.8. Store at room temperature.
6. 10 % SDS. Store at room temperature.
7. 10× Running buffer: 30.2 g Tris and 144.4 g Glycine. Dissolve in around 800 mL water, add 100 mL 10 % SDS, and then

complete to 1 L with water after all the powders totally dissolved. Store at room temperature.

8. 1× Running buffer: Dilute 10× running buffer with water. Store at room temperature.

2.6 Western Blot

1. Lysis buffer: 10 % Glycerol, 150 mM NaCl, 50 mM HEPES, 1 mM EDTA, 25 mM NaF, 10 μM $ZnCl_2$, 1 mM EGTA, and 1 % Triton in 500 mL water. Store at 4 °C.
2. Cocktail inhibitor. Prepare the 25× stock solution with water and store in aliquots at –20 °C.
3. Na_3VO_4. Prepare the stock solution (10 mg/mL) with water and store in aliquots at –20 °C.
4. Pierce BCA protein assay kit.
5. Polyvinyl difluoride (PVDF) membranes.
6. 10× Western blot transfer buffer: 30 g Tris and 150 g glycine; dissolve in around 800 mL water, add 100 mL 10 % SDS, and then complete to 1 L with water after all the powders totally dissolved. Store at room temperature.
7. 1× Western blot transfer buffer: 100 mL of 10× Western blot transfer buffer, 200 mL absolute EtOH, complete to 1 L with water. Store at room temperature.
8. 10× Tris buffered saline (TBS). 30.27 g Tris and 84.16 g NaCl (Sigma). Dissolve in around 800 mL of water. Adjust the pH of the solution to 7.4 after the powders totally dissolved, then complete to 1 L with water. Store at 4 °C.
9. 1× TBS: dilute 10× TBS with water. Store at room temperature.
10. Milk. Store at room temperature
11. Bovine Serum Albumin. Store at 4 °C.
12. Specific antibodies. Store in aliquots at –20 °C.
13. SuperSignal West Pico Chemiluminescent Substrate (Pierce Thermo Scientific).
14. Tabletop processor CURIX60 (AGFA).

3 Methods

Several types of SiRNA molecules including conventional SiRNA, sticky SiRNA (sSiRNA), and Dicer-substrate SiRNA (dSiRNA) are designed to downregulate Hsp27 in the prostate cancer model [11, 13, 17] (Fig. 2). Hsp27, an ATP-dependent molecular chaperone, is a vital regulator of cell survival and plays an important role in drug resistance [18], thus emerging as an attractive novel target for treating drug-resistant cancers such as castration-resistant prostate cancer [19–21]. Hsp27 conventional SiRNA is synthetic

a Conventional siRNA

5'-GCUGCAAAAUCCGAUGAGAC dTdT-3'
3'-dTdT CGACGUUUUAGGCUACUCUG-5'

b Dicer substrate siRNA (dsiRNA)

5'-GCUGCAAAAUCCGAUGAGACUGCdCdG-3'
3'-UUCGACGUUUUAGGCUACUCUGACGGC-5'

c Sticky siRNA (ssiRNA)

5'-GCUGCAAAAUCCGAUGAGAC dAdAdAdAdA-3'
3'-dTdTdTdTdT CGACGUUUUAGGCUACUCUG-5'

Fig. 2 Different types of siRNA molecules targeting heat shock protein Hsp27: (**a**) conventional siRNA, (**b**) dicer substrate siRNA (dsiRNA), and (**c**) sticky siRNA (ssiRNA)

double-stranded RNA molecules bearing 19 nucleotides and two dTdT overhangs at the 3′-ends (Fig. 2a), which is designed to mimic the Dicer cleavage product and therefore bypass Dicer processing to directly enter RISC for RNAi pathway [13]. Different from conventional SiRNA, Hsp27 Dicer-substrate SiRNA (denoted as dsiRNA thereafter) has a sense strand with the five nucleotide sequence UGCdCdG appended to the 3′-end, and for the antisense strand an ACGGC at the 5′-end and a short UU unit at the 3′-end (Fig. 2b) [17]. This dsiRNA bears an asymmetric blunt end and an overhang at the 3′-end and can be optimally processed by Dicer, hence yielding higher and more durable RNAi potency. The sticky SiRNA (denoted as ssiRNA thereafter) has complementary A_5/T_5 3′ overhangs (Fig. 2c) instead of dTdT overhangs in conventional SiRNA. Through the A_5/T_5 3′ overhangs, these sticky SiRNA molecules can self-assemble into "gene-like" longer double-stranded RNA molecules, which provide better cooperativity and stronger interactive forces than the individual SiRNA with the dendrimer vector in order to ensure efficient delivery [11].

Indeed, the flexible TEA-core PAMAM dendrimers can interact efficiently with all the above mentioned three different Hsp27 SiRNAs via electrostatic interaction (Subheading 3.1) to form stable nanoparticles (Subheading 3.2), which protected SiRNA from degradation (Subheading 3.3) and promoted efficient cell uptake (Subheading 3.5), leading to effective gene silencing at both mRNA (Subheading 3.6) and protein level (Subheading 3.7) in prostate cancer PC-3 cells. The most effective dendrimer for the delivery of Hsp27 SiRNA is generation 7 dendrimer (denoted as $\mathbf{G_7}$ thereafter), whereas for Hsp27 dsiRNA and Hsp27 ssiRNA, even lower generation 5 dendrimer (denoted as $\mathbf{G_5}$ thereafter) is sufficient for effective delivery. We present below the detailed protocols for our dendrimer-based SiRNA delivery.

3.1 Gel Shift Assays of siRNA/Dendrimer Complexes

1. Prepare a 1.2 % agarose gel: Add 0.6 g agarose into 50 mL 0.5×TBE buffer. Microwave the agarose mixture for however long it takes to melt completely. Let it cool on the benchtop for 5 or so minutes. Add 5 μL of EB (10 mg/mL) into the above solution and homogenize by shaking gently. Then pouring it into casting tray in the gel boxes sideways gently avoid the bubble and insert a comb into the slots in the tray. The gel should polymerize in about 30 min. Then put the gel carefully remove the comb for sample loading
2. Dilute dendrimers with 50 mM Tris–HCl buffer, pH 7.4 and SiRNAs with RNase-free water to an appropriate concentration, respectively. Store at 4 °C.
3. Mix 2 μL of the SiRNAs aqueous solution (100 ng/μL) with 2 μL of the desired amount of dendrimer G_5 or G_7 solution at various N/P ratios gently and equilibrate at 37 °C for 30 min (*see* **Note 6**). Add 1 μL loading dye to each sample before loading. The final quantity of the SiRNAs per well is 200 ng.
4. Load the above each sample in each well and subject to electrophoresis in 1.2 % agarose gel. The SiRNAs bands are stained by EB and then detected by an EASY CCD camera (type 429 K) (*see* **Note 4**). An example result is shown in Fig. 3.

3.2 Size and Zeta Potential Measurement of siRNA/Dendrimer Complexes

1. Gently mix 100 μL of the SiRNA aqueous solution (2 μM) with 100 μL of the desired amount of dendrimer $\mathbf{G_5}$ or $\mathbf{G_7}$ aqueous solution at N/P ratio 10 in the presence of sonication for 10 s. The final concentration of the SiRNA is 1 μM.
2. Incubate the above 200 μL mixture at 37 °C for 30 min for complexes formation.
3. Measure the size and zeta potential by using Zetasizer Nano-ZS (Malvern) with a He-Ne ion laser of 633 nm. An example result is shown in Table 1.

3.3 Stability of siRNA/Dendrimer Complex Against RNaseA

1. Prepare two 1.2 % agarose gels: Add 1.2 g agarose into 100 mL 0.5× TBE buffer (*see* Subheading 3.1, **step 1**).
2. Mix certain amount of SiRNAs (100 ng/μL) (1.4 μg of SiRNA, 1.8 μg of dsiRNA, 2.2 μg of ssiRNA) and equal volume of the indicated amounts of dendrimer G_5 or G_7 at N/P ratio 10 and keep them at 37 °C for 30 min.
3. Incubate SiRNAs alone or the complexes in the presence of 0.01 μg/μL RNase A for certain time at 37 °C.
4. Withdraw aliquots (4 μL) of the corresponding solution after certain time incubation; add to 1.5 μL 1 % SDS solution on the ice (*see* **Note 7**). The final quantity of the SiRNAs per well is 200 ng. Add 1 μL loading dye to each sample before loading.

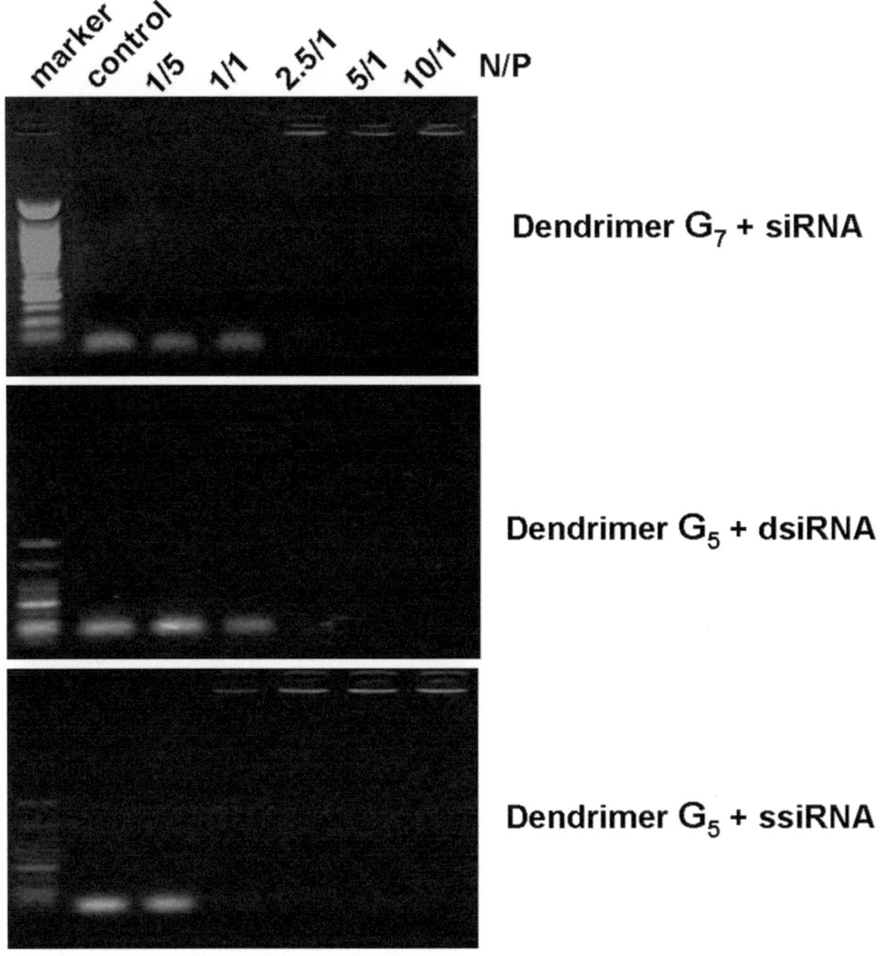

Fig. 3 Gel retardation of different siRNA molecules including siRNA, dicer substrate siRNA (dsiRNA), and sticky siRNA (ssiRNA) by dendrimer $\mathbf{G_7}$ or $\mathbf{G_5}$. Fig. 3a adapted with permission from ChemMedChem. 2009, 4:1302–1310, Copyright 2014 Wiley. Fig. 3b adapted from Nanomedicine NBM 2014, 10:1627–1636, Copyright 2014, with permission from Elsevier

Table 1
Size and zeta potential of nanoparticles formed by dendrimers with different siRNA molecules

Nanoparticles	Size (nm)	Zeta potential (mV)
Dendrimer $\mathbf{G_7}$/siRNA	105 ± 5.0	21 ± 1.2
Dendrimer $\mathbf{G_5}$/dsiRNA	97 ± 7.1	30 ± 1.4
Dendrimer $\mathbf{G_5}$/ssiRNA	110 ± 3.8	35 ± 2.7

5. Load the above each sample in each well and subject to electrophoresis in 1.2 % agarose gel. The SiRNA bands are stained by ethidium bromide and then detected by a Herolab EASY CCD camera (type 429 K) (Herolab) (*see* **Note 4**). An example result is shown in Fig. 4.

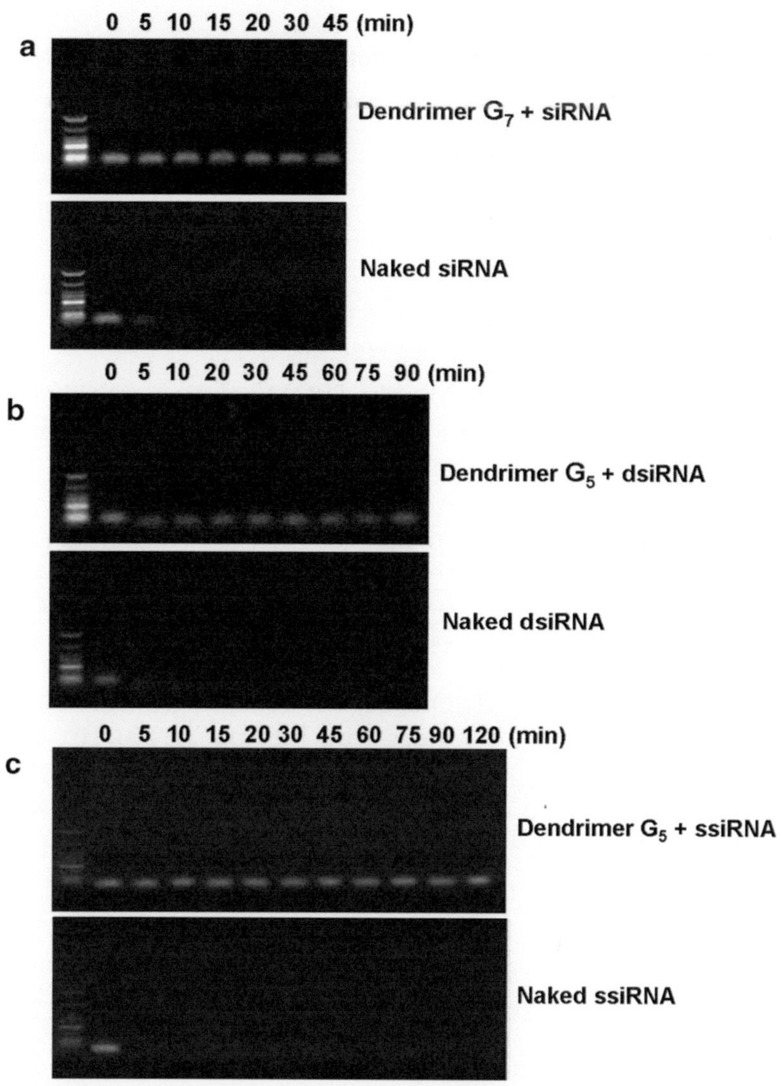

Fig. 4 Protection of different siRNA molecules against RNase digestion by dendrimer $\mathbf{G_7}$ or $\mathbf{G_5}$. (**a**) siRNA/dendrimer $\mathbf{G_7}$, (**b**) dsiRNA/dendrimer $\mathbf{G_5}$, (**c**) ssiRNA/dendrimer $\mathbf{G_5}$. Adapted with permission from ChemMedChem. 2009, 4: 1302-1310, Copyright 2014 Wiley; Nanomedicine NBM 2014, 10: 1627–1636, Copyright 2014, with permission from Elsevier; Mol. Pharmaceutics 2012; 9: 470–481, Copyright 2012 American Chemical Society

3.4 Preparation of siRNA/Dendrimer Complexes for Transfection

The following is an example when 4-well glass chamber slides (Labtek, Nunc) (500 μL for each chamber) is used for transfection.

1. Dilute SiRNA stock solution to 500 nM and dendrimer $\mathbf{G_5}$ or $\mathbf{G_7}$ stock solution to desired concentration at N/P ratio 10 in the volume of 50 μL with Opti-MEM transfection medium, respectively.
2. Gently vortex the above solution for 10 s and equilibrate for 10 min at room temperature.

Table 2
The volume of siRNA and dendrimer solution for transfection in each type of transfection plates

Type of plates	Total volume (mL)	siRNA solution (μL)	Dendrimer solution (μL)	Additional volume (mL)
4-Well glass chamber slides	0.5	50	50	0.4
6 cm dish	2.0	200	200	1.6

3. Mix the SiRNA solution with dendrimer solution and gently vortex for 10 s. Then equilibrate 30 min for complexes formation at room temperature.
4. Add another 400 μL Opti-MEM to complete to 500 μL for transfection. The final concentration of SiRNAs is 50 nM and the N/P ratio is 10. The volume of SiRNA and dendrimer solution changes according to the type of transfection plates (Table 2).

3.5 Cell Uptake by Confocal Microscopy

1. One day before assay, grow PC-3 cells in 4-well glass chamber slides with seeding at 10×10^4 cells/chamber in 1 mL of DMEM medium plus 10 % FBS.
2. Prepare the SiRNA/dendrimer complexes (500 μL for each chamber) as described in Subheading 3.4 in the dark by using fluorescen-labeled SiRNAs (*see* **Note 8**).
3. Before transfection, wash cells with 1 mL of prewarmed PBS. Then incubate with 500 μL of SiRNA/dendrimer complexes in Opti-MEM transfection medium.
4. After 4-h incubation at 37 °C, wash the cells with cold PBS for three times and stain them with 10 μg/mL Hoechst 34580 (nuclear dye for live cells) for 5–10 min at room temperature. Then wash the cells again with cold PBS three times and reserve cells in PBS for observation.
5. Visualize the cells with A Zeiss LSM 510 Meta laser scanning confocal microscope equipped with inverted Zeiss Axiovert 200 M stand. Images are acquired using LSM 510 software (Carl Zeiss GmbH). An example result is shown in Fig. 5.

3.6 Analysis of mRNA Expression by Quantitative Real-Time (qRT)-PCR

1. One day before assay, grow PC-3 cells in 6 cm dish with seeding at 1.5×10^5 cells/chamber in 4 mL of DMEM medium plus 10 % FBS.
2. Prepare the SiRNA/dendrimer complexes (2 mL for each 6 cm dish) as described in Subheading 3.4.
3. Before transfection, wash cells with 4 mL of prewarmed PBS. Then incubate with 2 mL of SiRNA/dendrimer complexes in Opti-MEM transfection medium.

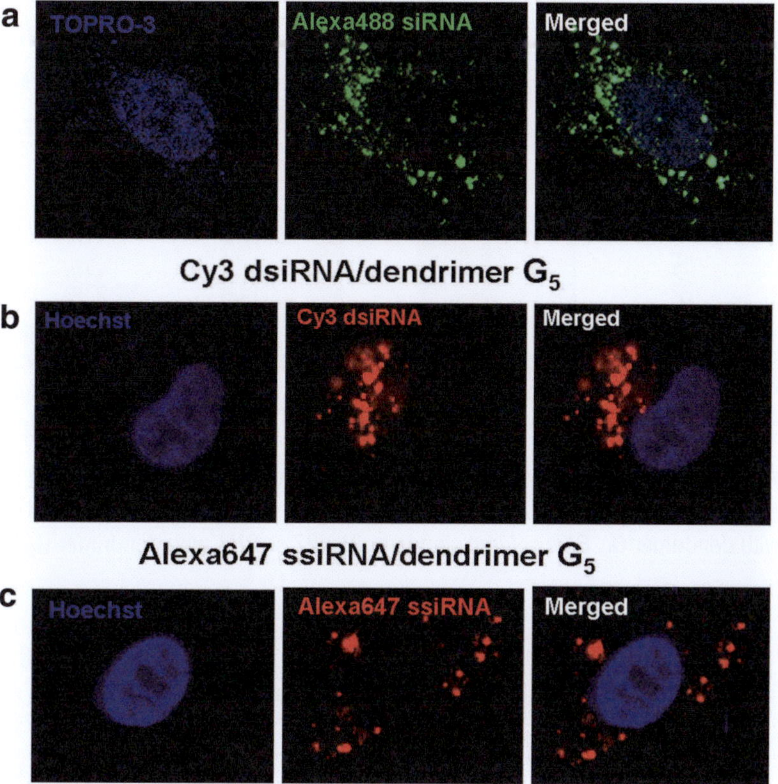

Fig. 5 Cell uptake of fluorescent-labeled siRNA/dendrimer complexes on PC-3 cells analyzed by live-cell confocal microscopy. (**a**) Alexa488 siRNA/dendrimer $\mathbf{G_7}$, (**b**) Cy3 dsiRNA/dendrimer $\mathbf{G_5}$, (**c**) Alexa647 ssiRNA/dendrimer $\mathbf{G_5}$. Adapted with permission from ChemMedChem. 2009, 4: 1302–1310, Copyright 2014 Wiley; Nanomedicine NBM 2014, 10: 1627–1636, Copyright 2014, with permission from Elsevier; Mol. Pharmaceutics 2012; 9: 470–481, Copyright 2012 American Chemical Society

4. After 8 h of incubation at 37 °C, change the transfection medium with complexes into fresh DMEM medium plus 10 % FBS.
5. After 48 h of treatment, pellet cells and isolate total RNAs using TRIzol method or AllPrep DNA/RNA/Protein Mini Kit by following the procedure offered by suppliers (*see* **Note 5**).
6. Reverse-transcribe RNA into cDNA by using ImProm-II™ Reverse Transcription system (*see* **Note 9**): Mix 1 μg RNAs with 0.5 μg oligo dT primer and complete to the final volume of 5 μL with RNase-free water. Heat the above mixture at 70 °C for 5 min and put 4 °C for 5 min. After that, add 15 μL mixture consisting of the following reagents to the product to achieve the final volume of 20 μL: 1 μL ImProm-II™ reverse transcriptase, 0.5 μL of RNase inhibitor, 1 μL of 10 mM dNTP mix, 4.8 μL of 25 mM $MgCl_2$, 4 μL of 5×RT buffer, 3.7 μL of RNase-free

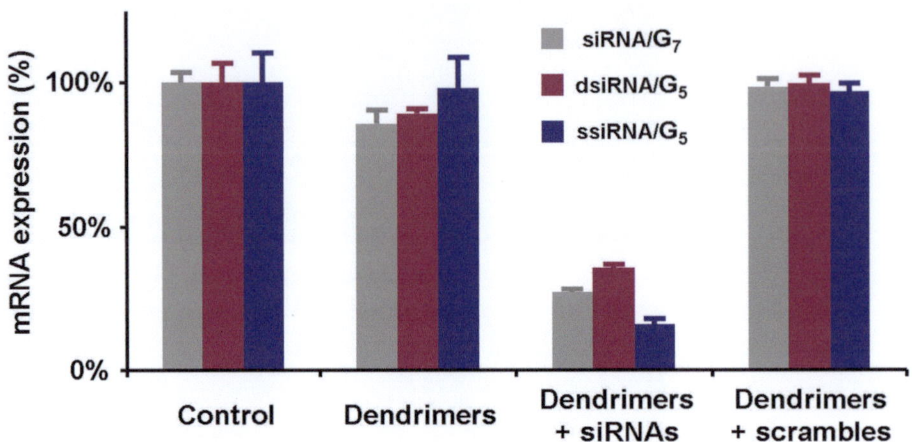

Fig. 6 Different siRNAs delivered by dendrimers downregulate targeted gene Hsp27 expression at mRNA level on PC-3 cells. *Grey column* stands for Hsp27 siRNA with dendrimer $\mathbf{G_7}$. *Violet column* stands for Hsp27 dsiRNA with dendrimer $\mathbf{G_5}$. *Blue column* stands for Hsp27 ssiRNA with dendrimer $\mathbf{G_5}$

water. Incubate the reaction (Total volume of 20 μL) at 25 °C for 5 min and 42 °C for 1 h. Heat the mixture at 70 °C for 15 min to inactivate reverse transcriptase and chill on ice or put at 4 °C. The cDNA is ready for qRT-PCR analysis.

7. Analyze the expression of the target genes by quantitative RT-PCR using 2× SYBR Premix Ex Taq and specific primer sets at a final concentration of 200 nM (Triplex assay) (*see* **Notes 9** and **10**). Use 18S expression as an internal control for normalization of the qPCR data. An example result is shown in Fig. 6.

3.7 Analysis of Protein Expression by Western Blot

1. Perform in vitro transfection of SiRNA/dendrimer complexes as described above (*see* Subheading 3.6, **steps 1–4**).
2. After 3 days of treatment, pellet cells and extract total protein using lysis buffer containing cocktail inhibitor (1×) and Na_3VO_4 (10 μg/mL).
3. Add 50 μL lysis buffer containing cocktail inhibitor and Na_3VO_4 to resuspend cell pellet of each dish and incubate for 30 min at 4 °C.
4. Centrifuge at 15,871 ×*g* for 45–60 min at 4 °C and collect the lysate.
5. Quantify the protein in the lysate using Pierce BCA protein assay kit.
6. Prepare a 10 % SDS polyacrylamide gel: Mix the following reagents to prepare a 20 mL running gel: 9.3 mL water, 5.2 mL 1.5 M Tris–HCl, pH 8.8, 5.0 mL 40 % acrylamide/bis-acrylamide solution, 200 μL 10 % SDS, 200 μL 10 % APS, and 8 μL TEMED. Cast the gel within a 7.25 cm × 10 cm × 1.5 mm gel cassette. Allow space for stacking gel and gently overlay within

isopropanol. The stacking gel should be polymerized in about 45–60 min. Mix the following reagents to make a 12 mL stacking gel: 8.9 mL MilliQ water, 1.5 mL 1.5 M Tris–HCl, pH 6.8, 1.5 mL 40 % acrylamide/bis-acrylamide solution, 120 μL 10 % SDS, 120 μL 10 % APS, and 12 μL TEMED. Insert a 15-well gel comb immediately without introducing air bubbles. The stacking gel should be polymerized in about 30–60 min.

7. Migrate samples containing equal amounts of protein (15 μg) from lysates in 10 % SDS polyacrylamide gel.
8. Carefully remove the comb and rinse the wells with 1× running buffer.
9. Denature samples containing equal amounts of protein (15 μg) from lysates at 95 °C for 5 min and chill the samples on the ice for 10 min.
10. Load the samples in each well and connect to a power supply. Electrophorese the gel first at 80 V for 10 min then at 120 V for 60 min.
11. Electrophoretic transfer the protein from the SDS-PAGE gel to the PVDF membrane using Mini Trans-Blot® Electrophoretic Transfer Cell (BIO-RAD) at 30 V overnight at room temperature or at 90 V for 1.5 h at 4 °C according to the manufacturer's protocols.
12. Immunoblot the membrane: Block nonspecific binding by incubating membrane in TBS (1×) with 5 % milk for 60 min at room temperature. Wash with 1× TBS two times (5 min/time).
13. Incubate the blocked membrane in specific primary antibodies and dilute in 1× TBS with 1 % milk and 1 % BSA for 1–2 h at room temperature or overnight at 4 °C (*see* **Note 11**). Wash with TBS (1×) three times (5 min/time).
14. Incubate the membrane for 1 h at room temperature with horseradish peroxidase (HRP)-conjugated secondary antibodies and dilute in 1× TBS with 1 % milk and 1 % BSA (*see* **Note 12**). Wash with 1× TBS three times (5 min/time).
15. Incubate membrane in SuperSignal West Pico Chemiluminescent Substrate according to the data sheet.
16. Visualize specific proteins using tabletop processor CURIX60 in the darkroom. An example result is shown in Fig. 7.

4 Notes

1. The structurally flexible TEA-core PAMAM dendrimers have been synthesized as previously reported [8, 10]. Generation 5 (**G_5**) bears 96 terminal amines and the molecular weight is 21,727 Da. Generation 7 (**G_7**) bears 384 terminal amines and the molecular weight is 87,489 Da.

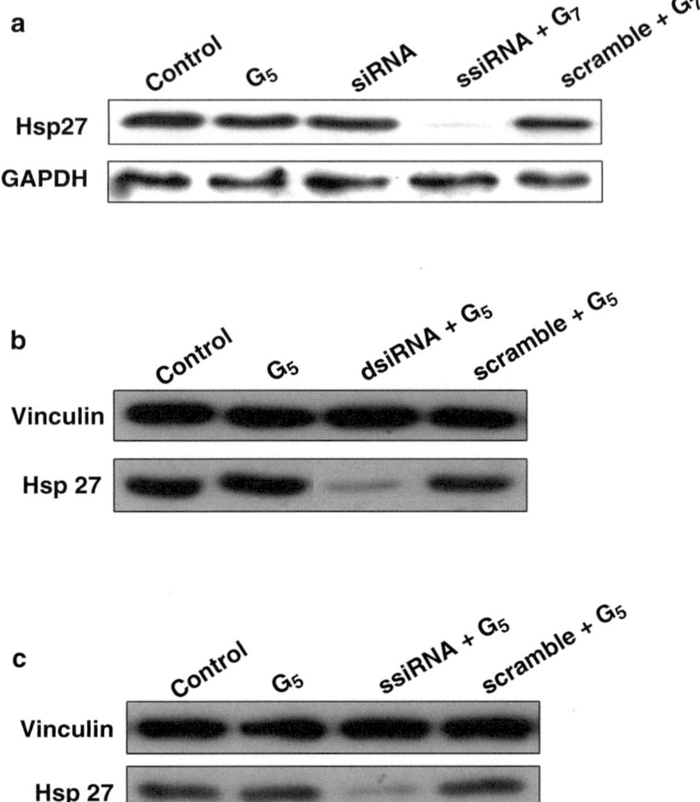

Fig. 7 Different siRNAs delivered by dendrimers downregulate targeted gene Hsp27 expression at protein level on PC-3 cells. (**a**) Hsp27 siRNA with dendrimer $\mathbf{G_7}$. (**b**) Hsp27 dsiRNA with dendrimer $\mathbf{G_5}$. (**c**) Hsp27 ssiRNA with dendrimer $\mathbf{G_5}$. Fig. 7a and 7c adapted with permission from ChemMedChem. 2009, 4: 1302–1310, Copyright 2014 Wiley; Mol. Pharmaceutics 2012; 9: 470–481, Copyright 2012 American Chemical Society

2. Unless stated otherwise, all chemicals purchase from Sigma-Aldrich and all solution should be prepared in water that has a resistivity of 18.2 MΩ-cm and autoclaved. This standard is referred to as "water" in the text.
3. Dendrimers aliquots should be prepared in the volume to be freeze/thawed less than five times.
4. Ethidium bromide is a potent mutagen. Ethidium bromide solution must be handled with extreme caution and decontaminated prior to disposal.
5. TRIzol Reagent contains phenol (toxic and corrosive) and guanidine isothiocyanate (an irritant), and may be a health hazard if not handled properly. Always work with TRIzol Reagent in a fume hood, and always wear a lab coat, gloves, and safety glasses.

6. N/P refers to [total terminal amines in dendrimer]/[phosphates in SiRNA].
7. Aliquots should be put immediately on the ice after withdrawn from the mixture to inactivate the RNase A. Meanwhile, the presence of SDS can also inactivate the RNase A and displace the SiRNA from the SiRNA/dendrimer complexes.
8. All the fluorescent-labeled SiRNAs are sensitive to light and the exposure to light should be limited from this step until the end of this subheading.
9. Allow all the needed reagents in the kit to reach room temperature before use.
10. qRT-PCR primer sequences as follows: Hsp27 primer F: 5′-TCCCTGGATGTCAACCACTTC-3′ and primer R: 5′-TCTCCACCACGCCATCCT-3′; 18S primer F: 5′-CTACCACATCCAAGGAAGGC-3′ and primer R: 5′-TTTTCGTCACTACCTCCCCG-3′.
11. Rabbit anti-human Hsp27 polyclonal antibody: Dilute to 1:5000 and incubate at room temperature for 1.5–2 h. Rabbit anti-human GAPDH polyclonal antibody is diluted to 1:2500 and incubated overnight at 4 °C. Mouse anti-human vinculin monoclonal antibody is diluted to 1:2000 and incubated at room temperature for 1.5–2 h.
12. Secondary antibodies: Horseradish peroxidase (HRP)-conjugated anti-rabbit or anti-mouse monoclonal antibody. Dilute to 1:5000 and incubate at room temperature for 1 h.

Acknowledgments

This work was supported by the international ERA-Net EURONANOMED European Research project DENANORNA, PACA Canceropôle, INCa, Aix-Marseille Université, CNRS, INSERM, China Scholarship Council (XL), Association pour la Recherche sur les Tumeurs de la Prostate (XL), Association Française contre les Myopathies (XL), and under the auspice of European COST Action TD0802 "Dendrimers in Biomedical Applications."

References

1. de Fougerolles A, Vornlocher HP, Maraganore J, Lieberman J (2007) Interfering with disease: a progress report on siRNA-based therapeutics. Nat Rev Drug Discov 6:443–453
2. Castanotto D, Rossi JJ (2009) The promises and pitfalls of RNA-interference-based therapeutics. Nature 457:426–433
3. Whitehead KA, Langer R, Anderson DG (2009) Knocking down barriers: advances in siRNA delivery. Nat Rev Drug Discov 8:129–138
4. Kanasty RL, Whitehead KA, Vegas AJ, Anderson DG (2012) Action and reaction: the biological response to siRNA and its delivery vehicles. Mol Ther 20:513–524
5. Couto LB, High KA (2010) Viral vector-mediated RNA interference. Curr Opin Pharmacol 10:534–542

6. Vögtle F, Richardt G, Werner N (2009) Dendrimer chemistry: concepts, syntheses, properties, applications. Wiley-VCH, Weinheim, p 354
7. Liu X, Rocchi P, Peng L (2012) Dendrimers as non-viral vectors for siRNA delivery. New J Chem 36:256–263
8. Wu J, Zhou J, Qu F, Bao P, Zhang Y, Peng L (2005) Polycationic dendrimers interact with RNA molecules: polyamine dendrimers inhibit the catalytic activity of Candida ribozymes. Chem Commun 313–315
9. Behr JP (1997) The proton sponge: a trick to enter cells the viruses did not exploit. Chimia 51:34–36
10. Liu X, Wu J, Yammine M, Zhou J, Posocco P, Viel S, Liu C, Ziarelli F, Fermeglia M, Pricl S, Victorero G, Nguyen C, Erbacher P, Behr JP, Peng L (2011) Structurally flexible triethanolamine core PAMAM dendrimers are effective nanovectors for DNA transfection in vitro and in vivo to the mouse thymus. Bioconjug Chem 22:2461–2473
11. Liu X, Liu C, Laurini E, Posocco P, Pricl S, Qu F, Rocchi P, Peng L (2012) Efficient delivery of sticky siRNA and potent gene silencing in a prostate cancer model using a generation 5 triethanolamine-core PAMAM dendrimer. Mol Pharm 9:470–481
12. Zhou J, Wu J, Hafdi N, Behr JP, Erbacher P, Peng L (2006) PAMAM dendrimers for efficient siRNA delivery and potent gene silencing. Chem Commun 2362–2364
13. Liu XX, Rocchi P, Qu FQ, Zheng SQ, Liang ZC, Gleave M, Iovanna J, Peng L (2009) PAMAM dendrimers mediate siRNA delivery to target Hsp27 and produce potent antiproliferative effects on prostate cancer cells. ChemMedChem 4:1302–1310
14. Zhou J, Neff CP, Liu X, Zhang J, Li H, Smith DD, Swiderski P, Aboellail T, Huang Y, Du Q, Liang Z, Peng L, Akkina R, Rossi JJ (2011) Systemic administration of combinatorial dsiRNAs via nanoparticles efficiently suppresses HIV-1 infection in humanized mice. Mol Ther 19:2228–2238
15. Liu C, Liu X, Rocchi P, Qu F, Iovanna JL, Peng L (2014) Arginine-terminated generation 4 PAMAM dendrimer as an effective nanovector for functional siRNA delivery in vitro and in vivo. Bioconjug Chem 25:521–532
16. Liu X, Liu C, Catapano CV, Peng L, Zhou J, Rocchi P (2014) Structurally flexible triethanolamine-core poly(amidoamine) dendrimers as effective nanovectors to deliver RNAi-based therapeutics. Biotechnol Adv 32:844–852
17. Liu X, Liu C, Chen C, Bentobji M, Cheillan FA, Piana JT, Qu F, Rocchi P, Peng L (2014) Targeted delivery of Dicer-substrate siRNAs using a dual targeting peptide decorated dendrimer delivery system. Nanomedicine. doi:10.1016/j.nano.2014.1005.1008
18. Acunzo J, Katsogiannou M, Rocchi P (2012) Small heat shock proteins HSP27 (HspB1), alphaB-crystallin (HspB5) and HSP22 (HspB8) as regulators of cell death. Int J Biochem Cell Biol 44:1622–1631
19. Rocchi P, So A, Kojima S, Signaevsky M, Beraldi E, Fazli L, Hurtado-Coll A, Yamanaka K, Gleave M (2004) Heat shock protein 27 increases after androgen ablation and plays a cytoprotective role in hormone-refractory prostate cancer. Cancer Res 64:6595–6602
20. Cornford PA, Dodson AR, Parsons KF, Desmond AD, Woolfenden A, Fordham M, Neoptolemos JP, Ke Y, Foster CS (2000) Heat shock protein expression independently predicts clinical outcome in prostate cancer. Cancer Res 60:7099–7105
21. Zoubeidi A, Gleave M (2012) Small heat shock proteins in cancer therapy and prognosis. Int J Biochem Cell Biol 44:1646–1656

Chapter 12

Chitosan Nanoparticles for SiRNA Delivery In Vitro

Héloïse Ragelle, Kevin Vanvarenberg, Gaëlle Vandermeulen, and Véronique Préat

Abstract

RNA interference, the process in which small interfering RNAs (SiRNAs) silence a specific gene and thus inhibit the associated protein, has opened new doors for the treatment of a wide range of diseases. However, efficient delivery of SiRNAs remains a challenge, especially due to their instability in biological environments and their inability to cross cell membranes. To protect and deliver SiRNAs to mammalian cells, a variety of polymeric nanocarriers have been developed. Among them, the polysaccharide chitosan has generated great interests. This derivative of natural chitin is biodegradable and biocompatible, and can complex SiRNAs into nanoparticles on account of its positive charges. However, chitosan presents some limitations that need to be taken into account when designing chitosan/SiRNA nanoparticles. Here, we describe a method to prepare SiRNA/chitosan nanoparticles with high gene silencing efficiency and low cytotoxicity by using the ionic gelation technique.

Key words Chitosan, SiRNA, Nanoparticles, Ionic gelation, RNA interference, Gene silencing, Nucleic acid delivery, PEGylated chitosan

1 Introduction

Chitosan is one of the most studied polymers in non-viral SiRNA delivery, on account of several unique attributes: (1) its polycationic nature under slightly acidic conditions allows for complexation of SiRNA into nanoparticles (NPs) via a fast, easy, and gentle process [1]; (2) chitosan is biodegradable and biocompatible, which is critical to in vivo administration [2, 3]; and (3) its versatile chemical structure enables functionalization to impart the polymer with new properties and/or to enhance its performance [4]. The molecular characteristics of chitosan and formulation parameters are both decisive for the gene silencing effect of SiRNA NPs and need to be optimized. In particular, the molecular weight (MW) and the degree of deacetylation (DD) (defined as the percentage of deacetylated primary amine groups) are two important features that influence siRNA NP formation and efficacy. On account of its

Kato Shum and John Rossi (eds.), *SiRNA Delivery Methods: Methods and Protocols*, Methods in Molecular Biology, vol. 1364, DOI 10.1007/978-1-4939-3112-5_12, © Springer Science+Business Media New York 2016

stiff molecular topology and short length, SiRNA needs a high number of positive charges to remain efficiently bound. Therefore, chitosans with DD equal or higher than 80 % are typically used to form SiRNA NPs. Regarding the MW, it was described that chitosan molecules that were 5–10 times the length of SiRNA, i.e., chitosan of MW from 65 to 170 kDa, were most suitable for NP formation [5]. Indeed, the MW should be high enough to efficiently complex SiRNA into stable NPs, but low enough to allow SiRNA unpacking and release into the cells. Two main methods for chitosan/siRNA NP formulation are described in the literature: simple complexation where SiRNA is mixed with chitosan alone and ionic gelation where a cross-linker agent, e.g., sodium tripolyphosphate, is added. Generally, ionic gelation is preferred since the NPs show higher stability compared to the simple complexation method [6]. Finally, an important parameter that needs to be determined experimentally is the positive/negative charge ratio. This ratio is decisive for the formation of NPs of suitable size.

While chitosan has demonstrated significant promise for in vitro SiRNA delivery, some limitations remain for in vivo administration. First, chitosan suffers from low water solubility at pH values higher than 6.5, due to partial protonation of the primary amino groups, which can weaken SiRNA binding and affect the stability of the NPs [7]. In addition, the transfection efficiency mediated by chitosan is restricted by its poor buffering capacity and inability to trigger endosomal escape [8, 9]. To address these limitations, a number of chitosan derivatives as well as formulation improvements have been described. Among them, the use of poly(ethylene glycol) (PEG) grafted chitosan (C_{PEG}) efficiently improves polymer solubility as well as the NP colloidal stability and the addition of endosomal disrupting agents, such as primary amine group-rich compounds: poly(ethylene imine) has shown to increase transfection efficiency [10].

In this chapter, we describe a method to formulate efficient chitosan/siRNA NPs by ionic gelation. The formulation was optimized in terms of chitosan characteristics, formulation process and addition of excipients in order to obtain NPs of 100–200 nm diameter that display high in vitro gene silencing efficiency in multiple cell culture models.

2 Materials

2.1 Nanoparticle Preparation

1. Chitosan derived from crustaceans: MW 190–310 kDa, DD 75–85 %.
2. Isocyanate terminated methoxy-poly(ethylene)glycol (mPEG-ISC): Mn = 1000 g/mol.
3. Dimethylformamide (DMF).

4. Hydrazine, anhydrous.
5. Phthalic anhydride.
6. siRNAs, purified by HPLC: Luc SiRNA (MW 13,300 g/mol): sense 5′-CUUACGCUGAGUACUUCGAtt-3′; control (ctrl) SiRNA (MW 13,300 g/mol): sense 5′-AUAGUGCAA CGAUGAGCUCtt-3′ and EGFP SiRNA (MW 13,490 g/ mol): sense 5′-pACCCUGAAGUUCAUCUGCAcc-3′. Lower case letters = deoxyribonucleotides and p = phosphate residues.
7. Sodium tripolyphosphate (TPP).
8. Poly(ethylene imine) (PEI), branched, 25 kDa.
9. Sodium acetate and acetic acid.
10. Transfection reagent: INTERFERin®, Polyplus transfection™.
11. Zetasizer ZS Malvern Instruments Ltd.
12. Zeta potential cuvette, Malvern Instruments Ltd.
13. 1 % Sodium dodecyl sulfate (SDS) solution: Weigh 1 g of SDS and add 99 ml of water.
14. 0.2 M Sodium acetate, pH 4.5: Add 11.5 ml of 3 M sodium acetate and 3.7 ml of 17.5 M acetic acid 17.5 M into 484.8 ml of RNase-free water.
15. 1 mg/ml Chitosan solution, pH 5.8: Weigh 40 mg of chitosan in a conical tube. Add 40 ml of 0.2 M sodium acetate buffer to obtain a final chitosan concentration of 1 mg/ml. Vortex until chitosan is completely dissolved. Adjust the pH to 5.8 using NaOH 10 M solution. Filter the solution using 0.45 μm syringe filter (*see* **Note 1**).
16. 1 mg/ml C_{PEG} solution: Weigh 40 mg of C_{PEG} in a conical tube. Add 40 ml of RNAse-free water. Vortex until C_{PEG} is completely dissolved. Filter the solution using 0.45 μm syringe filter.
17. 1 mg/ml PEI solution: Weigh PEI in a conical tube using a plastic pipette, add RNase-free water to obtain 1 mg/ml final solution. Sonicate the solution. Adjust the pH to 7.4 using acetic acid 1.75 M. Filter the solution using 0.22 μm syringe filter.
18. 1 mg/ml TPP solution: weigh 40 mg of TPP in a conical tube, add 40 ml of RNase-free water, vortex. Filter the solution using a syringe filter 0.22 μm (*see* **Note 2**).
19. 50 μM SiRNA solution in RNase-free water: Briefly centrifuge the tubes to ensure that the dried SiRNA is at the bottom of the tube. Add RNase-free water to obtain a final SiRNA concentration of 50 μM, gently vortex and make aliquots of 200 μl. Store the SiRNA aliquots at −20 °C.

2.2 Materials for Cell Culture

1. B16F10 luc cells: Murine melanoma cell line B16F10 stably expressing firefly luciferase.
2. H1299 EGFP obtained from [11]: Human non-small-cell lung carcinoma cell line H1299 expressing the destabilized enhanced green fluorescence protein.
3. Complete cell culture medium for B16F10 luc cells: Minimum essential Medium (MEM) with Glutamax supplemented with 10 % fetal bovine serum (FBS), and 1 % 10,000 U/ml penicillin-streptomycin.
4. Complete cell culture medium for H1299 EGFP cells: RPMI 1640 supplemented with 10 % FBS and 1 % 10,000 U/ml penicillin-streptomycin.
5. Cell culture plates: Black 96-well plates and transparent 12-well.
6. Phosphate-buffered saline (PBS).
7. Luminometer, GloMax Promega.
8. Fluorescence-activated cell sorting (FACS), FACSCalibur BD.

3 Methods

3.1 Preparation of PEGylated Chitosan (C_{PEG})

PEG chains were grafted onto the hydroxyl groups of chitosan as described in detail in [10].

1. To protect the amine groups of chitosan, transfer 1 g of chitosan into a glass tube containing 5 ml of anhydrous DMF, add 2.4 g of phthalic anhydride and stir at 130°C for 7 h.
2. Rapidly pour the chitosan–phthalimide solution in a water/ice mixture under vigorous stirring.
3. Filter, lyophilize, and store the chitosan–phthalimide under N_2 at 6 °C until further use.
4. To functionalize with PEG, dissolve 500 mg of chitosan–phthalimide in 25 ml DMF in a round bottom flask, then add 0.48 g mPEG-ISC, and let the reaction go overnight at room temperature.
5. To deprotect the amino groups, add 0.8 ml of hydrazine, increase the temperature to 110 °C, and let the deprotection reaction go for 3 h.
6. Dry the C_{PEG} under vacuum, solubilize the collected residues into sodium acetate buffer and remove the phthalimide derivatives by filtration.
7. Adjust the pH of the polymer solution to 10, dialyze against water until neutral pH and lyophilize.

3.2 Nanoparticle Formulation

1. In an Eppendorf tube, mix 125 μl (125 μg) of TPP and 14.8 μl (10 μg) of EGFP SiRNA or 15 μl (10 μg) of luc or control SiRNA. This is Eppendorf A.
2. In another Eppendorf tube, mix 250 μl (250 μg) of chitosan, 41.6 μl (41.6 μg) of C_{PEG} and 50 μl (50 μg) of PEI. This is Eppendorf B.
3. Add all solution in Eppendorf A to Eppendorf B. Mix vigorously by pipetting and vortex for 30 s.
4. Leave the nanoparticle mix for 1 h at room temperature.
5. Add RNase-free water to the NPs up to a volume of 1 ml.
6. Use as is or centrifuge the mix for 30 min at 17,860 × *g*, 25 °C, and suspend the NPs into 1 ml RNase-free water.
7. For each assay, prepare a formulation containing a luc or EGFP SiRNA and another one containing ctrl SiRNA (*see* **Notes 3–5**).

3.3 Nanoparticle Characterization

1. Size measurements using dynamic light scattering (e.g., Nanosizer ZS): Transfer the diluted NP suspension to a cuvette adapted to the equipment. Let the suspension equilibrate for 2 min at 25 °C before performing the measurements (*see* Table 1).
2. Zeta potential measurements using electrophoretic mobility: Transfer the diluted NP suspension to a zeta potential cuvette. Let the suspension equilibrate for 2 min at 25 °C before performing the measurements (*see* Table 1).
3. Encapsulation efficiency (EE): After centrifugation of the NP suspension (17,860 × *g*, 30 min), transfer 100 μl of the supernatant to a 96-well plate and mix with 100 μl of a diluted solution of Oligreen® (Life technologies, 1:200 in buffer TE 1×). Repeat this in triplicate. Incubate for 5 min in dark and measure the fluorescence (wavelengths: excitation 480 nm and emission 520 nm). The amount of SiRNA in the supernatant ($\text{siRNA}_{\text{supernatant}}$) is determined from a standard curve (recommended concentration range: 0.5 μg/ml to 0.1 ng/ml) and the encapsulation efficiency is calculated using the following equation where SiRNA is the initial quantity of $\text{SiRNA}_{\text{NPs}}$ is the initial quantity of SiRNA in the nanoparticles:

$$\text{EE\%} = 100 - \left(\begin{array}{r} (100 \times \text{siRNA amount in the supernatant}) \\ / \text{ siRNA initial amount in the NPs} \end{array} \right).$$

Table 1

Typical physicochemical characteristics of the chitosan-based NPs loaded with siRNA ($n \geq 3$)

Size (nm)	PdI	Zeta potential (mV)	EE (%)
190 ± 30	0.25 ± 0.05	30 ± 6	100 ± 0.5

3.4 Determination of In Vitro Gene Silencing Using Luciferase System

1. The day before the experiment, plate B16F10 luc cells on black or white 96-well plates at a cell density of 8000 cells per well in complete cell culture medium.
2. Prepare 1 ml NPs/medium mix: add 135 μl NPs to 865 μl of cell medium in an Eppendorf tube. Mix. Add 100 μl of the mix to the wells to obtain a final concentration of 100 nM SiRNA ($n = 8$).
3. As a positive control, mix 1 μl INTERFERin® with 0.2 μl SiRNA (final concentration of 100 nM) following the manufacturer's protocol.
4. Incubate for 4 h at 37 °C.
5. After 4 h, aspirate the medium and add 100 μl of fresh complete medium/well. Incubate at 37 °C for 48 h.
6. At 48 h post-transfection, add 100 μl of ONE-Glo™ (Promega) reagent/well.
7. Measure the luminescence within 30 min.
8. To determine the percentage of luciferase inhibition, use the following equation: $\%\text{inhibition} = 100 - ((100 \times \text{RLU}_{\text{luc}}) / \text{RLU}_{\text{ctl}})$ where RLU_{luc} is the mean of relative light unit (RLU) for luc SiRNA and RLU_{ctl} is the mean of RLU for control SiRNA (Fig. 1).

3.5 Determination of In Vitro Gene Silencing Using EGFP System

1. The day before the experiment, plate H1299 EGFP cells on a 12-well plate at a cell density of 10^5 cells/well in complete cell culture medium.
2. Add 135 μl of NPs and 865 μl of complete cell culture medium (final SiRNA concentration of 100 nM) and incubate for 4 h.

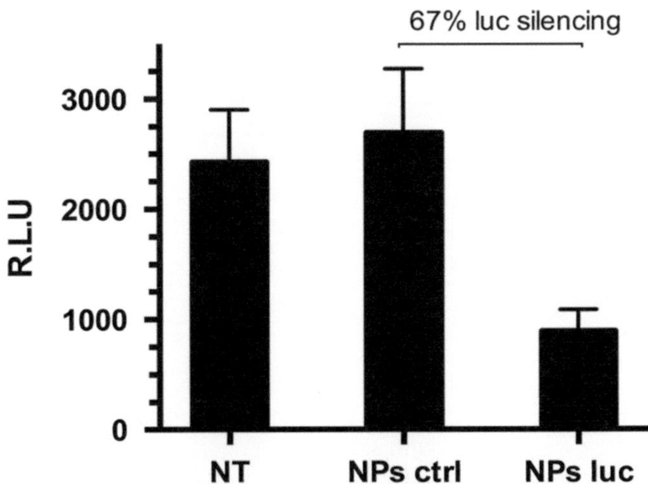

Fig. 1 Gene silencing efficiency of NPs on B16F10 luc cells after 4 h incubation at 200 nM. The luminescence expressed in relative light units (RLU) was measured at 48 h ($n = 8$). Data represent mean ± SD. *NT* non-treated cells

As a negative control, prepare the same NPs loaded with the ctrl SiRNA. As a positive control, use INTERFERin® transfection reagent.

3. After 4 h, remove the medium and add 1 ml of fresh medium/well.
4. After 48 h, wash the cells with 1 ml PBS, and add 200 μl trypsin for 5 min.
5. Neutralize trypsin with 800 μl of medium, transfer the cell suspension to an Eppendorf tube, and centrifuge at 500 × *g* for 5 min.
6. Add 300 μl of PBS and transfer to a tube for FACS analysis.
7. For each measurement, perform the FACS analysis on 10,000 cells.
8. Using FlowJo software, determine the FL1 median (median of the EGFP peak) for each fluorescence histogram.
9. Calculate the percentage of EGFP silencing using the following equation:

$$\%\,\text{EGFP silencing} = 100 - \left(\text{FL1 median}_{\text{NP EGFP}} \times 100 \,/\, \text{FL1 median}_{\text{NP CTRL}}\right).$$

4 Notes

1. Throughout all of the experiments, strict RNase-free conditions must be applied: use RNase-free tubes, gloves, tips with filter, and a dedicated pipette set. Before the experiment, spray the bench and the gloves with a solution of 1 % SDS.
2. The chitosan, PEI, and TPP solutions can be stored for up to 1 month at 4 °C. After this time, variations of the NP characteristics (larger size, high PdI) might be observed.
3. It is recommended to let the chitosan, PEI, and TPP solutions stabilize at room temperature for 1 h before making the NPs.
4. To optimize the NP characteristics, it is recommended to set the SiRNA and the chitosan amounts constant and to vary the amount of TPP.
5. It is recommended to prepare the NPs just before performing the in vitro tests.

Acknowledgement

This work was supported by the Walloon Region (BioWin project TARGETUM).

References

1. Mao S, Sun W, Kissel T (2010) Chitosan-based formulations for delivery of DNA and siRNA. Adv Drug Deliv Rev 62:12–27
2. Baldrick P (2010) The safety of chitosan as a pharmaceutical excipient. Regul Toxicol Pharmacol 56:290–299
3. Kean T, Thanou M (2010) Biodegradation, biodistribution and toxicity of chitosan. Adv Drug Deliv Rev 62:3–11
4. Raemdonck K, Martens TF, Braeckmans K, Demeester J, De Smedt SC (2013) Polysaccharide-based nucleic acid nanoformulations. Adv Drug Deliv Rev 65:1123–1147
5. Liu X, Howard KA, Dong M, Andersen MO, Rahbek UL, Johnsen MG, Hansen OC, Besenbacher F, Kjems J (2007) The influence of polymeric properties on chitosan/siRNA nanoparticle formulation and gene silencing. Biomaterials 28:1280–1288
6. Katas H, Alpar HO (2006) Development and characterisation of chitosan nanoparticles for siRNA delivery. J Control Release 115: 216–225
7. Ragelle H, Vandermeulen G, Preat V (2013) Chitosan-based siRNA delivery systems. J Control Release 172:207–218
8. Kim TH, Kim SI, Akaike T, Cho CS (2005) Synergistic effect of poly(ethylenimine) on the transfection efficiency of galactosylated chitosan/DNA complexes. J Control Release 105:354–366
9. Lai WF, Lin MC (2009) Nucleic acid delivery with chitosan and its derivatives. J Control Release 134:158–168
10. Ragelle H, Riva R, Vandermeulen G, Naeye B, Pourcelle V, Le Duff CS, D'Haese C, Nysten B, Braeckmans K, De Smedt SC, Jerome C, Preat V (2014) Chitosan nanoparticles for siRNA delivery: optimizing formulation to increase stability and efficiency. J Control Release 176:54–63
11. Jensen DM, Cun D, Maltesen MJ, Frokjaer S, Nielsen HM, Foged C (2010) Spray drying of siRNA-containing PLGA nanoparticles intended for inhalation. J Control Release 142:138–145

Chapter 13

Non-Covalently Functionalized of Single-Walled Carbon Nanotubes by DSPE-PEG-PEI for SiRNA Delivery

King Sun Siu, Yujuan Zhang, Xiufen Zheng, James Koropatnick, and Wei-Ping Min

Abstract

The expression of a gene can be specifically downregulated by small interfering RNA (SiRNA). Modified carbon nanotubes (CNT) can be used to protect SiRNA and facilitate its entry into cells. Regardless of that, simple and efficient functionalization of CNT is lacking. Effective SiRNA delivery can be carried out using non-covalently functionalized CNT, where non-covalent (versus covalent) functionalization is simpler and more expeditious. Non-covalently functionalized single walled carbon nanotubes (SWCNT) that include a lipopolymer are described here. Polyethylenimine (PEI) conjugated to 1,2-distearoyl-sn-glycero-3-phosphoethanolamine-*N*-[amino(polyethylene glycol)-2000] (DSPE-PEG) was generated and the products used to disperse CNT to form DSPE-PEG-PEI/CNT (DGI/C), an agent capable of facilitating SiRNA delivery to cells in vitro and organs and cells in vivo.

Key words SiRNA delivery, RNAi, Gene silencing, Carbon nanotubes, Polyethylenimine, PEI, Polyethylene glycol, PEG

1 Introduction

Carbon nanotubes (CNT) have been used for gene delivery and found to enhance efficiency of transfection into mammalian cells [1–11]. Nevertheless, pristine CNT (p-CNT) are not soluble in most bio-compatible liquids unless functionalized by the addition of covalently bound moieties [12]. Non-covalently functionalized CNT for effective delivery of SiRNA has also been developed. Some of the methods described in the literature for non-covalent functionalization [10, 13, 14] are based on the method of Kam et al. [15] and involve use of a lipopolymer DSPE-PEG for CNT dispersion. On the other hand, some methods are based on aromatic π-π stacking interaction [4, 16]. The lipopolymer method exploits the hydrophobic tail to interact hydrophobically with CNT, while PEG helps to disperse the whole complex in water [15] and acts as a site for covalent conjugation of SiRNA to the CNT through disulphide

Kato Shum and John Rossi (eds.), *SiRNA Delivery Methods: Methods and Protocols*, Methods in Molecular Biology, vol. 1364, DOI 10.1007/978-1-4939-3112-5_13,

bonding. This approach is relatively simple to carry out, highly efficient, and has been used by Liu et al. to deliver SiRNA to T cells [10]. This method has been further developed by Cai et al. as an efficient magnetic CNT "spear" for in vitro transfection of nucleic acids into T and B cells [13, 14]. In this strategy, non-covalently dispersed CNT was conjugated to plasmid DNA (pDNA) directly, or condensed with pDNA by addition of DSPE-PEG-poly(l-lysine). CNT for delivery of nucleic acids in this fashion were associated with nickel to create magnetic constructs for purification.

SiRNA is a powerful tool for research and is an attractive molecule for treatment of diseases with known targets, including cancer [17, 18]. Small molecules or biologics (including antibodies) are limited in their capacity to target "undruggable" protein molecules (i.e., proteins in which targeted regions are shared with non-target molecules or structures, leading to undesirable off-target toxicity; or where targeted regions are unavailable for interaction with targeting molecules because the overall structure of proteins limits availability for interaction). Such "undruggable" targets can be susceptible to SiRNA, which targets the mRNA encoding the target proteins and reduces the target protein level itself by reducing or ablating transcription of that protein [19, 20]. However, the delivery of nucleic acids into mammalian cells, a process called transfection, remains a challenge; the size, lipophilicity, polyanionic nature, and susceptibility of SiRNA to degradation by RNase limits application of SiRNA technology to disease therapy. An efficient delivery system is necessary to apply SiRNA to treatment of human disease.

Poly(ethylenimine) (PEI) is a cationic polymer that has been used extensively for nucleic acid delivery [21–23]. There are two kinds of PEI: branched PEI [24] and linear PEI [25]. They are water soluble, basic, and positively charged at physiologic pH. The ratio of commercially available branched PEI has a primary, secondary, and tertiary amine ratio of 1:1:1, as suggested by von Harpe et al. [26]. PEI (M_w 800 kDa) was introduced for delivery of pDNA by Boussif et al. in 1995 [21]. According to Abdallah et al., PEI can form a polyplex with pDNA that, upon entry into cells in vivo, produces transgene expression levels comparable to those obtained using lentiviral or adenoviral vectors introduced into brain cells in an animal model in vivo via direct brain injection [27]. Use of 25 kDa (M_w 25,000, M_n 10,000) and higher molecular weight PEI in PEI/nucleic acid complexes promotes high transfection efficiency [28]. However, the toxicity of complexes using high molecular weight PEI is also much higher than complexes generated using lower molecular weight PEI [29, 30]. Regardless of that limitation, PEI can condense SiRNA and facilitate endosomal escape [31], a desirable characteristic to allow SiRNAs to act in the interior of cells. It has been shown that PEI-conjugated CNT can increase transfection efficiency [2].

To promote the release of SiRNA from CNT in the interior of cells, non-covalent binding and stabilization of SiRNA by cationic charge is preferable to covalent binding. The chemical procedure to non-covalently associate CNT with SiRNA is relatively simple. To combine the advantageous SiRNA transfection properties conferred by PEI and the properties of DSPE-PEG (which enable non-covalent functionalization and dispersion of CNT) [10, 32, 33], DSPE-PEG-PEI (DGI) was prepared and used to non-covalently functionalize CNT, in a modification of the method of Liu et al. [10]. The materials and methods pertinent to that preparation are described below.

2 Materials

Prepare all solutions for SiRNA using nucleases-free water and nuclease-free pipette tips. Prepare synthetic mixtures using ultrapure water, which is purified by deionized water and analytical grade reagents. Prepare and store all reagents at room temperature (unless indicated otherwise). Diligently follow all waste disposal regulations when disposing waste materials. Store the DGI and DGI/C at −80 °C after lyophilization. When measuring volumes of liquids in high accuracy, allow solutions to equilibrate to room temperature for 30 min prior to measurement of volumes.

2.1 Synthesis and Characterization of Lipopolymer

1. Polyethylenmine (PEI, M_w 25,000, M_n 10,000) (*see* **Note 1**).
2. *N*-(3-Dimethylaminopropyl)-*N'*-ethylcarbodiimide hydrochloride (EDC) (*see* **Note 2**).
3. *N*-Hydroxysuccinimide (NHS).
4. 1,2-distearoyl-sn-glycero-3-phosphoethanolamine-*N*-[carboxy (polyethylene glycol)-2000] (ammonium salt) (DSPE-PEG-COOH) (*see* **Note 3**).
5. Dimethyl sulfoxide (DMSO).
6. 0.1 M 2-(*N*-morpholino)ethanesulfonic acid (MES), pH 5.5 buffer: Dissolve 1.952 g of MES in 90 mL of water and adjusting pH to 5.5, then adjusting the volume to 100 mL with water.
7. Regenerated cellulose dialysis membrane: 50 kDa molecular weight cut-off (MWCO).
8. Deuterated water (D_2O).
9. NMR tube.

2.2 Functionalization and Characterization of DGI/C

1. Purified single-walled carbon nanotubes (95 %, Nano-C) (*see* **Note 4**).
2. Vacuum filtration with stainless steel screen (filter cylinder, base, screen, and vacuum flask).
3. 0.22 μm nylon filter.

4. Centrifugal filter unit, 100 kDa MWCO.
5. Carbon-coated copper grid for TEM.
6. Reverse tweezers.
7. 50× TAE buffer: add 242 g of Tris–OH, 57.1 mL glacial acetic acid, 100 mL of 500 mM EDTA to 800 mL water; adjust pH to 8.3 and then adjust the volume to 1 L.
8. Agarose.
9. 6× DNA loading buffer.

2.3 Materials and siRNA for In Vitro Transfection and In Vivo Delivery

1. SiRNA: Double-stranded SiRNAs Silencer® Cy™3 Labeled GAPDH SiRNA siGAPDH and Luciferase GL2 Duplex (siScramble) were used in our experiment.
2. Nuclease-free water.
3. RNase-free pipette tips.
4. 10 % glucose solution (*see* **Note 5**).

3 Methods

Carry out all procedures at room temperature (25 °C) unless otherwise specified.

3.1 Synthesis of 1,2-Distearoyl-sn-Glycero-3-Phosphoethanolamine-N-Poly(Ethylene Glycol)-Poly (Ethylenimine) (DGI)

1. Dissolve 1 g of PEI in 50 mL of MES buffer and mix well.
2. Add 200 μL (0.5 mg, 0.175 μmol) of 2.5 mg/mL DSPE-PEG-COOH to a glass round bottom flask.
3. Dry the lipopolymer in chloroform with purging with compressed air.
4. Weigh out 1 mg (5.2 μmol) of EDC.
5. Add 5 mL of DMSO to the dried polymer and then add the EDC.
6. Put a magnetic stir bar into the solution and stir the mixture with magnetic stirring for 15 min.
7. Put desired amount of PEI solution and MES buffer into a container and mix well (*see* **Note 6**).
8. Add the PEI solution into the DSPE-PEG-COOH solution (*see* Fig. 1).
9. Cap the flask with a rubber or glass stopper.
10. Allow the reaction to proceed overnight.
11. Dialyze the reaction mixture using a dialysis membrane (MWCO 50 kDa) against deionized water for 48 h. Change the water every 4–6 h during dialysis.
12. Lyophilize the product for 2 days (*see* **Note 7**).

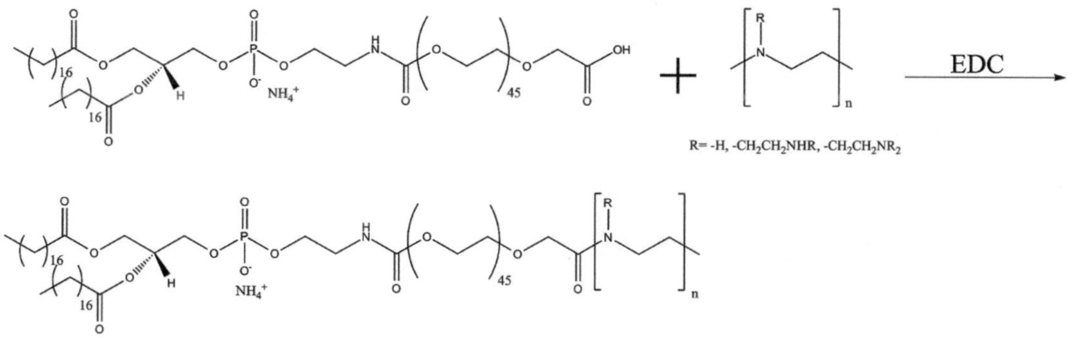

Fig. 1 Synthesis of DGI. Reaction scheme of DGI. DSPE-PEG-COOH was activated with EDC and then PEI was added for the conjugation

13. Analyze the product by ^{1}H-NMR (*see* Subheading 3.2).
14. Optional analysis: size exclusion chromatography (*see* **Note 8**).

3.2 Nuclear Magnetic Resonance (^{1}H-NMR) of DGI

1. Dissolve 0.2 mg of the product from Subheading 3.1 in 0.5 mL of deuterated water for ^{1}H NMR analysis.
2. Transfer the polymer solution to an NMR tube.
3. Record the ^{1}H-NMR at room temperature and use D_2O as the reference solution.
4. Set the acquisition delay to 5 s.
5. Record the spectrum for 128 scans.
6. Calculate the ratio between the peaks of PEI to PEG (*see* Fig. 2) to determine if the synthesis was successful (*see* **Note 9**).

3.3 Non-covalent Functionalization of SWCNT by DGI

1. Weigh out 5 mg of DGI in a 50 mL polyethylene centrifuge tube (*see* **Note 10**).
2. Add 20 mL of water to dissolve the polymer (*see* **Note 11**).
3. Add 5 mg of SWCNT into the polymer solution (*see* **Note 12**).
4. Sonicate the solution for 1 h at 60 °C. Vortex the solution every 15 min.
5. Remove the undissolved CNT by vacuum filtration through a 0.22 μ nylon filter (*see* **Note 13**).
6. Collect the filtrate (*see* **Note 14**).
7. Pour the filtrate into a centrifugal filter unit and remove the unbound polymer by repeated centrifugation.
8. Use 15 mL of water to wash the DGI/C. Repeat 3 times.
9. Take out the concentrated DGI/C carefully by pipette into a centrifugation tube.

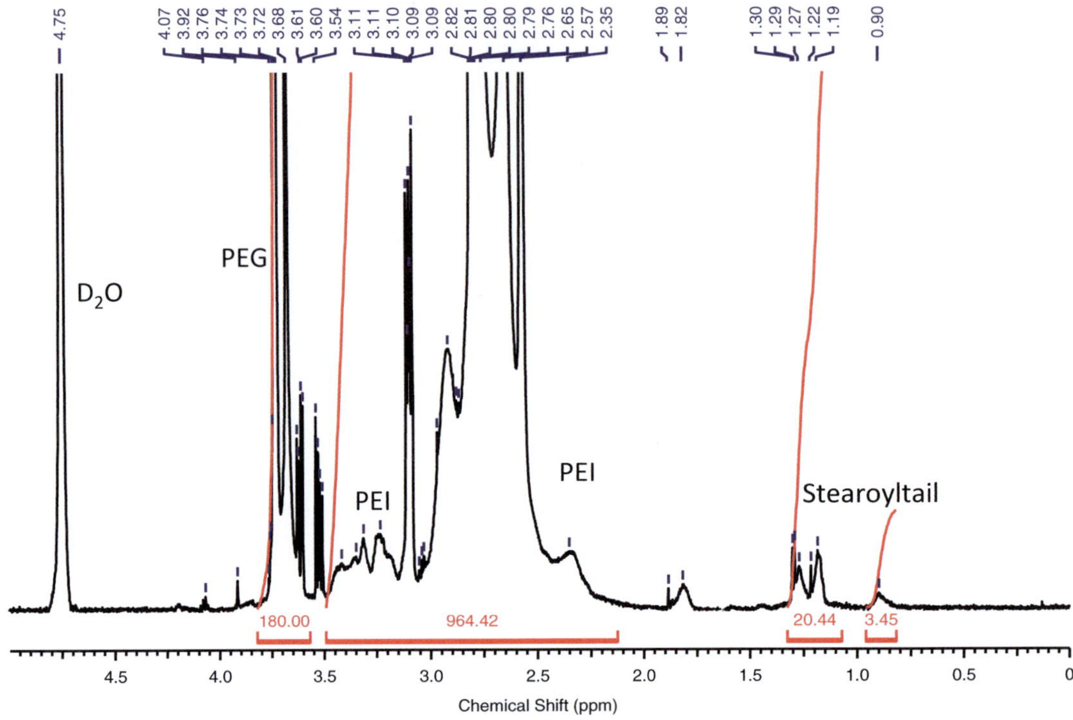

Fig. 2 ^{1}H-NMR of DGI. The polymers were dissolved in D_2O and the acquisition delay was 5 s

10. Remove the CNT residue by repeated centrifugation. Repeat this process 3–5 times if the DGI/C is to be used in an in vivo study.
11. Lyophilize the DGI/C solution (*see* **Note 15**).
12. Weigh out the dried DGI/C and store it at −80 °C.
13. Analyze the DGI/C/SiRNA by gel shift assay (*see* Subheading 3.4).
14. Analyze the DGI/C by transmission electronic microscope (TEM) (*see* Subheading 3.5).
15. Optional: Analyze the DGI/C/SiRNA by zeta potential measurement (*see* **Note 16**).

3.4 Gel Shift Assay

This assay is to assess whether the dispersed CNT (DGI/C) is able to condense the SiRNA.

1. Dilute the SiRNA solution to 0.5 μg/μL using sterile deionized water.
2. Dissolve the DGI/C solutions and dilute to 1 μg/μL (*see* **Note 17**).
3. Prepare seven tubes of SiRNA solution containing 0.5 μg each (i.e., 1 μL).

Table 1
Composition of DSPE-PEG-PEI in weight ratio calculated by ^{1}H-NMR and molecular weight by SEC

	PEI:DSPE-PEG ratio (feed)	PEI:PEG ratio (^{1}H-NMR)	M_n	M_w	PDI (M_n/M_w)
DGI 5	6.6:1	5.0:1	13,440	21,904	1.63
DGI 9	12.1:1	8.9:1	13,245	21,256	1.60
DGI 18[a]	11.1:1	17.1:1	14,645	24,249	1.66

[a]DGI 18 was made by EDC/NHS coupling

4. Prepare 1 μg/μL DGI/C solutions in six additional tubes, and prepare a blank tube without DGI/C.
5. Prepare the DGI/C/SiRNA complexes by mixing equal volumes of DGI/C with SiRNA, or prepare mixtures as shown in Table 1 (*see* **Note 18**).
6. When preparing DGI/C/SiRNA complexes as described in **step 5**, add DGI/C solutions into SiRNA solutions and mix well (*see* **Note 19**).
7. Incubate the DGI/C/SiRNA solutions for 30 min.
8. Prepare a 0.8 % agarose gel in TAE buffer (w/v) (*see* **Note 20**).
9. Mix loading dye with DGI/C/SiRNA solutions using 2 μL of 6× loading buffer (*see* **Note 21**).
10. Put the gel into the gel tank and load the samples into the wells.
11. Perform electrophoresis for 15 min.
12. Visualize the SiRNA using a UV illuminator and photograph the gels.

3.5 Transmission Electron Microscope (TEM)

This technique is to assess whether the dispersed CNT (DGI/C) is singly dispersed.

1. Use reverse tweezers to hold the edge of the copper grid.
2. Put the tweezers with the copper grid on a flat surface. Place the front (shiny side) of the grid facing upward.
3. Add 20 μL of 10 μg/μL DGI/C solution onto the copper grid.
4. Incubate the solution for 10 min (*see* **Note 22**).
5. Remove the solution by putting a small piece of tissue paper at the edge of the copper grid without touching the grid.
6. Dry the copper grid in air overnight (Table 2).

3.6 In Vitro siRNA Transfection with DGI/C/ siRNA

1. Dilute 2 μg of SiRNA into 10 μL (*see* **Note 23**).
2. Dilute 10 μg of DGI/C into 10 μL.

Table 2
A reference table for mixing solution for gel shift assay

Well	1	2	3	4	5	6	7
DGI/C:siRNA ratio (w/w)	0	0.5	1	1.5	2	2.5	3
Vol. of siRNA solution (μL)	1	1	1	1	1	1	1
Vol. of water (μL)	4	4	4	4	4	4	4
Vol. of DGI/C solution (μL)	0	0.25	0.5	0.75	1	1.25	1.5
Vol. of water (μL)	5	4.75	4.5	4.25	4	3.75	3.5

3. Put the DGI/C solution into SiRNA solution and mix well.
4. Incubate for 30 min.
5. These solutions can be added into tissue culture wells containing mammalian cells, to transfect SiRNA into those cells.

3.7 In Vivo siRNA Delivery Using DGI/C/siRNA

1. Dilute 200 μg of Cy3-labeled SiRNA into 100 μL (*see* **Note 24**).
2. Dilute 600 μg of DGI/C into 100 μL RNase-free water.
3. Add DGI/C solution into SiRNA solution and incubate for 30 min.
4. Add 200 μL of 10 % glucose solution and mix well.
5. The solution is ready for injection into mice (20 g body weight).

4 Notes

1. PEI is hygroscopic and it should be stored under nitrogen. If not, it will take up water and weights will be inaccurate.
2. If stored at –20 °C, EDC should be allowed to return to room temperature before opening the bottle to prevent hydrolysis.
3. DSPE-PEG-COOH is available as a solution in chloroform or as a dry powder. It is important to keep in mind that chloroform is a volatile solvent and can vaporize, even when stored under freezing conditions. The amount of DSPE-PEG-COOH will not be accurate if vaporization occurs.
4. CNT of greater purity can be obtained by prolonging the sonication time.
5. Sterile 10 % glucose can be made by dissolving 1 g of glucose in 10 mL of water followed by filtration through a sterile 0.22 μ filter.
6. The amount of PEI in MES buffer for DGI 5 and DGI 9 are 165 μL and 302 μL, respectively. The amount of MES buffer

added should be 4.84 mL and 4.68 mL, respectively. Reference amounts of PEI are as follows: for DGI 18, use 100 μL of DSPE-PEG-COOH (2.5 mg/mL; 0.09 μmol), 0.50 mg (2.6 μmol) of EDC, and 0.20 mg (0.18 μmol) of NHS. It is easier to add an aliquot of dissolved EDC/NHS into the DSPE-PEG-COOH solution. The reaction is prone to errors as a consequence of scale-up effects and conditions may have to be optimized.

7. If the volume of the diluted polymer solution is too great, purging the solution with compressed air overnight may reduce the volume. DSPE-PEG-PEI might accelerate trans-amination at high temperature and the use of a rotary evaporator should be avoided. Trans-amination might lead to products with higher molecular, which is not ideal for transfection into cells and tissues.
8. Size Exclusion Chromatography (SEC) is to determine if there were any unreacted DSPE-PEG or PEI in the DGI, it can also be used to determine if the conjugation was successful by comparing with the PEI alone. Use 0.2 M ammonium acetate/acetic acid (pH 5.3) as mobile phase, flow rate 1 mL/min, at room temperature for 35 min per run. Prepare the samples at a concentration of 10 mg/mL in water, filter through 0.2 μm Supor membrane filters, and inject with a 100 μL volume loop. Calibration curves should be generated using PEO/PEG standards. Molecular weights can be calculated using Empower 3 software (Waters). The following standard molecular weights were used in our calculations: 615, 1010, 3930, 12,140, 20,000, 31,380, 71,700, 106,500.
9. NMR is an exceptionally useful technique to understand the structure of a molecule or the ratios of two polymers in a co-polymer. It is also a specialized instrument which requires training for successful, safe, and accurate use. Follow guidelines for use of the instrument in your workplace and ask for help if you did not know how to proceed. The chemical shift of the peaks from δ 2.5–3.5 were from PEI, the presence of the peaks from δ 3.0–3.5 are characteristic of successful conjugation, and the peak at δ 3.68 was from PEG. The integration of the PEG peak was set to 180 (corresponding to 45 repeat units) and then the PEI:PEG weight ratio was calculated using the following equation:

$$\text{Weight ratio of PEI : PEG} = \frac{\left(\left(\text{Integral of PEI peak} - 2 \text{ for DSPE methylene}\right) / 4\text{H per PEI monomer}\right) \times 43\text{ g / mol}}{\left(\text{Integral of PEG peak} / 4\text{H per PEG monomer}\right) \times 44\text{ g / mol}}$$

The ^{1}H NMR of PEI can be used for comparison.

10. DGI is a sticky polymer. Rather than weighing it out using weight paper, weigh it out in a plastic tube and add, by pipette, the amount of water necessary to obtain the desired solution.
11. Vortex or sonication may help dissolve the polymer. Store the polymer in −80 °C to preserve its integrity. If the polymer comes out of solution it should not be used to functionalize CNTs.
12. Static electricity can be troublesome when weighing out SWCNT using weighing paper. That can be overcome by using aluminum foil as a substitute for the weighing paper.
13. Follow the filtration setup as provided by the supplier. The setup order used by us was: filter cylinder, filter membrane, screen, base, and vacuum flask. Check if there is leakage in the system. Stop pour DGI/C solution into the system if it leaks. Wrapping the connection between the funnel cylinder and the base tightly with plastic wrap or parafilm may help to prevent leakage. Keep in mind that some filter membranes must be wetted with methanol or ethanol before adding water.
14. "Filtrate" (the aqueous portion) refers to the liquid after filtration inside the vacuum flask. The filtrate contains water-soluble DGI/C.
15. If a small tube of the DGI/C is lyophilized, wrap the tube with parafilm and pierce it the parafilm with a needle to create an exit point for vapor. This will prevent the sample from escaping from the tube during lyophilization and pressure equalization.
16. Zeta potential is a measure to determine colloidal stability of the DGI/C/SiRNA complexes. The colloidal complexes were stabilized by the repulsive charge on the surface (to prevent flocculation) and higher the magnitude of the zeta potential (negative or positive), the more likely the colloid to be dispersed. The DGI/C/SiRNA prepared should be positively charged in order to prevent precipitation and facilitate cellular uptake. To determine zeta potential, prepare the DGI/C/SiRNA complexes using 5 μg of SiRNA dissolve in 1 mL of deionized water and desired amount of DGI/C (eg. For DGI/C:SiRNA ratio of 2:1, mix 10 μg of DGI/C in 10 μL and 5 μg of SiRNA in 10 μL an incubate at room temperature for 30 min). Follow the instruction of the zeta potential machine. Use the same source of water to clean and rinse the flow cell or disposable cell.
17. Diluted DGI/C should always be stored at −80 °C. Always have a concentrated solution or most of the aliquot stored at −80 °C. Alternatively, it can be stored in the dried form.

18. DGI/C is able to complex with SiRNA at a ratio of 1:1 (w/w) and 2:1 for DGI/C 18. If it cannot inhibit the SiRNA migration effectively at all ratios (the migration distance of naked SiRNA and DGI/C/SiRNA complexes are the same), the DGI/C cannot effectively interact with SiRNA.
19. Tap the tube with your finger to mix the solution.
20. TAE buffer can be made by diluting the 50× stock solution into 1× TAE buffer. The gel can be made by adding 0.8 g of agarose into 100 mL of 1× TAE buffer. Heat the agarose/buffer mixture to boiling (microwave for 1.5–2 min, depending on the microwave power) but do not allow to boil for more than a few seconds. Wait until the solution has cooled to 50 °C and then add 10 μL of EtBr solution. Mix the solution well and pour it into the gel tray and insert the comb. Cool to room temperature (approximately 25 °C) until completely solid.
21. 15–20 min of electrophoresis run time is usually sufficient to determine whether RNA is associated with complexes.
22. The conditions depend on the concentration of DGI/C and on the CNT. Optimization might be necessary if a different source or purity of CNT was used.
23. Optimize the ratio and amount of DGI/C and SiRNA by first determining the amount of toxicity exerted by 48 h of exposure of the cells of interest to DGI/C using a trypan blue dye exclusion assay (do *not* use MTT assay because CNT interact with MTT) or propidium iodide staining. A minimum cell viability of 80 % should be obtained. Use a ratio of DGI/C:SiRNA ranging from 2:1 to 10:1. If transfection efficacies were low, use a higher DGI/C:SiRNA ratio and/or a larger amount of SiRNA. To avoid unacceptable levels of toxicity, the lowest DGI/C:SiRNA ratio and least amount of SiRNA consistent with adequate knockdown of target mRNA is recommended.
24. Cy3-labeled SiRNA is recommended to visualize SiRNA localization in various organs after in vivo administration. Optimize the amount of DGI/C and SiRNA if necessary. The total volume can be reduced. Glucose solution instead of PBS or saline should be used.

References

1. Gao LZ, Nie L, Wang TH, Qin YJ, Guo ZX, Yang DL, Yan XY (2006) Carbon nanotube delivery of the GFP gene into mammalian cells. Chembiochem 7:239–242
2. Liu Y, Wu DC, Zhang WD, Jiang X, He CB, Chung TS, Goh SH, Leong KW (2005) Polyethylenimine-grafted multiwalled carbon nanotubes for secure noncovalent immobilization and efficient delivery of DNA. Angew Chem Int Ed Engl 44:4782–4785
3. Zhang Z, Yang X, Zhang Y, Zeng B, Wang S, Zhu T, Roden RB, Chen Y, Yang R (2006) Delivery of telomerase reverse transcriptase small interfering RNA in complex with positively

charged single-walled carbon nanotubes suppresses tumor growth. Clin Cancer Res 12:4933–4939

4. Varkouhi AK, Foillard S, Lammers T, Schiffelers RM, Doris E, Hennink WE, Storm G (2011) siRNA delivery with functionalized carbon nanotubes. Int J Pharm 416:419–425
5. Foillard S, Zuber G, Doris E (2011) Polyethylenimine-carbon nanotube nanohybrids for siRNA-mediated gene silencing at cellular level. Nanoscale 3:1461–1464
6. Ladeira MS, Andrade VA, Gomes ER, Aguiar CJ, Moraes ER, Soares JS, Silva EE, Lacerda RG, Ladeira LO, Jorio A, Lima P, Leite MF, Resende RR, Guatimosim S (2010) Highly efficient siRNA delivery system into human and murine cells using single-wall carbon nanotubes. Nanotechnology 21:385101
7. Al-Jamal KT, Toma FM, Yilmazer A, Ali-Boucetta H, Nunes A, Herrero MA, Tian B, Eddaoudi A, Al-Jamal WT, Bianco A, Prato M, Kostarelo K (2010) Enhanced cellular internalization and gene silencing with a series of cationic dendron-multiwalled carbon nanotube: siRNA complexes. FASEB J 24:4354–4365
8. Podesta JE, Al-Jamal KT, Herrero MA, Tian B, Ali-Boucetta H, Hegde V, Bianco A, Prato M, Kostarelos K (2009) Antitumor activity and prolonged survival by carbon-nanotube-mediated therapeutic siRNA silencing in a human lung xenograft model. Small 5:1176–1185
9. Herrero MA, Toma FM, Al-Jamal KT, Kostarelos K, Bianco A, Da Ros T, Bano F, Casalis L, Scoles G, Prato M (2009) Synthesis and characterization of a carbon nanotube-dendron series for efficient siRNA delivery. J Am Chem Soc 131:9843–9848
10. Liu Z, Winters M, Holodniy M, Dai H (2007) siRNA delivery into human T cells and primary cells with carbon-nanotube transporters. Angew Chem Int Ed Engl 46:2023–2027
11. Zhang C, Tang N, Liu X, Liang W, Xu W, Torchilin VP (2006) siRNA-containing liposomes modified with polyarginine effectively silence the targeted gene. J Control Release 112:229–239
12. O'Connell MJ (2006) Carbon nanotubes: properties and applications. Taylor & Francis, Boca Raton
13. Cai D, Mataraza JM, Qin ZH, Huang Z, Huang J, Chiles TC, Carnahan D, Kempa K, Ren Z (2005) Highly efficient molecular delivery into mammalian cells using carbon nanotube spearing. Nat Methods 2:449–454
14. Cai D, Doughty CA, Potocky TB, Dufort FJ, Huang Z, Blair D, Kempa K, Ren ZF, Chiles TC (2007) Carbon nanotube-mediated delivery of nucleic acids does not result in non-specific activation of B lymphocytes. Nanotechnology 18
15. Kam NW, Liu Z, Dai H (2005) Functionalization of carbon nanotubes via cleavable disulfide bonds for efficient intracellular delivery of siRNA and potent gene silencing. J Am Chem Soc 127:12492–12493
16. Chen Y, Yu L, Feng XZ, Hou S, Liu Y (2009) Construction, DNA wrapping and cleavage of a carbon nanotube-polypseudorotaxane conjugate. Chem Commun (Camb) 4106–4108
17. Mercer KE, Pritchard CA (2003) Raf proteins and cancer: B-Raf is identified as a mutational target. Biochim Biophys Acta 1653:25–40
18. Leicht DT, Balan V, Kaplun A, Singh-Gupta V, Kaplun L, Dobson M, Tzivion G (2007) Raf kinases: function, regulation and role in human cancer. Biochim Biophys Acta 1773: 1196–1212
19. Seyhan AA (2011) RNAi: a potential new class of therapeutic for human genetic disease. Hum Genet 130:583–605
20. Perrimon N, Ni JQ, Perkins L (2010) In vivo RNAi: today and tomorrow. Cold Spring Harb Perspect Biol 2:a003640
21. Boussif O, LezoualC'H F, Zanta MA, Mergny MD, Scherman D, Demeneix B, Behr JP (1995) A versatile vector for gene and oligonucleotide transfer into cells in culture and in vivo: polyethylenimine. Proc Natl Acad Sci U S A 92:7297–7301
22. Grayson AC, Doody AM, Putnam D (2006) Biophysical and structural characterization of polyethylenimine-mediated siRNA delivery in vitro. Pharm Res 23:1868–1876
23. Zintchenko A, Philipp A, Dehshahri A, Wagner E (2008) Simple modifications of branched PEI lead to highly efficient siRNA carriers with low toxicity. Bioconjug Chem 19: 1448–1455
24. Heinrich U, Walter H (1939) Polymerization of ethylene imines. IG Farbenindustrie AG, USA
25. Milton C, Nummy WR (1957) Preparation of polyimines from 2-oxazolidone. Arnold Hoffman & Co Inc., USA
26. von Harpe A, Petersen H, Li Y, Kissel T (2000) Characterization of commercially available and synthesized polyethylenimines for gene delivery. J Control Release 69:309–322
27. Abdallah B, Hassan A, Benoist C, Goula D, Behr JP, Demeneix BA (1996) A powerful nonviral vector for in vivo gene transfer into the adult mammalian brain: polyethylenimine. Hum Gene Ther 7:1947–1954

28. Godbey WT, Wu KK, Mikos AG (1999) Size matters: molecular weight affects the efficiency of poly(ethylenimine) as a gene delivery vehicle. J Biomed Mater Res 45:268–275
29. Tang GP, Yang Z, Zhou J (2006) Poly (ethylenimine)-grafted-poly[(aspartic acid)-co-lysine], a potential non-viral vector for DNA delivery. J Biomater Sci Polym Ed 17: 461–480
30. Kunath K, von Harpe A, Fischer D, Petersen H, Bickel U, Voigt K, Kissel T (2003) Low-molecular-weight polyethylenimine as a non-viral vector for DNA delivery: comparison of physicochemical properties, transfection efficiency and in vivo distribution with high-molecular-weight polyethylenimine. J Control Release 89:113–125
31. Kichler A, Leborgne C, Coeytaux E, Danos O (2001) Polyethylenimine-mediated gene delivery: a mechanistic study. J Gene Med 3:135–144
32. Kam NW, Liu Z, Dai H (2006) Carbon nanotubes as intracellular transporters for proteins and DNA: an investigation of the uptake mechanism and pathway. Angew Chem Int Ed Engl 45:577–581
33. Kam NW, O'Connell M, Wisdom JA, Dai H (2005) Carbon nanotubes as multifunctional biological transporters and near-infrared agents for selective cancer cell destruction. Proc Natl Acad Sci U S A 102:11600–11605

Chapter 14

SiRNA In Vivo-Targeted Delivery to Murine Dendritic Cells by Oral Administration of Recombinant Yeast

Kun Xu, Zhongtian Liu, Long Zhang, Tingting Zhang, and Zhiying Zhang

Abstract

SiRNA therapeutics promise a future where any target in the transcriptome could be potentially addressed. However, the delivery of SiRNAs and targeting of particular cell types or organs are major challenges. A novel, efficient, and safe delivery system for promising the introduction of SiRNAs into particular cell types within living organisms is of great significance. Our previous studies have proved that recombinant protein (MSTN) and exogenous gene (EGFP) as vaccines, and furthermore functional CD40 shRNA expression can be delivered into dendritic cells (DCs) in mouse by oral administration of recombinant yeast (*Saccharomyces cerevisiae*). Here, we describe the details of the promising and innovative approach based on oral administration of recombinant yeast that allows in vivo-targeted delivery of functional SiRNA to murine intestinal DCs.

Key words SiRNA, Targeted delivery, Dendritic cells, Oral administration, Yeast, *S. cerevisiae*

1 Introduction

RNA interference (RNAi) is known as a natural gene silencing process in eukaryotic cells mediated by one kind of short RNAs, named as small interfering RNAs (SiRNAs) [1, 2]. Short hairpin RNAs (shRNAs) are the precursors of SiRNAs, which are synthesized in the nucleus and processed to be SiRNAs in cytoplasm [3]. RNAi functions to knock down the expression of a target gene with high specificity at the transcriptional level [4–7], where any target in the transcriptome including previously "undruggable" targets could be potentially addressed. shRNA/SiRNA mediated gene regulation has been emerged as a commonly used technique for gene manipulation. However, delivery of these small RNAs and targeting of particular cell types or organs are still major challenges. A novel, efficien, and safe delivery system for the introduction of SiRNAs into particular cell types within living organisms is of great significance.

Kato Shum and John Rossi (eds.), *SiRNA Delivery Methods: Methods and Protocols*, Methods in Molecular Biology, vol. 1364, DOI 10.1007/978-1-4939-3112-5_14,

Proteins have been proved to be successfully delivered into dendritic cells (DCs) by yeast [8–11] without being degraded in the digestive environment [12]. Also, our previous study demonstrated that recombinant protein (MSTN) [13, 14] and exogenous functional gene (EGFP) [15] can be in vivo delivered into murine DCs by oral administration of recombinant yeast (*Saccharomyces cerevisiae*, *S. cerevisiae*). We hypothesize that we can regulate the expression of particular genes in DCs by targeted delivery of functional SiRNAs with oral administration of recombinant yeast harboring correspondingly designed shRNA expression vectors.

DCs function as professional antigen-presenting cells (APCs) both in innate and adaptive immune responses [16–19]. We chose to knock down the expression of CD40, which is an activation receptor [20] related to the maturation of DCs [21], and the interference may influence immune responses [6, 22–26] and contribute to the immunotherapy for allergy [27], autoimmune renal disease [28], and inhibition of arthritic disease [29]. On the other hand, previous reports suggested that enhanced suppression could be achieved by introducing a small nuclear RNA (snRNA) leader sequence after the U6 promoter and flanking the target shRNA design with an endogenous miRNA (miR30) sequence [30–33]. To test our hypothesis for SiRNA in vivo targeted delivery, we generated miR30-based shRNA expression vectors for CD40 targeting, and results from mice with oral administration of the recombinant yeast demonstrated that functional CD40 shRNA/SiRNAs were successfully introduced to DCs [34].

The principle for our yeast-mediated SiRNA targeted delivery system is diagrammatically shown in Fig. 1. Briefly, after the oral administration, recombinant yeast is recognized and swallowed by DCs; the shRNA expression vector is transported from yeast to the nucleus of DCs and transcribed into pri-shRNA; pri-shRNA is subsequently processed into pre-shRNA by the RNase III enzyme Drosha; with the help of Exportin-5, pre-shRNA is transported to the cytoplasm; pre-shRNA is further processed into mature SiRNA by Dicer; SiRNA functions with Ago/RISC leading to target mRNA cleavage and degradation.

The yeast-mediated SiRNA targeted delivery system promises an innovative and safe approach for in vivo targeted delivery of functional SiRNAs to intestinal DCs in mouse with a time schedule (Fig. 2) simply including several procedures as follows: construction of shRNA and reporter vectors and confirmation of shRNA/SiRNA functional expression (3.1/3.2), preparation and oral administration of recombinant yeast (3.3/3.4), isolation of intestinal DCs and detection of target gene expression (3.5/3.6).

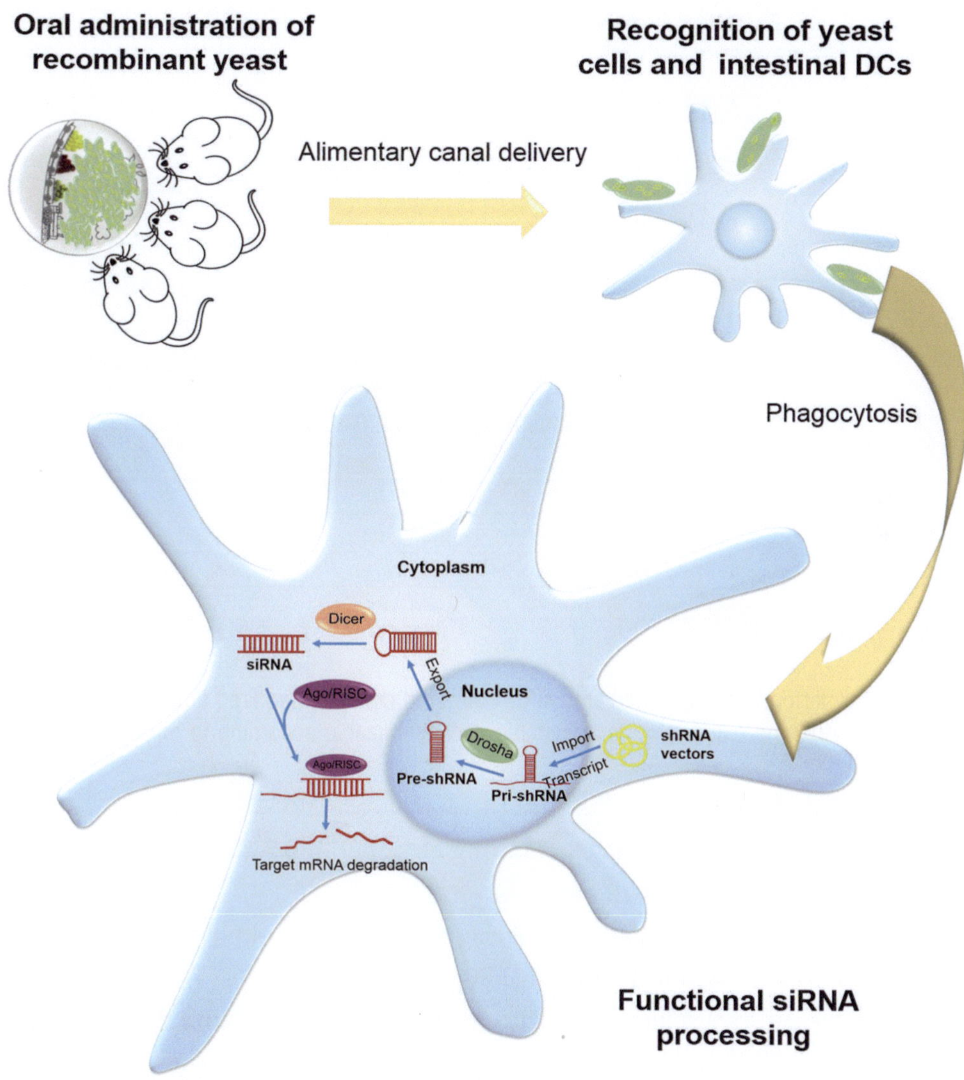

Fig. 1 Principle of yeast-mediated siRNA targeted delivery. After the oral administration, the shRNA expression vector is transported from yeast to DCs, transcribed, and processed into mature siRNA leading to target mRNA cleavage and degradation

2 Materials

2.1 Equipment

1. Nanodrop 2000 micro-spectrophotometer.
2. Horizontal and vertical electrophoresis systems.
3. Fluorescence microscope.
4. Surgical scissors and knife; cytometer; stainless-steel sieve (200 and 400 mesh).

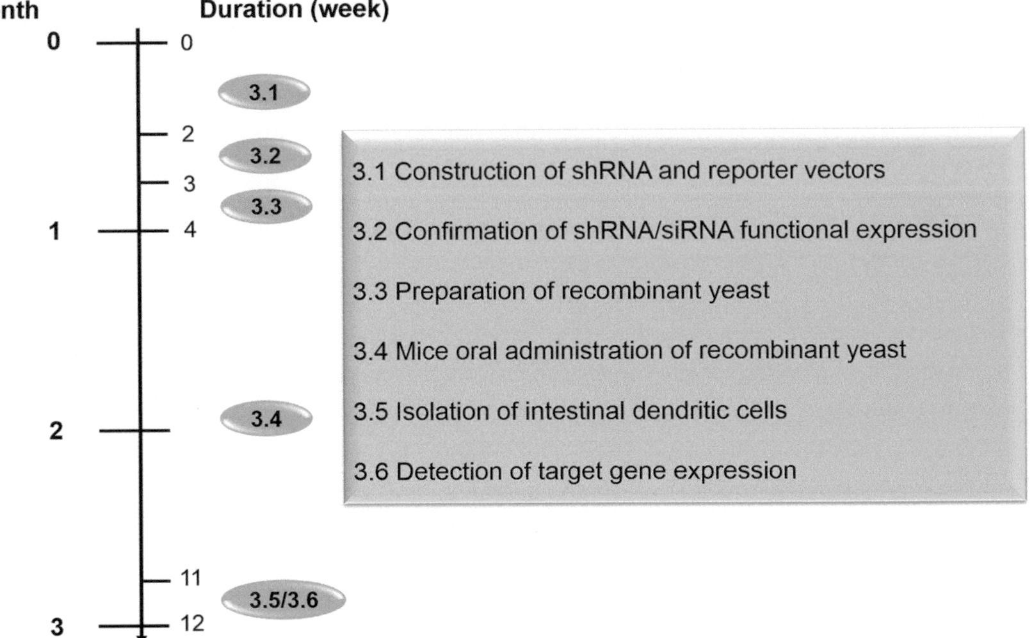

Fig. 2 Timeline for performing yeast-mediated siRNA targeted delivery. Required time for each procedure is just estimated for reference, and may be adjustable and optimizable according to the detailed experiment design

5. Blood cell counting chamber; light microscope.
6. MACS separator.
7. Trans-Blot cell (*see* **Note 1**).
8. Horizontal shaking table.
9. ChemiDoc XRS imaging system (*see* **Note 2**).

2.2 Plasmids, Cells, and Mice

1. pRS426, a typical shuttle vector containing origins of replication as well as selectable markers for both *E. coli* (oriE, ampR) and *S. cerevisiae* (2 μm, URA3) (*see* **Note 3**).
2. pRS426-CMV-GFP-polyA, a vector based on pRS426, containing a T2A-EGFP open reading frame (ORF) driven by the CMV promoter (*see* **Note 4**).
3. *Escherichia coli* DH5α competent cells (*see* **Note 5**).
4. *Saccharomyces cerevisiae* JMY1 (MATα *his3-Δ1 trp1-289rad1-Δ ura3-52*) (*see* **Note 6**).
5. Human 293T cells (*see* **Note 7**).
6. C57BL/6 mice aged 7–10 weeks (*see* **Note 8**).

2.3 Growth Media

1. Culture *E. coli* DH5α cells at 37 °C with standard LB medium. Select and culture DH5α cells harboring related plasmid using LB supplemented with 100 μg/mL ampicillin.

2. Culture *S. cerevisiae* JMY1 cells at 30 °C with standard YPDA medium. Select and culture JMY1 cells harboring shRNA expression or the parental vector with SD/-Ura medium.
3. Routinely maintain human 293T cells in DMEM medium supplemented with 10 % fetal bovine serum (FBS) at 37 °C with 5 % (v/v) CO_2.

2.4 Construction of shRNA Expression and Reporter Vectors

1. Agarose; 10 mg/mL EB.
2. Standard PCR reagents (Taq; dNTP; PCR buffer).
3. Human 293 T and mouse NIH 3 T3 genome DNAs (*see* **Note 9**).
4. 1× TBE: Dilute the 10× 890 mM Tris-boric acid, pH 8.3, 20 mM EDTA 1:10 with distilled water.
5. Restriction enzyme*Kpn*I, *Sal*I, *Sac*I, *Eco*RI, and *Xho*I with buffer supplied.
6. T4 DNA ligase with buffer supplied.
7. PCR primers for hU6L27: hU6-F (5′-TTGggtaccCCCGAGTCCAACACCCGTGGG-3′, *Kpn*I), hU6-R (5′-CTAgtcgacTAGTATATGTGCTGCCGAAGCG-3′, *Sal*I).
8. miR30 sequence synthesized (*see* **Note 10**):
   5′-(*Sal*I)gtcgacTAGGGATAACAGGGTAATTGTTTGAATGAGGCTTCAGTACTTTACAGAATCGTTGCCTGCACATCTTGGAAACACTTGCTGGGATTACTTCTTCAGGTTAACCCAACAGAAGGctcgag(*Xho*I)AAGGTATATGCTGTTGACAGTGAGCGCCgaattc(*Eco*RI)CTGGTTCAAGGGGCTACTTTAGGAGCAATTATCTTGTTTACTAAAACTGAATACCTTGCTATCTCTTTGATACATTTTTACAAAGCTGAATTAAAATGGTATAAATTAAATCACTTTTTTgagctc (*Sac*I)-3′.
9. PCR primers for miR30-shRNA (Take CD40shRNA309 as an example [34], *see* **Note 11**):
   CD40shRNA309-R1,
   5′-TACATCTGTGGCTTCACTAGATTGGGTTCACAGTGTCTGTTCGCTCACTGTCAACAGCA-3′,
   CD40shRNA309-R2,
   5′-TTCCGAGGCAGTAGGCACACAGACACTGTGAACCCAATCTACATCTGTGGCTTCACTA-3′,
   EcoRI-R, 5′-CTTgaa**ttcCGAGGCAGTAGGCA**-3′ *Eco*RI.
10. DNA gel extraction kit.
11. Reagents for LB medium.
12. 100 mg/mL ampicillin. Store at −20 °C.
13. Test tubes; 90 mm plates; sterile 1.5 mL tubes.

14. *E. coli* plasmid mini preparation kit.
15. Sequencing primers: M13F (−47), 5′-CGCCAGGGTTTTC CCAGTCACGAC-3′; M13R (−48), 5′-AGCGGATAACA ATTTCACACAGGA-3′.

2.5 Confirmation of shRNA/siRNA Functional Expression

1. Cell culture 6-well plates; sterile 1.5 mL tubes.
2. DMEM medium supplemented with 10 % FBS.
3. *Sofast* transfection reagent (*see* **Note 12**).
4. Total RNA and cDNA preparation kits (*see* **Note 13**).
5. Standard PCR reagents.
6. PCR primers: miR30.F, 5′-AGAATCGTTGCCTGCACATC-3′; miR30.R, 5′-GAGATAGCAAGGTATTCAG-3′

2.6 Preparation of Recombinant Yeast

1. YPDA medium; SD/-Ura medium; LB/Amp plates and liquid medium.
2. Test tubes; 90 mm plates; 250 mL/2 L Erlenmeyer flasks; sterile 1.5 mL/50 mL tubes.
3. 10× Lithium acetate stock: 1 M lithium acetate in distilled water. Store at 4 °C.
4. PEG stock: 50 % w/v PEG in distilled water. Store at 4 °C (*see* **Note 14**).
5. SS-DNA: 2 mg/mL Salmon sperm DNA. Store at −20 °C (*see* **Note 15**).
6. Transformation mix (360 μL in total per transformation): 240 μL PEG (50 % w/v), 36 μL 1.0 M lithium acetate, 50 μL SS-DNA (2.0 mg/mL), X μL plasmid DNA, 34-X μL sterile ddH_2O.
7. Yeast plasmid mini preparation kit.
8. *E. coli* plasmid mini preparation kit.
9. Restriction enzyme*Kpn*I, *Eco*RI and *Xho*I with buffer supplied.
10. Reagents for agarose gel electrophoresis.
11. PBS Buffer: 137 mM sodium chloride, 2.7 mM potassium chloride, 10 mM sodium hydrogen phosphate, 2 mM potassium hydrogen phosphate.

2.7 Isolation of Intestinal Dendritic Cells

1. Sterile 1.5 mL/50 mL tubes.
2. PBS buffer; FBS.
3. Digestive juice I: PBS added 1 mM EDTA, 1 mM dithiothreitol, and 10 % FBS.
4. Digestive juice II: PBS added 100 U/L collagenase IV and 10 % FBS.

5. Buffer III: PBS added 2 mM EDTA and 0.5 % bovine serum albumin (BSA).
6. MACS magnetic cell sorting reagent kit.

2.8 Detection of Target Gene Expression (Western Blotting)

1. Sterile 1.5 mL tubes.
2. RIPA lysis buffer: 100 Mm PMSF Stock, 6× SDS loading buffer, stripping buffer (*see* **Note 16**).
3. 1×Tris-glycine running buffer: Dilute the 5× stock (125 mM Tris; 1.25 M glycine; 0.5 % w/v SDS) 1:5 with distilled water.
4. Transfer buffer: 48 mM Tris; 39 mM glycine; 0.037 % w/v SDS; 20 % v/v methanol.
5. TBST buffer: 20 mM Tris–HCl, 150 mM sodium chloride, 0.05 % v/v Tween20.
6. Blocking buffer: 5 % skim milk in TBST.
7. Primary antibodies (*see* **Note 17**): Rabbit anti-mouse CD40 polyclonal antibodies, rabbit anti-mouse β-actin polyclonal antibodies.
8. Secondary antibody: HRP-conjugated goat anti-rabbit antibody.
9. Immobilon™ Western Chemiluminescent HRP Substrate (*see* **Note 18**).

3 Methods

3.1 Construction of shRNA and Reporter Vectors

To establish the yeast-mediated SiRNA delivery system, firstly the human U6 promoter sequence along with a 27 nt snRNA leader and the miR30 sequence with intent restriction enzyme sites are introduced into the backbone vector pRS426, a typical shuttle vector for both *E. coli* and *S. cerevisiae*, to construct an enhanced shRNA expression vector (*see* **Note 19**). Secondly, the gene of interest (GoI)-EGFP cassette driven by CMV promoter for reporter mRNA fusion is designed and introduced with the same backbone vector. All the plasmid vectors are constructed simply by standard double digestion and ligation cloning method (Fig. 3). The following procedure describes the details for construction of the shRNA expression and EGFP reporter vectors.

1. Construction of pRS426-hU6L27: Perform standard PCR reaction (Template 293 T genome DNA; primers hU6-F/R) to amplify the hU6 promoter and the leader sequence. Isolate the resulting products from agarose gel (1 % agarose in 1× TBE, running buffer: 1× TBE). Perform standard double-restriction digestion for the hU6L27 fragment and pRS426 with *Kpn*I/*Sal*I, and isolate the intent fragments from agarose gel. Perform standard ligation reaction, transform *E. coli*

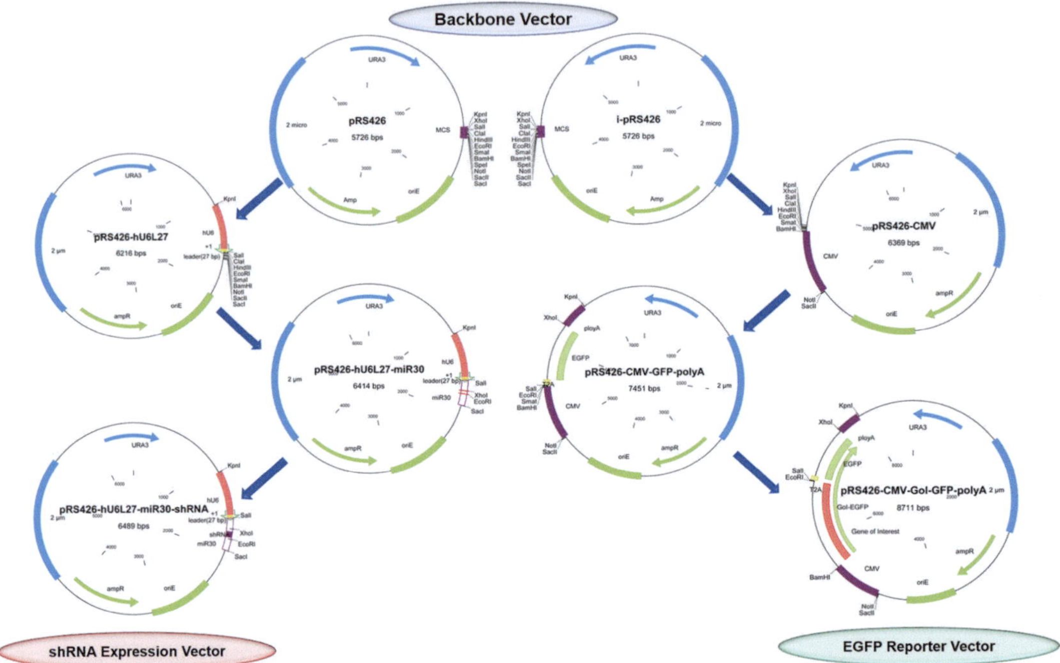

Fig. 3 Schematic drawing for the construction of shRNA and reporter vectors. For shRNA expression vector, the human U6 promoter sequence along with a 27 nt snRNA leader and the flanking miR30 sequence were introduced to enhance the suppression. For the reporter vector, the gene of interest (GoI) were designed to be transcribed along with EGFP as a single-reporter mRNA fusion

DH5α competent cells, and select positive colonies on LB/Amp plates. Culture positive colonies, harvest plasmid DNA, double digest with *Kpn*I/*Sal*I, and run agarose gel to confirm positive clones. Finally, sequence with M13F (−47) primer to confirm the sequence.

2. Construction of pRS426-hU6L27-miR30: Synthesize miR30 sequence directly with *Xho*I/*Eco*RI sites introduced (*see* **Note 10**) and insert it into the *Sal*I/*Sac*I sites of pRS426-hU6L27 similarly by standard double digestion and ligation cloning method. Confirm positive clones by double digestion with *Sal*I/*Sac*I and sequencing with M13R (−48) primer.

3. Construction of pRS426-hU6L27-miR30-shRNA: To construct the miR30-shRNA expression vector, a method of three-step sequential PCR was developed (*see* **Note 11**). Generally, use plasmid pRS426-hU6-miR30 as template for the first step PCR, and the gel purified product from the preceding PCR step for the second and the third steps. The primers used in each step were: hU6-F/shRNA-R1, hU6-F/shRNA-R2 and hU6-F/EcoRI-R, respectively. Insert the final product into the *Kpn*I/*Eco*RI sites of pRS426-hU6L27-miR30 also by standard double digestion and ligation cloning method.

Confirm positive clones by double digestion with *Xho*I/*Eco*RI and sequencing with M13R (−48) primer.

4. Construction of the reporter vector (*see* **Note 4**): Perform standard PCR reactions for CMV promoter, T2A-EGFP-polyA and GoI fragments with pEGFP-C1, pLenti-EF1α-Blank, and NIH 3 T3 genome DNA as templates to construct pRS426-CMV, pRS426-CMV-GFP-polyA, and pRS426-CMV-GoI-GFP-polyA, respectively (Fig. 3). Also construct these plasmids by standard double digestion and ligation cloning method, and confirm positive clones by double digestion and sequencing.

3.2 Confirmation of shRNA/siRNA Functional Expression

To confirm the shRNA/SiRNA functional expression before oral administration of recombinant yeast, the EGFP reporter vector pRS426-CMV-GoI-GFP-polyA is suggested to be constructed, and used to transfect 293 T cells with the shRNA expression vector pRS426-hU6L27-miR30-shRNA or the control parental vector pRS426-hU6L27-miR30. Check the intensity of EGFP fluorescence to confirm indirectly the shRNA/SiRNA functional expression and suppression on GoI-EGFP reporter mRNA fusion (*see* **Note 20**). Conduct RT-PCR to confirm miR30-shRNA expression directly (*see* **Note 21**).

1. Prepare and co-transfect 293 T cells with the EGFP reporter vector and shRNA expression or the control parental vector within 6-well plate using *Sofast* transfection reagent according to the manual. Check and compare the fluorescence between experimental and control groups 24 h after transfection by a fluorescent microscope.
2. Harvest transfected 293 T cells 24 h after transfection (*see* **Note 22**). Prepare total RNA and subsequently cDNA with corresponding reagent kits according to standard manuals (*see* **Note 23**).
3. Perform standard PCR reaction using the cDNA as template with primers miR30.F/R. Check the PCR products with agarose gel (1.5 % agarose). The PCR fragment for miR30-shRNA should be longer (68 bp) with shRNA inserted than the control (Fig. 4a).

3.3 Preparation of Recombinant Yeast

After confirming the miR30-shRNA functional expression, the shRNA expression vector and the control parental vector are used to transform the yeast strain JMY1, separately. Positive colonies are picked, confirmed, and cultured to prepare yeast transformants for oral administration.

1. Transformation of yeast cells: The yeast strain JMY1 was transformed using LiAc method (*see* **Note 24**) as described by Gietz et al. [35–37]. Briefly, inoculate a single colony of JMY1 in 2 mL YPDA with shaking overnight at 30 °C; 1:50 inoculate

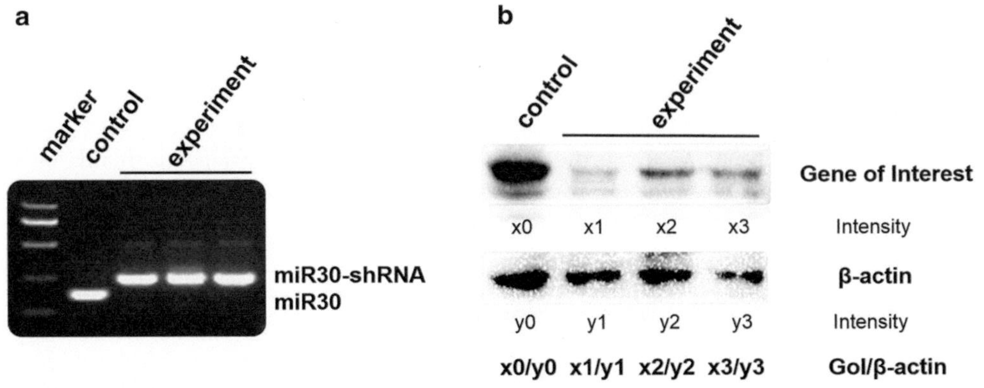

Fig. 4 Detection for the expression of shRNA and gene of interest. (**a**) RT-PCR detection for the shRNA expression. (**b**) Western blotting for the expression of gene of interest (GoI)

50 mL YPDA and incubate at 30 °C on a shaker at 200 rpm until $OD_{600}=0.5$ (*see* **Note 25**); harvest the culture and wash twice with sterile water and once with 100 mM LiAc; resuspend the cells in the 360 μL transformation mix (*see* **Note 26**) and heat shock in a water bath at 42 °C for 45 min; centrifuge and resuspend the cells in 200–1000 μL sterile water (*see* **Note 27**); plate 200 μL on SD/-Ura plates and incubate for at least 3 days at 30 °C.

2. Pick and incubate positive colonies in SD/-Ura liquid medium with shaking overnight at 30 °C and extract yeast plasmid DNA with corresponding reagent kit according to standard manual, the remaining culture was stored at 4 °C directly or at −20 °C with 20 % glycerol for the subsequent inoculation.
3. Transform *E. coli* DH5α competent cells with the yeast plasmid extractions. Plate on LB/Amp plates and incubate overnight at 37 °C.
4. Pick and incubate positive colonies in LB/Amp liquid medium with shaking overnight at 37 °C, and extract plasmid DNA with plasmid mini preparation kit according to standard manual.
5. Double digest with *Kpn*I/*Eco*RI, *Xho*I/*Eco*RI and run agarose gel to confirm the plasmids recovered from yeast transformant colonies as the shRNA expression or control parental vector.
6. Large scale preparation of recombinant yeast: Inoculate 1 mL the remaining culture (**step 2**) of confirmed single transformant colony in 50 mL SD/-Ura liquid medium with shaking overnight at 30 °C; 1:50 inoculate 1 L SD/-Ura liquid medium until $OD_{600}=1.0$ (*see* **Note 28**); harvest the culture with centrifuge at 3000 × *g* for 5 min; wash with sterile water for twice; resuspend cells in 40 mL PBS buffer (*see* **Note 29**); aliquot into 1 mL in sterile 1.5 mL tubes; and store at −20 °C until oral administration.

3.4 MiceOral Administration of Recombinant Yeast

1. Mice and diets: Five female C57BL/six mice aged 10 weeks are suggested to be randomly allocated to one experiment or control group. House under standard conditions of room temperature and dark–light cycles with plenty of food and water.
2. Oral administration of yeast (*see* **Note 30**): Feed the experiment group with yeast transformant harboring the shRNA expression vector and the control group with that harboring the control parental vector. The oral administration will be performed for 50 days with about 5×10^8 yeast cells (400 μL PBS Stock) per mouse every 3 days (*see* **Note 31**).

3.5 Isolation of Intestinal Dendritic Cells

To prove directly that the shRNA/SiRNA expression has been delivered to and functioned in the mouse intestinal DCs after the oral administration, the intestinal DCs of the experiment and the control mice are isolated separately (*see* **Note 32**) and subjected to western blotting for the detection of target protein expression.

1. Sacrifice the mice of the experiment and the control groups (by cervical dislocation), separate out the small intestines and colons and wash with 4 °C precooling PBS containing 10 % fetal bovine serum (*see* **Note 33**).
2. Cut the samples into tiny tissues (*see* **Note 34**) and digest with 20 mL digestive juice I with shaking at 155 rpm, 37 °C, for 40 min.
3. Centrifuge for digested fragments at 4 °C $1500 \times g$ for 5 min and wash with precooling PBS for two times.
4. Further digest with 20 mL digestive juice II with shaking at 155 rpm, 37 °C, for 40 min (*see* **Note 35**).
5. Filter through a stainless-steel sieve for digested cells and wash with 5 mL precooling PBS for two times (*see* **Note 36**).
6. Resuspend the cells with 400 μL Buffer III (per 10^8 cells), add 100 μL anti-CD11c-conjugated magnetic cell sorting microbeads, and incubate at 4 °C for 30 min with gentle shaking.
7. Add 10 mL precooling Buffer III (per 10^8 cells), mix gently, centrifuge at 4 °C $600 \times g$ for 10 min, discard the supernatant completely, and resuspend the cells and microbeads with 500 μL Buffer III.
8. Wash MACS Separation Column with 500 μL Buffer III, and settle it in the magnetic field of MACS Separator.
9. Add the cell suspension to MACS Separation Column, which will adsorb the microbeads and conjugated cells when the suspension flow through by gravity. Wash the column with 500 μL Buffer III for two times.
10. Remove the MACS Separator, and wash the column with 200 μL Buffer III to collect the sorted DCs.

3.6 Detection of Target Gene Expression

1. Centrifuge for purified DCs at 4 °C 1500 × *g* for 10 min, add 100 μL RIPA Lysis Buffer with 1 mM PMSF and incubate on ice for 30 min with gentle mix every 3 min (*see* **Note 37**).
2. Centrifuge at 4 °C 12000 × *g* for 10 min, transfer the supernatants into new 1.5 mL tubes, add 20 μL 6× SDS loading buffer, mix well, and treat in boiling water for 10 min (*see* **Note 38**).
3. Perform standard SDS-PAGE in 1×Tris-glycine running buffer and standard electrotransfer in transfer buffer (*see* **Note 39**).
4. Block the transferred membrane with 5 % skim milk in TBST with gentle shaking at room temperature for 1 h.
5. 1:1000–5000 dilute the primary antibody (against the GoI, e.g., rabbit anti-mouse CD40 polyclonal antibodies, *see* **Note 17**) with 5 % skim milk in TBST, incubate the membrane at room temperature for 1–2 h, and wash with TBST for 3–5 times with gentle shaking for 10 min each time.
6. 1:5000–10000 dilute (*see* **Note 40**) the secondary antibody (HRP-conjugated goat anti-rabbit antibody) with 5 % skim milk in TBST, incubate the membrane at room temperature for 1–2 h, and wash with TBST for 5–6 times with gentle shaking for 10 min each time.
7. Detect reactive target protein by Bio-Rad ChemiDoc XRS imaging system with Millipore Immobilon™ Western Chemiluminescent HRP Substrate.
8. Wash membrane once for 10 min with TBST, incubate the membrane in Stripping Buffer with gentle shaking for 15 min, and then wash the stripped membrane at least three times for 8 min each time.
9. Repeat the blocking, antibody incubating, and imaging (**step 4–7**) with primary rabbit anti-mouse β-actin polyclonal antibodies for detection of the reference β-actin protein (*see* **Note 41**).
10. Analyze intensity of different protein bands using the software Quantity One. The GoI/β-actin ratios are calculated to assess the relative expression of Gene of Interest (Fig. 4b).

4 Notes

1. Trans-blot cells or semidry trans-blot instrument from other manufacturers are alternative.
2. Imaging systems from other manufacturers are alternative. The imaging system should be capable for general agarose gel, polyacrylamide gel, and western blot chemiluminescent imaging.
3. For the construction of shRNA expression and the parental vectors, if pRS426 is not available, you may use another yeast shuttle vector of your choice.

4. The vector was constructed previously for another research project "Functional gene in vivo targeted delivery by yeast" in our lab. Here, we just use it to construct the EGFP reporter vector pRS426-CMV-GoI-GFP-polyA, transcribing the GoI-T2A-EGFP reporter mRNA fusion (For the construction procedure, *see* Subheading 3.1). The EGFP reporter vector need not be a yeast shuttle vector, for example, you can use pEGFP-C1 with GoI-T2A inserted in front of the EGFP sequence within one ORF. T2A was introduced because a direct GoI-EGFP protein fusion often diminish the fluorescence seriously. Besides, the GoI-T2A-EGFP reporter mRNA fusion must be designed within one ORF, if not available, an EGFP-T2A-GoI mRNA fusion is alternative.
5. For cloning and construction of related plasmid vectors; you can also use another standard strain. You may purchase or prepare the competent cells of your choice, and store them in a −80 °C refrigerator before usage.
6. Use as vehicle to deliver the particular shRNA/SiRNA. JMY1 is the private nomenclature of the strain in our lab; if not available, you may use another standard strain of your choice and be aware to check the uptake of this strain by the particular DCs in advance.
7. Human 293 T were chosen as host cell to confirm the shRNA/SiRNA functional expression (*see* Subheading 3.2) because of the easy transfection. However, if available, mouse cell lines are suggested, such as NIH 3 T3 and C2C12.
8. Laboratory animals for in vivo delivery experiment; to make it precisely, the mice should all be male or female. Kunming mice and other species (such as fish and chickens) are also available in our similar projects.
9. The genome DNAs can be prepared with corresponding cells using genomic DNA extraction kit or routine reagents and protocols.
10. The miR30 sequence are flanked with *Sal*I/*Sac*I sites for cloning. You may choose other sites of your choice. The synthesized DNA fragment may be provided within a plasmid by the commercial manufacturer (e.g., we use pUC57-miR30 provided by Genscript, Nanjing, China).
11. The three-step sequential PCR was designed to introduce the arrangement "Droshasite-shRNA left arm-miR30 Loop-shRNA right arm-Drosha site" orderly. Replace the underlined CD40shRNA309 sequences within the primers by the sequences of your shRNAs in order to construct delivery vectors against your favorite gene (GoI).
12. You may use other transfection reagents of your choice, such as Lipofectamine 2000 (Invitrogen) and X-tremeGENE HP (Roche).

13. To guarantee the quality of total RNA and cDNA, we suggest to use corresponding reagent kits. However, you can use routine reagents and protocols if you like.
14. For PEG stock solution, you may choose PEG3350, PEG4000, PEG6000, or PEG8000 of your choice.
15. Denature SS-DNA in boiling water for 10 min, quickly chill and keep on ice before use.
16. You may use those reagents provided by other manufacturers or prepare by yourself according to routine protocols.
17. Choose appropriate primary antibodies against your target gene. Beware that endogenous mouse antibodies may appear on the membrane. If possible, avoid to use murine primary antibodies and anti-mouse secondary antibodies.
18. You may use chemiluminescent HRP substrate provided by other manufacturers of your choice.
19. The intent for introducing the snRNA leader and miR30 sequence was to construct an enhanced shRNA expression vector. However, whether a routine shRNA expression vector is capable for the in vivo targeted delivery remains to be explored.
20. To quantify and further confirm the functional suppression on GoI-EGFP reporter mRNA fusion, you may count EGFP intense positive cells by flow cytometer and conduct RT-PCR or western blotting to check the GoI or EGFP expression.
21. To further verify the shRNA/SiRNA expression and processing, you may conduct Northern blotting to distinguish the pri-shRNA, pre-shRNA and SiRNA (Fig. 3) here in the transfect 293 T cells or subsequently the isolated DCs.
22. You may harvest the cells 24–72 h after transfection.
23. For cDNA preparation the reverse transcription should be performed with a specific primer located at 3′ end of the introduced miR30 sequence.
24. There are different versions for the yeast transformation protocol by LiAc method. As we need not to require high transformation efficiency, you may choose any available protocol for the transformation.
25. For the yeast transformation, OD_{600} between 0.4 and 0.8 will do.
26. Prepare transformation mix before use. To make the cell pellet completely mixed, vortex vigorously and this usually takes about 1 min. As alternative, you may resuspend cell pellet with 84 μL SS-DNA/Plasmid DNA mix, add 36 μL 1.0 M LiAc mix quickly but gently, and then add 240 μL 50 % PEG and vortex.
27. The volume of resuspending sterile water should be estimated according to previous transformation efficiency to make sure

to generate suitable number of single colonies on the selecting SD/-Ura plates.

28. For oral administration, OD_{600} between 1.0 and 1.6 will do.
29. Usually, the yeast cells are considered as 5×10^7/mL when OD_{600} at 1.0. We suggest to resuspend the yeast cells from 1 L culture with 40 mL PBS. Thus, one mouse will be given 400 μL PBS Stock (about 5×10^8 cells) once. However, to make it precise, you may count the cells with blood cell counting chamber under a light microscope before resuspending.
30. When feeding with yeast, remove water more than 6 h before oral administration. Use Eppendorf micropipettor with 200 μL tips to aliquot the thawed yeast PBS stock, just put the tip in front of the mouth of the thirsty mice, and feed drop by drop. The mice will enjoy the feeding pleasantly. If the mice do not like the yeast PBS stock, maybe you have to mix the yeast cells into their diets.
31. We fed 2 mL (1 mL/tube) yeast PBS Stock for one group (5 mice, 400 μL each) once, and the cell number was set at about 5×10^8/mouse. The feeding period for our previous "Yeast-mediated vaccine" projects was 45 days. Here we suggest 50 days to ensure the delivery. However, the feeding dosage and period further remain to be investigated.
32. All steps should be carried out on ice or at 4 °C. 2 or more mice can be used for one treatment to ensure to get enough samples.
33. Remove mesenterium and fat from the small intestines and colons as possible.
34. Cut the samples into tiny tissues as small as possible.
35. Make sure to digest the samples sufficiently in the two steps of digestion, which will help with the subsequent filtering with stainless-steel sieve and the yield of DCs.
36. Filter first with 200 mesh stainless-steel sieve and then 400 mesh to remove insufficient digested tissues. This will help with the subsequent sorting with MACS Separation Column. Count the cells with blood cell counting chamber before resuspending with Buffer III.
37. To make the lysis sufficient, short treatment with ultrasonic may be given every 3–5 min.
38. After lysis, do centrifuge to discard the gelatinous precipitate, which contains the genome DNA and will disturb the subsequent gel loading.
39. Select suitable separating gel and optimize the transfer time for your target protein.
40. The dilution factor depends on the quality of your antibodies.
41. You may use other reference gene (e.g., GAPDH or β-tubulin) and choose corresponding antibodies of your choice.

Acknowledgments

The authors would like to thank the colleagues in Professor Zhang's lab for their excellent technical assistance and helpful collaboration. We are grateful to financial support from China's Ministry of Agriculture (948 Program 2013-Z27), China's Ministry of Science and Technology (National Science and Technology Major Project 2014ZX0801009B and 973 Program 2011CBA01002) and National Natural Science Foundation of China (NSFC 31172186).

References

1. Bartel DP (2009) MicroRNAs: target recognition and regulatory functions. Cell 136: 215–233
2. Shukla GC, Singh J, Barik S (2011) MicroRNAs: processing, maturation, target recognition and regulatory functions. Mol Cell Pharmacol 3:83–92
3. Burnett JC, Rossi JJ, Tiemann K (2011) Current progress of siRNA/shRNA therapeutics in clinical trials. Biotechnol J 6:1130–1146
4. Hannon GJ (2002) RNA interference. Nature 418:244–251
5. Paddison PJ, Hannon GJ (2002) RNA interference: the new somatic cell genetics? Cancer Cell 2:17–23
6. Karimi MH, Ebadi P, Pourfathollah AA, Soheili ZS, Samiee S, Ataee Z, Tabei SZ, Moazzeni SM (2009) Immune modulation through RNA interference-mediated silencing of CD40 in dendritic cells. Cell Immunol 259:74–81
7. Grimm D (2009) Small silencing RNAs: state-of-the-art. Adv Drug Deliv Rev 61:672–703
8. Stubbs AC, Martin KS, Coeshott C, Skaates SV, Kuritzkes DR, Bellgrau D, Franzusoff A, Duke RC, Wilson CC (2001) Whole recombinant yeast vaccine activates dendritic cells and elicits protective cell-mediated immunity. Nat Med 7:625–629
9. Haller AA, Lauer GM, King TH, Kemmler C, Fiolkoski V, Lu Y, Bellgrau D, Rodell TC, Apelian D, Franzusoff A, Duke RC (2007) Whole recombinant yeast-based immunotherapy induces potent T cell responses targeting HCV NS3 and Core proteins. Vaccine 25:1452–1463
10. Wansley EK, Chakraborty M, Hance KW, Bernstein MB, Boehm AL, Guo Z, Quick D, Franzusoff A, Greiner JW, Schlom J, Hodge JW (2008) Vaccination with a recombinant Saccharomyces cerevisiae expressing a tumor antigen breaks immune tolerance and elicits therapeutic antitumor responses. Clin Cancer Res 14:4316–4325
11. Bernstein MB, Chakraborty M, Wansley EK, Guo Z, Franzusoff A, Mostbock S, Sabzevari H, Schlom J, Hodge JW (2008) Recombinant Saccharomyces cerevisiae (yeast-CEA) as a potent activator of murine dendritic cells. Vaccine 26:509–521
12. Blanquet S, Meunier JP, Minekus M, Marol-Bonnin S, Alric M (2003) Recombinant Saccharomyces cerevisiae expressing P450 in artificial digestive systems: a model for biodetoxication in the human digestive environment. Appl Environ Microbiol 69:2884–2892
13. Begriche K, Levasseur PR, Zhang J, Rossi J, Skorupa D, Solt LA, Young B, Burris TP, Marks DL, Mynatt RL, Butler AA (2011) Genetic dissection of the functions of the melanocortin-3 receptor, a seven-transmembrane G-protein-coupled receptor, suggests roles for central and peripheral receptors in energy homeostasis. J Biol Chem 286:40771–40781
14. Zhang T, Sun L, Xin Y, Ma L, Zhang Y, Wang X, Xu K, Ren C, Zhang C, Chen Z, Yang H, Zhang Z (2012) A vaccine grade of yeast Saccharomyces cerevisiae expressing mammalian myostatin. BMC Biotechnol 12:97
15. Kiflmariam MG, Yang H, Zhang Z (2013) Gene delivery to dendritic cells by orally administered recombinant Saccharomyces cerevisiae in mice. Vaccine 31:1360–1363
16. Banchereau J, Steinman RM (1998) Dendritic cells and the control of immunity. Nature 392:245–252
17. Palucka K, Banchereau J (1999) Dendritic cells: a link between innate and adaptive immunity. J Clin Immunol 19:12–25
18. Qian C, Qian L, Yu Y, An H, Guo Z, Han Y, Chen Y, Bai Y, Wang Q, Cao X (2013) Fas

signal promotes the immunosuppressive function of regulatory dendritic cells via the ERK/beta-catenin pathway. J Biol Chem 288: 27825–27835

19. Fu C, Jiang A (2010) Generation of tolerogenic dendritic cells via the E-cadherin/beta-catenin-signaling pathway. Immunol Res 46:72–78
20. Lutgens E, Daemen MJ (2002) CD40-CD40L interactions in atherosclerosis. Trends Cardiovasc Med 12:27–32
21. Ma DY, Clark EA (2009) The role of CD40 and CD154/CD40L in dendritic cells. Semin Immunol 21:265–272
22. Zheng X, Vladau C, Zhang X, Suzuki M, Ichim TE, Zhang ZX, Li M, Carrier E, Garcia B, Jevnikar AM, Min WP (2009) A novel in vivo siRNA delivery system specifically targeting dendritic cells and silencing CD40 genes for immunomodulation. Blood 113:2646–2654
23. Suzuki M, Zheng X, Zhang X, Li M, Vladau C, Ichim TE, Sun H, Min LR, Garcia B, Min WP (2008) Novel vaccination for allergy through gene silencing of CD40 using small interfering RNA. J Immunol 180:8461–8469
24. Moore TM, Shirah WB, Khimenko PL, Paisley P, Lausch RN, Taylor AE (2002) Involvement of CD40-CD40L signaling in postischemic lung injury. Am J Physiol Lung Cell Mol Physiol 283:L1255–L1262
25. Taylor PA, Friedman TM, Korngold R, Noelle RJ, Blazar BR (2002) Tolerance induction of alloreactive T cells via ex vivo blockade of the CD40:CD40L costimulatory pathway results in the generation of a potent immune regulatory cell. Blood 99:4601–4609
26. Ripoll E, Merino A, Herrero-Fresneda I, Aran JM, Goma M, Bolanos N, de Ramon L, Bestard O, Cruzado JM, Grinyo JM, Torras J (2013) CD40 gene silencing reduces the progression of experimental lupus nephritis modulating local milieu and systemic mechanisms. PLoS One 8, e65068
27. Suzuki M, Zheng X, Zhang X, Ichim TE, Sun H, Kubo N, Beduhn M, Shunnar A, Garcia B, Min WP (2009) Inhibition of allergic responses by CD40 gene silencing. Allergy 64:387–397
28. Wang Y, Wang YM, Wang Y, Zheng G, Zhang GY, Zhou JJ, Tan TK, Cao Q, Hu M, Watson D, Wu H, Zheng D, Wang C, Lahoud MH, Caminschi I, Harris DC, Alexander SI (2013) DNA vaccine encoding CD40 targeted to dendritic cells in situ prevents the development of Heymann nephritis in rats. Kidney Int 83:223–232
29. Zheng X, Suzuki M, Zhang X, Ichim TE, Zhu F, Ling H, Shunnar A, Wang MH, Garcia B, Inman RD, Min WP (2010) RNAi-mediated CD40-CD154 interruption promotes tolerance in autoimmune arthritis. Arthritis Res Ther 12:R13
30. Silva JM, Li MZ, Chang K, Ge W, Golding MC, Rickles RJ, Siolas D, Hu G, Paddison PJ, Schlabach MR, Sheth N, Bradshaw J, Burchard J, Kulkarni A, Cavet G, Sachidanandam R, McCombie WR, Cleary MA, Elledge SJ, Hannon GJ (2005) Second-generation shRNA libraries covering the mouse and human genomes. Nat Genet 37:1281–1288
31. Paddison PJ, Silva JM, Conklin DS, Schlabach M, Li M, Aruleba S, Balija V, O'Shaughnessy A, Gnoj L, Scobie K, Chang K, Westbrook T, Cleary M, Sachidanandam R, McCombie WR, Elledge SJ, Hannon GJ (2004) A resource for large-scale RNA-interference-based screens in mammals. Nature 428:427–431
32. Paddison PJ, Caudy AA, Bernstein E, Hannon GJ, Conklin DS (2002) Short hairpin RNAs (shRNAs) induce sequence-specific silencing in mammalian cells. Genes Dev 16:948–958
33. Zeng Y, Wagner EJ, Cullen BR (2002) Both natural and designed micro RNAs can inhibit the expression of cognate mRNAs when expressed in human cells. Mol Cell 9:1327–1333
34. Zhang L, Zhang T, Wang L, Shao S, Chen Z, Zhang Z (2014) In vivo targeted delivery of CD40 shRNA to mouse intestinal dendritic cells by oral administration of recombinant Saccharomyces cerevisiae. Gene Ther 21: 709–714
35. Gietz RD (2014) Yeast transformation by the LiAc/SS carrier DNA/PEG method. Methods Mol Biol 1163:33–44
36. Gietz RD, Schiestl RH (2007) Quick and easy yeast transformation using the LiAc/SS carrier DNA/PEG method. Nat Protoc 2:35–37
37. Gietz RD, Woods RA (2006) Yeast transformation by the LiAc/SS carrier DNA/PEG method. Methods Mol Biol 313:107–120

Chapter 15

TLR9-Targeted SiRNA Delivery In Vivo

Dewan Md Sakib Hossain, Dayson Moreira, Qifang Zhang, Sergey Nechaev, Piotr Swiderski, and Marcin Kortylewski

Abstract

The SiRNA strategy is a potent and versatile method for modulating expression of any gene in various species for investigational or therapeutic purposes. Clinical translation of SiRNA-based approaches proved challenging, mainly due to the difficulty of targeted SiRNA delivery into cells of interest and the immunogenic side effects of oligonucleotide reagents. However, the intrinsic sensitivity of immune cells to nucleic acids can be utilized for the delivery of SiRNAs designed for the purpose of cancer immunotherapy. We have demonstrated that synthetic ligands for the intracellular receptor TLR9 can serve as targeting moiety for cell-specific delivery of SiRNAs. Chemically synthesized CpG-SiRNA conjugates are quickly internalized by TLR9-positive cells in the absence of transfection reagents, inducing target gene silencing. The CpG-SiRNA strategy allows for effective targeting of TLR9-positive cells in vivo after local or systemic administration of these oligonucleotides into mice.

Key words TLR9, SiRNA, CpG, Cancer, Myeloid cells, Leukemia, Oligonucleotides

1 Introduction

The discovery of RNA interference (RNAi) mediated by small-interfering RNA (SiRNA) created a unique opportunity for targeting almost any disease-related gene for therapeutic purposes [1–3]. The SiRNA-based therapies can overcome the challenges of non-enzymatic proteins, such as transcription factors. Preclinical studies in various disease models, including cancer, demonstrated in vivo efficacy of RNAi in rodents [2] and in nonhuman primates [4]. Recent, first-in-human clinical trial confirmed gene silencing after systemic delivery of SiRNA to cancer patients [5]. Broad application of SiRNAs is still limited by the lack of cell-specific delivery, insufficient intracellular uptake, and cytoplasmic release. The effective target gene silencing usually requires encapsulating SiRNA into chemical formulations that may contribute to toxicities and side effects [1, 2]. Previously, several strategies have been

Kato Shum and John Rossi (eds.), *SiRNA Delivery Methods: Methods and Protocols*, Methods in Molecular Biology, vol. 1364, DOI 10.1007/978-1-4939-3112-5_15, © Springer Science+Business Media New York 2016

implemented to enhance targeted cell delivery by conjugating SiRNAs with cholesterol, cationic peptides, cell-specific antibodies, RNA aptamers, or ligands for cell surface receptors [3, 6]. However, many of these strategies result in endosomal entrapment of SiRNAs with limited cytoplasmic release and thereby limited effect on target mRNAs [2].

We have recently developed an original method to deliver therapeutic SiRNAs into specific target cells by targeting endosomal receptor Toll-like receptor 9 (TLR9) [7]. TLR9-positive cells recognize and internalize single-stranded oligodeoxyribonucleotides containing an unmethylated CpG motif (CpG ODN) [8, 9]. The conjugates of CpG ODNs to Dicer substrate SiRNAs (CpG-SiRNAs) [10] are actively internalized by human and mouse TLR9-positive cells, without any transfection reagents [7, 11]. The CpG-SiRNA conjugates interact with Dicer after uptake into early endosomes (EE). Importantly, TLR9 plays essential role in facilitating the release of uncoupled SiRNA from EE and trafficking to RNA-induced silencing complexes (RISC) on the surface of endoplasmic reticulum (ER) [12, 13]. The CpG-SiRNA processing results in efficient target gene silencing in various human and mouse TLR9-positive cells in vivo. These include hematopoietic cells, such as dendritic cells (DCs), macrophages, and B cells, as well as in certain blood cancers, e.g., in acute myeloid leukemia (AML), multiple myeloma, or B cell lymphoma (BCL) [7, 11, 14]. The efficacy studies in mice demonstrated direct and/or immune-mediated antitumor activity of CpG-SiRNAs targeting various tumorigenic factors, including *STAT3, STAT5, RELA/p65, BCL2L1*, or *S1PR1* [11, 15, 16]. Overall, CpG-SiRNA strategy allows for a two-pronged targeting of both cancer cells and immune cells to augment therapeutic effects. In this chapter, we describe methods for the design and application of CpG-SiRNA conjugates to deliver SiRNA into specific target cells in mice, starting from the oligonucleotide synthesis to local/systemic delivery into TLR9-positive cells in vivo. Additionally, we discuss critical troubleshooting points for target cell selection and optimizing gene silencing efficiency of CpG-SiRNA conjugates.

2 Materials

2.1 CpG-siRNA Design and Synthesis

1. The CpG-SiRNA conjugates consist of Dicer substrate SiRNA (25/27mer) [10] linked to the class A CpG/D19 (type A) or CpG1668 (type B) for human- or mouse-optimized versions of the reagent, respectively.
2. The two strands of the CpG-SiRNA are synthesized separately using Akta OligoPilot100 (GE) as follows:
   - SiRNA(SS)—25mer sense strand with or without fluorochrome added to the 3′ end (*see* **Note 1**).

- 5′-NNNNNNNNNNNNNNNNNNNNNNNNNdNdN-3′
- N = ribonucleotide; dN = deoxyribonucleotide.
- CpG-SiRNA(AS)—27mer antisense strand.
- *Human version* 5′-GGGGGGGG-linker-3′.
- *Mouse version* 5′-TCCATGACGTTCCTGATGCT-linker-NNNNNNNNNNNNNNNNNNNNNNNNNNN-3′.
- The linker consists of 5 units of the C3 Spacer (Glen Research, Sterling, VA); underlined are phosphorothioated residues. The second nucleotide from the 3′ end of CpG-SiRNA(AS) strand can be 2′O-methyl-modified to improve stability.

3. Both oligonucleotides are aliquoted and stored as lyophilized pellets of 50 nmol each below –20 °C until use.

2.2 Tissue Digestion

1. 10× Collagenase D stock solution: Determine enzyme activity in Wunsch units and prepare 4000 U Collagenase D per 1 mL of HBSS with calcium and magnesium; before use dilute in HBSS to 1× solution for tissue digestion.
2. 10× DNase I stock solution: Dissolve 10 mg DNase I in 1 mL of HBSS, and before use dilute in HBSS to 1× solution for tissue digestion.
3. 10× ACS buffer: For 1 L, dissolve 89.9 g NH_4Cl, 10.0 g $KHCO_3$, and 370 mg EDTA in milli-Q water, then adjust to pH 7.3, and dilute to 1× buffer solution in water for erythrocyte lysis.
4. Histopaque®-1083 for isolating viable cells from mouse tissues.

2.3 7 M Urea/ Polyacrylamide (PAGE) Gel

1. 10× TBE: For 500 mL, dissolve 54 g Tris base, 27.6 g boric acid, and 10 mL 0.5 M EDTA, pH 8.0 in water, adjust to pH 8.3, and dilute to 1× solution for gel running and staining.
2. 8 M Urea: For 200 mL, dissolve 96.1 g urea in water, and filter the solution.
3. 7 M Urea/20 % acrylamide solution: For 200 mL, dissolve 88.2 g urea in 100 mL of 40 % acrylamide/bis-acrylamide solution (19:1), adjust volume to 200 mL using milli-Q water and filter the solution, *see* Table 1.
4. Ethidium bromide: Use 2 μL of ethidium bromide solution (10 mg/mL) in 25 mL of 1× TBE buffer for gel staining.
5. Double-stranded RNA marker.
6. Nuclease-free water: DNase/RNase-free water.

Table 1
Composition of 7 M urea/15 % PAGE

Reagent	For 10 mL
7 M Urea/20 % acrylamide solution	7.5 mL
8 M Urea	1.5 mL
10× TBE	1 mL
APS (10 %)	80 μL
TEMED	7 μL

2.4 Flow Cytometry Reagents

1. Staining buffer: 1× PBS, 2 % FBS and 0.1 % NaN_3.
2. Antibodies: Fluorescently labeled antibodies specific to surface markers on immune and cancer cells.
3. Fc Block: Anti-CD16/CD32 antibody to block unspecific binding by Fcγ receptors.
4. Viability dye: 7AAD.
5. Flow cytometer: BD Accuri C6.

2.5 Time-Lapse Confocal Microscopy Reagents, Instruments and Software

1. Culture dishes: 35/14 mm #1.5 glass bottom tissue culture dishes.
2. Microscope: cLSM510-Meta inverted confocal microscope (Zeiss, Thornwood, NY).
3. Acquisition software: LSM software v.4.2 SP1 (Zeiss).
4. Post-acquisition analysis software: LSM Image Browser v.4.2.0.121 (Zeiss).

2.6 In Vivo CpG-siRNA Delivery

1. Syringes: 1 mL U-100 insulin syringe with 30G needle to inject CpG-SiRNA conjugates diluted in 1× PBS in vivo.
2. Anesthetic agents: Anesthesia induction chamber including isoflurane and oxygen supply.
3. Mice: For in vivo experiments C57BL/6 and NOD/SCID/IL-2RγKO (NSG) mice were purchased from National Cancer Institute (Frederick, MD) and the Jackson Laboratory (Bar Harbor, ME), respectively.

3 Methods

3.1 Hybridization of CpG-siRNA Conjugates

1. Add 250 μL of DNase/RNase-free water to each of the two 50 nmol aliquots of CpG-SiRNA(AS) and SiRNA(SS), mix well at 37 °C for 5 min, and make sure that both oligonucleotides are completely dissolved.

2. Combine both strands in a new 1.5 mL sterile tube to make 100 μM stock solution of CpG-SiRNA conjugate (use amber tubes for fluorescently labeled reagents).
3. Place the tube at 80 °C for 1 min; promptly transfer to the 37 ° C water bath and incubate for 1 h.
4. Spin down and aliquot hybridized CpG-SiRNA conjugates into desired volumes for long-term storage at −80 °C.

3.2 Oligonucleotide Quality Control on the 7 M Urea/PAGE Gels

1. Assemble the gel plates in the gel casting chamber, prepare 7 M urea/15 % PAGE gel according to Table 1 and pour quickly between gel plates, insert comb, and allow polymerizing for 30 min.
2. Dismount the polymerized gel from casting chamber and assemble into gel apparatus according to manufacturer's instruction.
3. Fill the gel chamber with 1× TBE, carefully remove the comb, wash all wells and pre-run for 30 min.
4. Dilute oligonucleotides to 10 μM, use 2 μL of diluted samples to mix with 0.5 μL of 0.5 M EDTA, pH 8.0 and 3 μL of 2× RNA loading dye, load into the wells together with the RNA marker, and run the gel at 250 V for 1 h.
5. Remove the gel and carefully transfer into a dish containing 1× TBE plus ethidium bromide, stain for 15–20 min, and then examine the gel on the UV transilluminator (Fig. 1) (*see* **Note 2**).

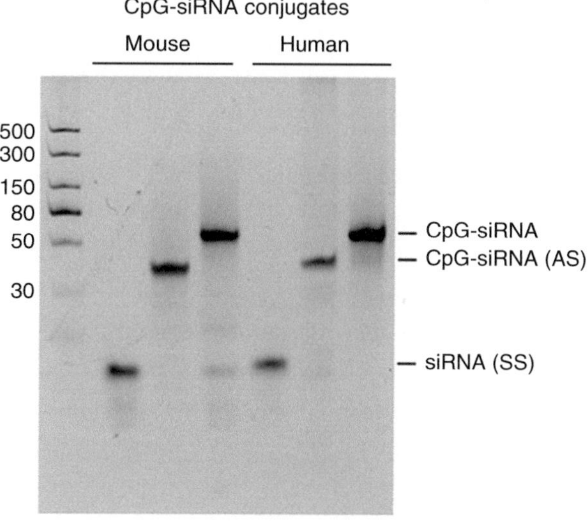

Fig. 1 Quality control for the hybridized CpG-siRNA constructs. Mouse and human versions of siRNA(SS), CpG-siRNA(AS), and CpG-siRNA conjugate were loaded on the 7 M urea/15 % PAGE gel. After electrophoresis gel was stained using ethidium bromide to visualize oligonucleotides compared to a dsRNA marker

3.3 The Cytofluorimetric Assessment of CpG-siRNA Internalization by Target Cells In Vitro

1. Culture the desired number of cells (usually 10^6 cells/mL) in the appropriate medium or, if using freshly isolated primary cells (*see* Subheadings 3.7–3.9), allow cells to adjust to the in vitro culture condition for 2–4 h (*see* **Note 3**).
2. Treat cells with various doses (10–500 nM) of fluorescently labeled CpG-SiRNA constructs, then collect cells at desired times to verify cellular uptake.
3. Collect cells, resuspend each sample in 100 μL of the staining buffer, add 1 μL of the Fc Block for 10 min, then stain cells with a combination of fluorescently labeled antibodies specific to immune cell markers for 20–30 min on ice, wash once with 2 ml of the staining buffer, and resuspend in the staining buffer plus 7AAD for exclusion of dead cells for the analysis using flow cytometry (Fig. 2a, b) (*see* **Note 4**).

3.4 Time-Lapse Confocal Microscopy to Verify Internalization of CpG-siRNA In Vitro

1. Plate and culture cells overnight at a desired density (usually 10^5 cells/well) in the appropriate medium using the 35/14 mm #1.5 glass-bottom tissue culture dishes.
2. Next day, wash cells twice in phenol red-free cell culture medium, then replace it with complete culture medium, and add fluorescently labeled CpG-SiRNA at 100–500 nM concentration.
3. Transfer culture dishes to the microscope chamber with environmental control (37 °C at 5 % CO_2), and incubate for desired times (0.25–48 h).
4. Acquire images at various time increments using cLSM510-Meta inverted confocal microscope, C-Apochromat 40×/1.2 water-immersed objective and LSM software v.4.2 SP1 (Zeiss).
5. Use LSM Image Browser v.4.2.0.121 for post-acquisition analysis (Zeiss) (Fig. 3) (*see* **Note 5**).

3.5 Local Peritumoral Delivery of CpG-siRNA to Target Cell Populations In Vivo

1. Prepare 7–10-week-old mice (C57BL/6 or NSG for syngeneic or xenotransplanted tumor models, respectively).
2. Inject tumor cells suspended in sterile PBS (usually $0.1–1 \times 10^6$ cells/100 μL) to generate subcutaneous (SC) tumor model.
3. After tumor size exceeds ~5 mm in diameter, prepare for injections of CpG-SiRNA.
4. Prepare solution of fluorescently-labeled CpG-SiRNA (e.g., CpG-SiRNA^{Cy3}) at 1–5 mg/kg (20–100 μg) in 100 μL of sterile PBS.
5. Use insulin syringe with 30G injection needle laid parallel to the skin and inserted flatly onto the tumor.

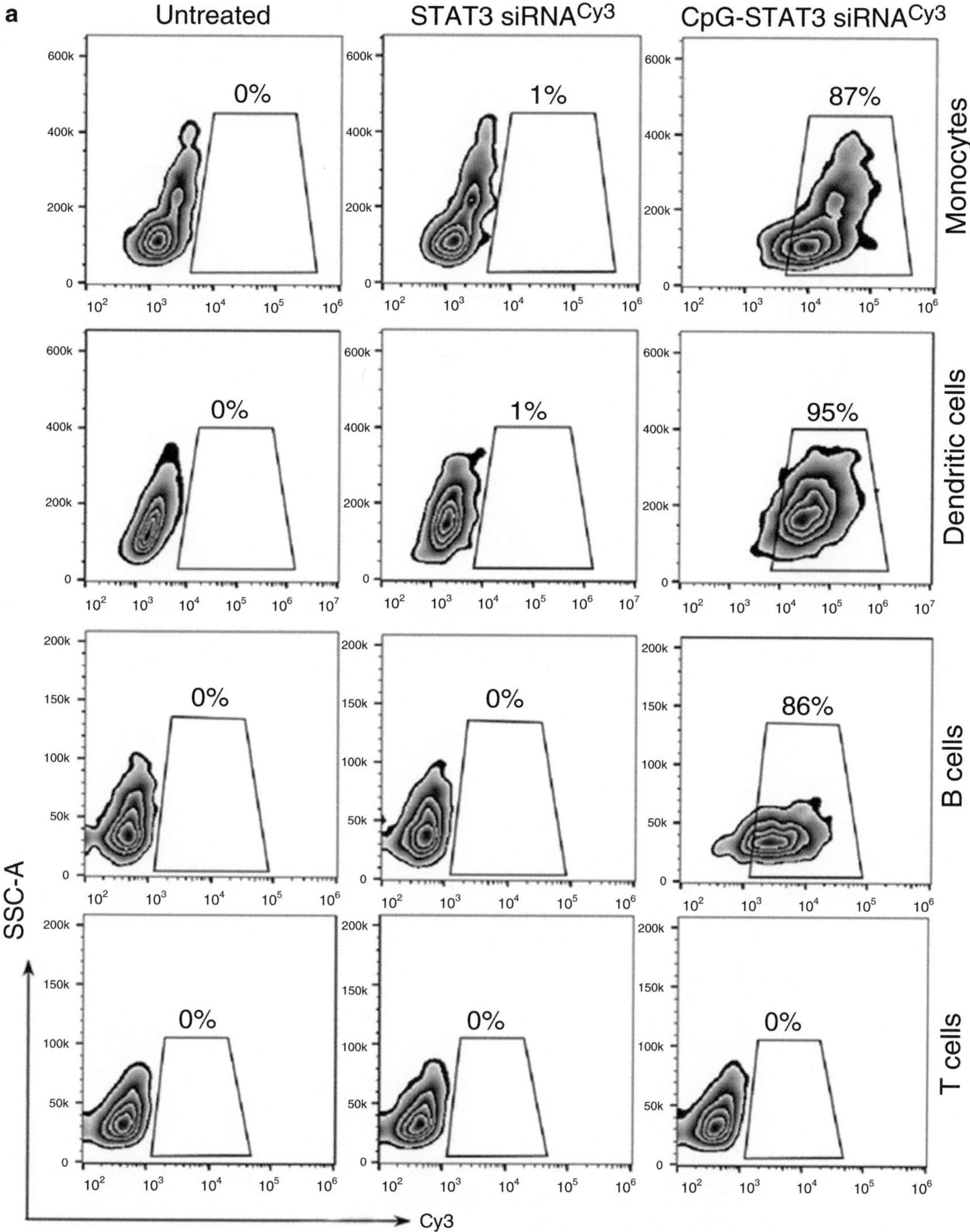

Fig. 2 Human immune and leukemic cells efficiently internalize CpG-siRNA conjugates in vitro. (**a**) Targeted delivery of CpG-*STAT3* siRNA into different primary human immune cells in vitro. Human PBMCs were incubated with fluorescently labeled CpG(A)-*STAT3* siRNA^{Cy3} or unconjugated *STAT3* siRNA^{Cy3} at 500 nM concentration for 1 h without any transfection reagents. Percentages of Cy3^{+} monocytes (CD14^{+}), plasmacytoid dendritic cells (BDCA2^{+}), B cells (CD19^{+}), and T cells (CD3^{+}) were assessed by flow cytometry. (**b**) CpG-siRNA internalization by human MM (KMS.11) and AML (MV4-11, KG1a) cells. Cells were incubated with 500 nM Cy3-labeled CpG-*STAT3* siRNA for 1 h. The percentage of Cy3^{+} leukemic cells was analyzed by flow cytometry

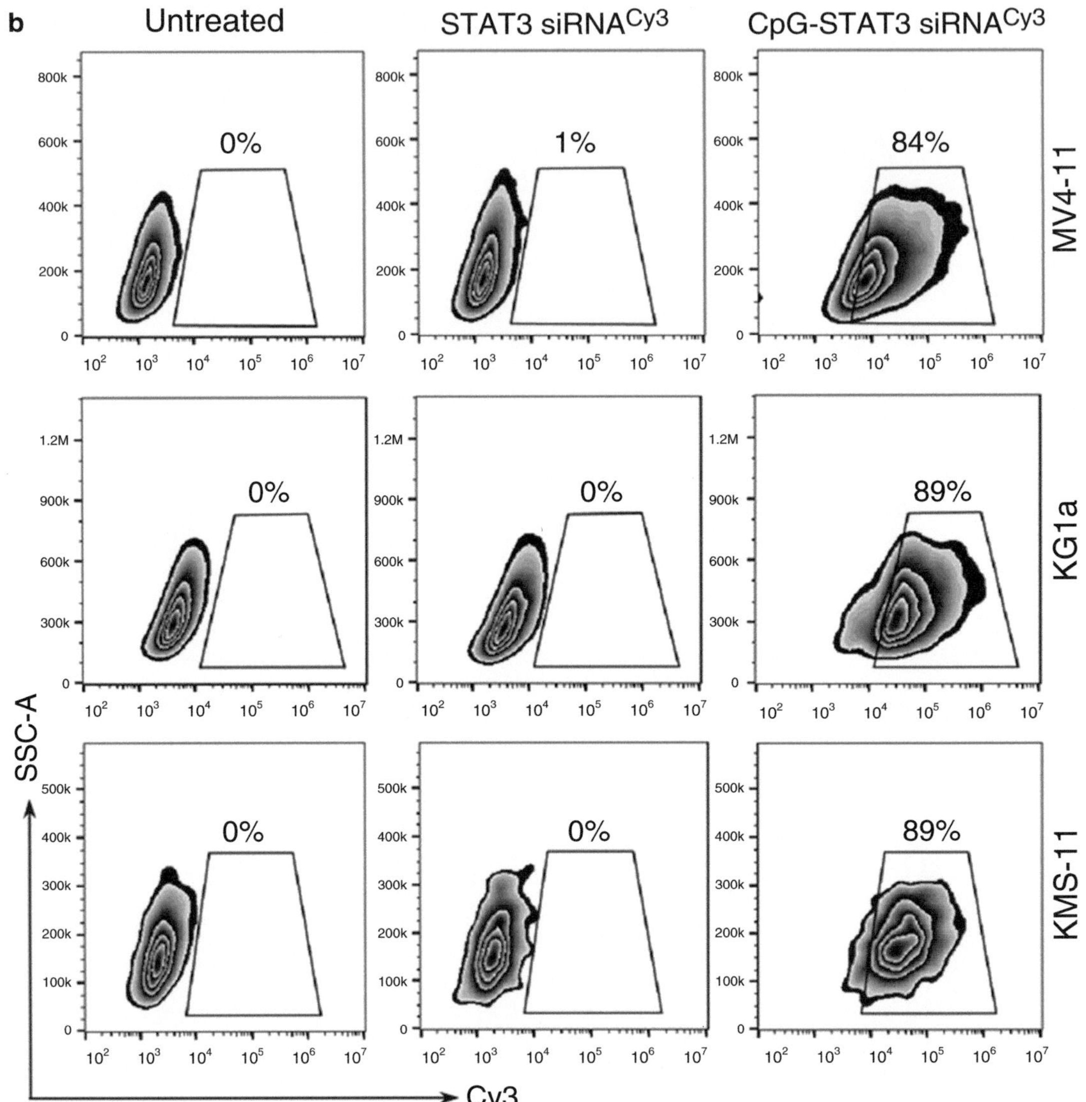

Fig. 2 (continued)

6. Slowly inject CpG-SiRNA solution in four different locations (~25 μL/location) without completely withdrawing the needle to reach maximum area without puncturing skin on the opposite side (*see* **Note 6**).
7. After the appropriate time (usually 1–24 h), euthanize mice, harvest tumors, tumor-draining, and peripheral lymph nodes, and prepare single-cell suspensions as described in Subheading 3.7 followed by the flow cytometric analysis of the oligonucleotide internalization as in Subheading 3.3.

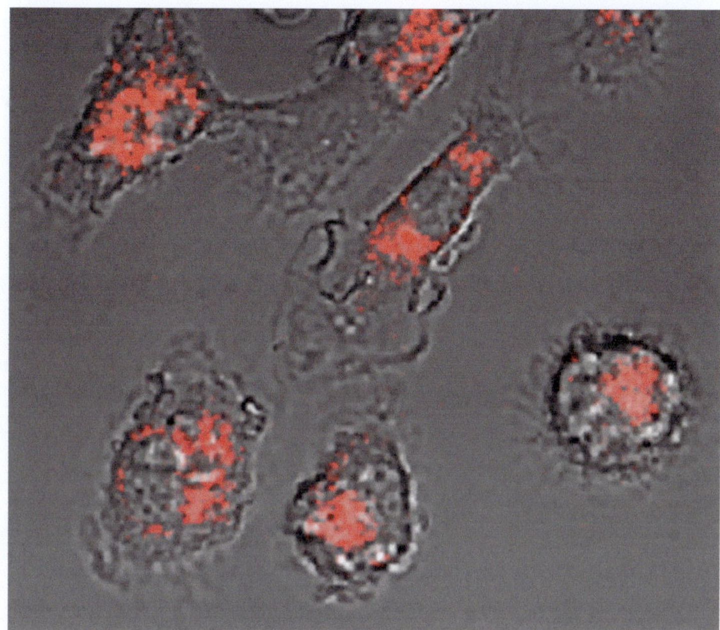

Fig. 3 Intracellular localization of CpG-siRNA after in vitro uptake. Representative images from time-lapse confocal microscopy of cultured primary bone-marrow derived macrophages incubated in the presence of 500 nM CpG-siRNA^{Cy3} for 0.5 h

3.6 Systemic Delivery of CpG-siRNA to Target Cell Populations In Vivo

1. Prepare 7–10-week-old mice (C57BL/6 or NSG for syngeneic or xenotransplanted tumor models, respectively) and anesthetize using isoflurane inhalation.
2. Prepare solution of fluorescently labeled CpG-SiRNA (e.g., CpG-SiRNA^{Cy3}) at 1–5 mg/kg in 200 μL of sterile PBS.
3. Position the anesthetized mouse on its side, use index finger and thumb to draw back skin of above and below the eye then carefully insert the 30G needle at ~45° angle at the corner of the eye, lateral to the medial canthus.
4. Slowly inject the CpG-SiRNA solution into the retrobulbar sinus then remove needle gently and apply ointment to reduce strain (*see* **Note** 7).
5. After the desired time (usually 1–24 h) euthanize mice, collect blood and harvest various organs (e.g., liver, spleen, kidney, bone marrow) to evaluate biodistribution of the fluorescently labeled oligonucleotides using as described in Subheadings 3.3, 3.7 and 3.8 (Fig. 4) (*see* **Notes 8** and **9**).

3.7 Isolation of Cells from Soft Tissues and Tumors

1. Transfer harvested tissues into Petri dishes and wash twice with HBSS.
2. Add 5–10 mL of 1× Collagenase D (400 U/mL) and 1× DNase I (1 mg/mL) and mince tissues into small cubes (~1–2 mm in diameter) using surgical scissors, and incubate for 30 min/37 °C.

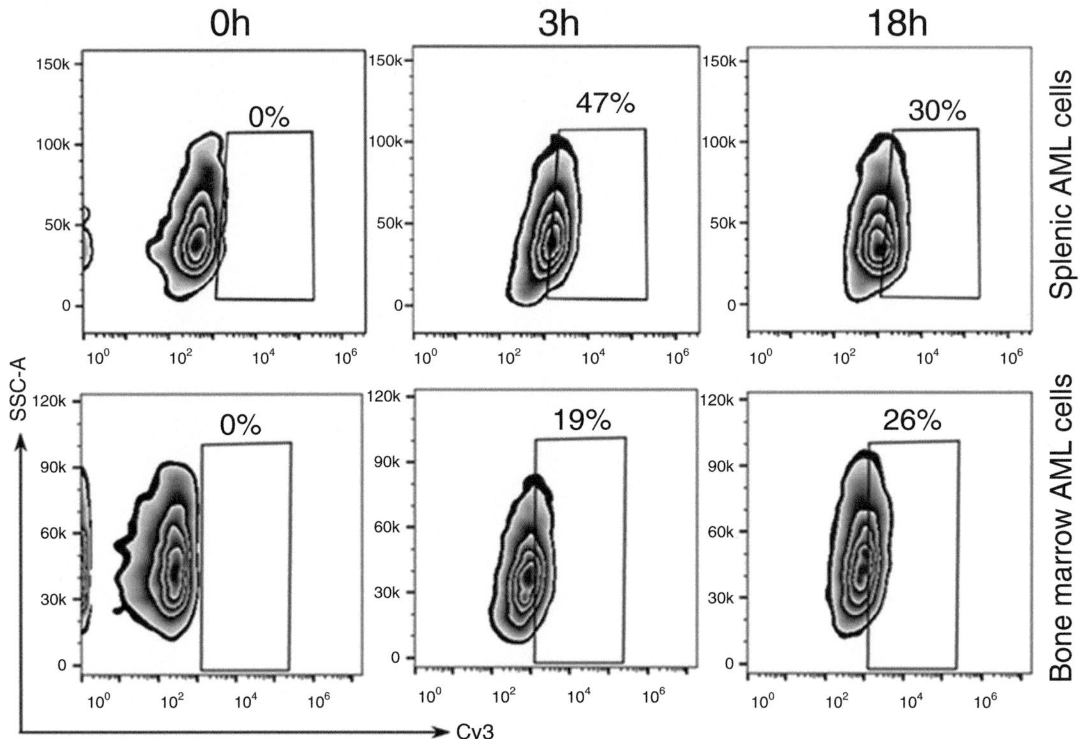

Fig. 4 CpG-*Stat3* siRNA is efficiently internalized by AML cells in vivo. C57BL/6 mice bearing *Cbfb-MYH11/Mpl+* AML were injected with a single dose of CpG-*Stat3* siRNA^{Cy3} (5 mg/kg). Uptake of fluorescently labeled siRNA was assessed using flow cytometry in AML cells isolated from spleen and bone marrow 3 or 18 h after injection

3. Add 10 mM EDTA (tissue culture grade) to stop enzymatic reaction and then pipette up and down to disperse tissues into single-cell suspension.
4. Filter cell suspensions through 70 μm cell strainers into 50 mL tubes, adjust volume to 20–30 mL with HBSS without calcium and magnesium and spin down at 500 × *g*/5 min/18 °C.
5. To lyse erythrocytes, add 5 mL of 1× ACS buffer to the cell pellet, resuspend, and incubate for 3 min/RT.
6. Add 25 mL HBSS without calcium and magnesium to stop the lysis, spin down, and resuspend cells into 10 mL HBSS without calcium and magnesium. To isolate viable cells, pipet 10 mL of cell suspension into 50 mL tube then carefully underlay with 5 mL of Histopaque®-1083 equilibrated to RT and centrifuge tubes at 400 × *g*/30 min/18 °C.
7. Collect viable cells from the interphase without touching the pellet, resuspend in HBSS without calcium and magnesium, spin down again, and then resuspend again for cell counting (Fig. 5) (*see* **Note 10**).

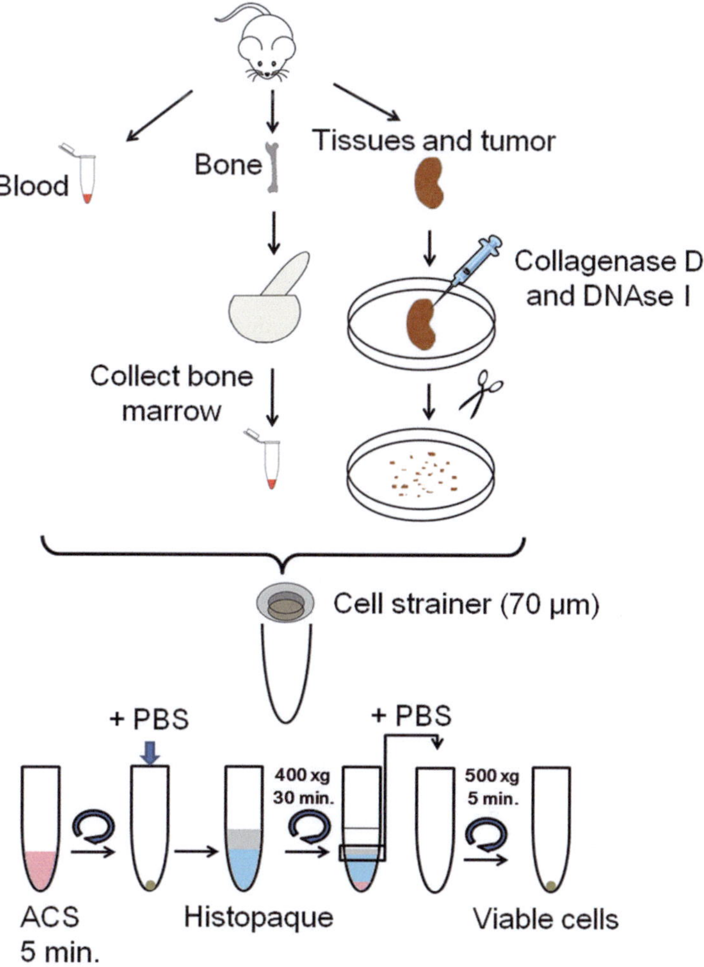

Fig. 5 Schematic of blood, bone, and soft tissue processing to assess organ and cellular biodistribution of CpG-siRNA conjugates

3.8 Isolation of Cells from the Bone Marrow

1. Euthanize mice, remove tibia and femur and place them in tubes with PBS.
2. Cleanly remove rests of muscles from bones, place them in a mortar and add 5 mL of PBS/3 % FBS, crush all bones using pestle for 1–2 min, and rinse with 1 mL of PBS/3 % FBS.
3. Aspirate the cell suspension up and down using 5 mL pipette, and then filter the content of the mortar through a 70 μm strainer into 50 mL tube and spin down at 500 × *g* for 5 min.
4. To lyse erythrocytes, add 5 mL of 1× ACS buffer to the cell pellet, resuspend, and incubate for 3 min/RT.
5. Add 25 mL PBS then spin down and re-suspend cells into 10 mL PBS.

6. To isolate viable cells, pipet 10 mL of cell suspension into 50 mL tube, then carefully underlay with 5 mL of Histopaque®-1083 equilibrated to RT, and centrifuge tubes at $400 \times g/30$ min/18 °C.
7. Collect viable cells from the interphase without touching the pellet, resuspend in HBSS without calcium and magnesium, spin down, and resuspend the cell pellet again for cell counting (Fig. 5).

3.9 Processing of Blood Cells

1. Transfer collected blood into 15 mL tube containing 3–4 mL ACS buffer to lyse blood cells.
2. Incubate for 5–10 min at RT and spin down $500 \times g$ for 5 min.
3. Wash cell pellet twice with PBS.
4. Count cell number and proceed to next experiment (Fig. 5).

4 Notes

1. The choice of fluorochrome and its position in the conjugate depends on the application. For uptake and biodistribution studies, it is best to label 3′ end of the SiRNA(SS) to prevent interference with the intracellular processing of the conjugate. For in vivo applications, more intensive fluorochromes are preferred, e.g., Cy3 or Alexa 488 rather than fluorescein.
2. Differences in concentrations or formation of secondary structures between CpG part of the molecule and the SiRNA(AS) could potentially interfere with proper hybridization of both oligonucleotides. Therefore, it is important to confirm the quality and the proper formation of hybridized duplex using gel electrophoresis.
3. For selection of optimal target cells, it is critical to confirm TLR9 expression. Cells lacking TLR9 internalize CpG-SiRNA but endosomal retention of the SiRNA is likely to prevent gene silencing effect [12].
4. Majority of TLR9-positive immune cells and malignant cells, such as leukemia and lymphoma, internalize the conjugate within 1–4 h at concentrations from 100 to 500 nM. Higher concentrations of the CpG-SiRNA usually do not improve uptake any further.
5. The purpose of testing conjugate uptake using confocal microscopy is to confirm that SiRNA is present intracellularly rather than associated with target cell surface. The exact localization of the construct cannot be distinguished by standard flow cytometry.

6. For peritumoral injections, it is important not to withdraw needle completely and to avoid additional puncturing of the skin which can result in leakage of the reagent. It is also advisable to hold the needle in position for few seconds after injecting the total 100 μL to allow for penetration of the solution deeper into tissues to minimize potential reflux.
7. In case of retroorbital injections, mice should not receive more than single injection per day. Additional injections are possible only when using alternate eyes with 1–2 days in between.
8. Efficient in vivo delivery of the SiRNA to target cells depends on their ability to quickly internalize CpG-SiRNA (within 0.5–1 h). The best results are usually achieved targeting normal and malignant myeloid cells in vitro and in vivo.
9. Conjugate degradation by serum nucleases as well as rapid kidney clearance limits CpG-SiRNA half-life in mouse circulation. Chemical modification of the oligonucleotide backbone and the conjugation of CpG-SiRNA to polyethylene glycol (PEG) can improve the pharmacokinetic profile of these conjugates.
10. To induce target gene silencing in vivo, mice should be treated using CpG-SiRNA conjugates for at least 2 days using daily injections of the reagent.

Acknowledgments

This work was supported by the Department of Defense grant number W81XWH-12-1-0132, Prostate Cancer Foundation, STOP CANCER Allison Tovo-Dwyer Memorial Career Development Award and by the National Cancer Institute of the National Institutes of Health under grant numbers R01CA155367 (to M.K.). The content is solely the responsibility of the authors and does not necessarily represent the official views of the NIH.

References

1. Snead NM, Rossi JJ (2012) RNA interference trigger variants: getting the most out of RNA for RNA interference-based therapeutics. Nucleic Acid Ther 22:139–146
2. Rettig GR, Behlke MA (2012) Progress toward in vivo use of siRNAs-II. Mol Ther 20:483–512
3. Davidson BL, McCray PB Jr (2011) Current prospects for RNA interference-based therapies. Nat Rev Genet 12:329–340
4. Zimmermann TS, Lee AC, Akinc A, Bramlage B, Bumcrot D, Fedoruk MN, Harborth J, Heyes JA, Jeffs LB, John M, Judge AD, Lam K, McClintock K, Nechev LV, Palmer LR, Racie T, Rohl I, Seiffert S, Shanmugam S, Sood V, Soutschek J, Toudjarska I, Wheat AJ, Yaworski E, Zedalis W, Koteliansky V, Manoharan M, Vornlocher HP, MacLachlan I (2006) RNAi-mediated gene silencing in non-human primates. Nature 441:111–114
5. Davis ME, Zuckerman JE, Choi CH, Seligson D, Tolcher A, Alabi CA, Yen Y, Heidel JD, Ribas A (2010) Evidence of RNAi in humans from systemically administered siRNA via targeted nanoparticles. Nature 464:1067–1070

6. Whitehead KA, Langer R, Anderson DG (2009) Knocking down barriers: advances in siRNA delivery. Nat Rev Drug Discov 8:129–138
7. Kortylewski M, Swiderski P, Herrmann A, Wang L, Kowolik C, Kujawski M, Lee H, Scuto A, Liu Y, Yang C, Deng J, Soifer HS, Raubitschek A, Forman S, Rossi JJ, Pardoll DM, Jove R, Yu H (2009) In vivo delivery of siRNA to immune cells by conjugation to a TLR9 agonist enhances antitumor immune responses. Nat Biotechnol 27:925–932
8. Krieg AM (2012) CpG still rocks! Update on an accidental drug. Nucleic Acid Ther 22:77–89
9. Kawai T, Akira S (2010) The role of pattern-recognition receptors in innate immunity: update on Toll-like receptors. Nat Immunol 11:373–384
10. Amarzguioui M, Lundberg P, Cantin E, Hagstrom J, Behlke MA, Rossi JJ (2006) Rational design and in vitro and in vivo delivery of Dicer substrate siRNA. Nat Protoc 1:508–517
11. Zhang Q, Hossain DM, Nechaev S, Kozlowska A, Zhang W, Liu Y, Kowolik CM, Swiderski P, Rossi JJ, Forman S, Pal S, Bhatia R, Raubitschek A, Yu H, Kortylewski M (2013) TLR9-mediated siRNA delivery for targeting of normal and malignant human hematopoietic cells in vivo. Blood 121:1304–1315
12. Nechaev S, Gao C, Moreira D, Swiderski P, Jozwiak A, Kowolik CM, Zhou J, Armstrong B, Raubitschek A, Rossi JJ, Kortylewski M (2013) Intracellular processing of immunostimulatory CpG-siRNA: toll-like receptor 9 facilitates siRNA dicing and endosomal escape. J Control Release 170:307–315
13. Stalder L, Heusermann W, Sokol L, Trojer D, Wirz J, Hean J, Fritzsche A, Aeschimann F, Pfanzagl V, Basselet P, Weiler J, Hintersteiner M, Morrissey DV, Meisner-Kober NC (2013) The rough endoplasmatic reticulum is a central nucleation site of siRNA-mediated RNA silencing. EMBO J 32:1115–1127
14. Hossain DM, Dos Santos C, Zhang Q, Kozlowska A, Liu H, Gao C, Moreira D, Swiderski P, Jozwiak A, Kline J, Forman S, Bhatia R, Kuo YH, Kortylewski M (2014) Leukemia cell-targeted STAT3 silencing and TLR9 triggering generate systemic antitumor immunity. Blood 123:15–25
15. Deng J, Liu Y, Lee H, Herrmann A, Zhang W, Zhang C, Shen S, Priceman SJ, Kujawski M, Pal SK, Raubitschek A, Hoon DS, Forman S, Figlin RA, Liu J, Jove R, Yu H (2012) S1PR1-STAT3 signaling is crucial for myeloid cell colonization at future metastatic sites. Cancer Cell 21:642–654
16. Moreira D, Zhang Q, Hossain DM, Nechaev S, Li H, Kowolik CM, D'Apuzzo M, Forman S, Jones J, Pal SK, Kortylewski M (2015) TLR9 signaling through NF-κB/RELA and STAT3 promotes tumor-propagating potential of prostate cancer cells. Oncotarget 6:17302–17313

Chapter 16

Aptamer-MiRNA Conjugates for Cancer Cell-Targeted Delivery

Carla L. Esposito, Silvia Catuogno, and Vittorio de Franciscis

Abstract

microRNAs (miRNAs) are short noncoding RNAs that effectively regulate the expression of a wide variety of genes. Increasing evidences have shown a fundamental role of miRNAs in cancer initiation and progression, thus indicating these molecules among the most promising for new approaches in cancer therapy. However, several hurdles limit the translation of miRNAs into the clinic. One of the most critical aspects is represented by the lack of a safe and reliable way to selectively target organs and tissues. Therefore, the development of cell-specific delivery means has become an essential step for the translation of miRNA-based therapeutics to clinic for cancer management. To this end aptamer-based approaches may provide efficient delivery tools for the selective accumulation of miRNA to target tumors, their intracellular uptake, processing, and functional silencing of target genes. In this chapter, we discuss the direct conjugation of miRNAs to aptamers against transmembrane receptors as innovative experimental approach for their selective delivery to cancer cells.

Key words microRNAs, Aptamers, Targeted delivery, Therapeutic RNA, Cancer

1 Introduction

Despite the advances in surgical techniques, radiotherapy, and chemotherapy, cancer remains one of the leading causes of death worldwide. Yet, cancer results from the progressive accumulation of multiple genetic and epigenetic alterations that confer the proliferative and invasive characteristics to tumor cells [1]. Thus, the development of gene-based approaches enabling to modulate gene expression in target cancer cells represents an important challenge for personalized cancer therapies to be effective. In this respect, the great potential of emerging new therapeutics based on nucleic acid compounds, including miRNAs and siRNAs, has recently attracted great interest.

MiRNAs are short noncoding RNAs processed from longer precursor molecules to produce an imperfect miRNA duplex of about 19–25 nucleotides in length [2]. In the cytoplasm, the

Kato Shum and John Rossi (eds.), *SiRNA Delivery Methods: Methods and Protocols*, Methods in Molecular Biology, vol. 1364, DOI 10.1007/978-1-4939-3112-5_16, © Springer Science+Business Media New York 2016

RNA-induced silencing complex (RISC) drives the strand of the duplex that directs the silencing (named guide strand) to the target messenger RNA (mRNA), while the complementary strand (passenger strand) is degraded. By recognizing specific sites at the 3′ untranslated region (3′ UTR) of the target genes, the miRNAs down regulate gene expression by a posttranscriptional mechanism that depends on the degree of complementarity of the guide strand to the target mRNA sequence [3] (Fig. 1a). Each miRNA can potentially target multiple mRNAs, and a single mRNA can be targeted by multiple miRNAs at its 3′ UTR. As such, miRNAs may act in concert to regulate gene expression and, in turn, to coordinately regulate key biological processes, including cell cycle, apoptosis, differentiation, motility and angiogenesis. Deregulation of miRNAs has been shown to be involved in the pathogenesis of several diseases, including cardiovascular, cancer, and metabolic diseases [4].

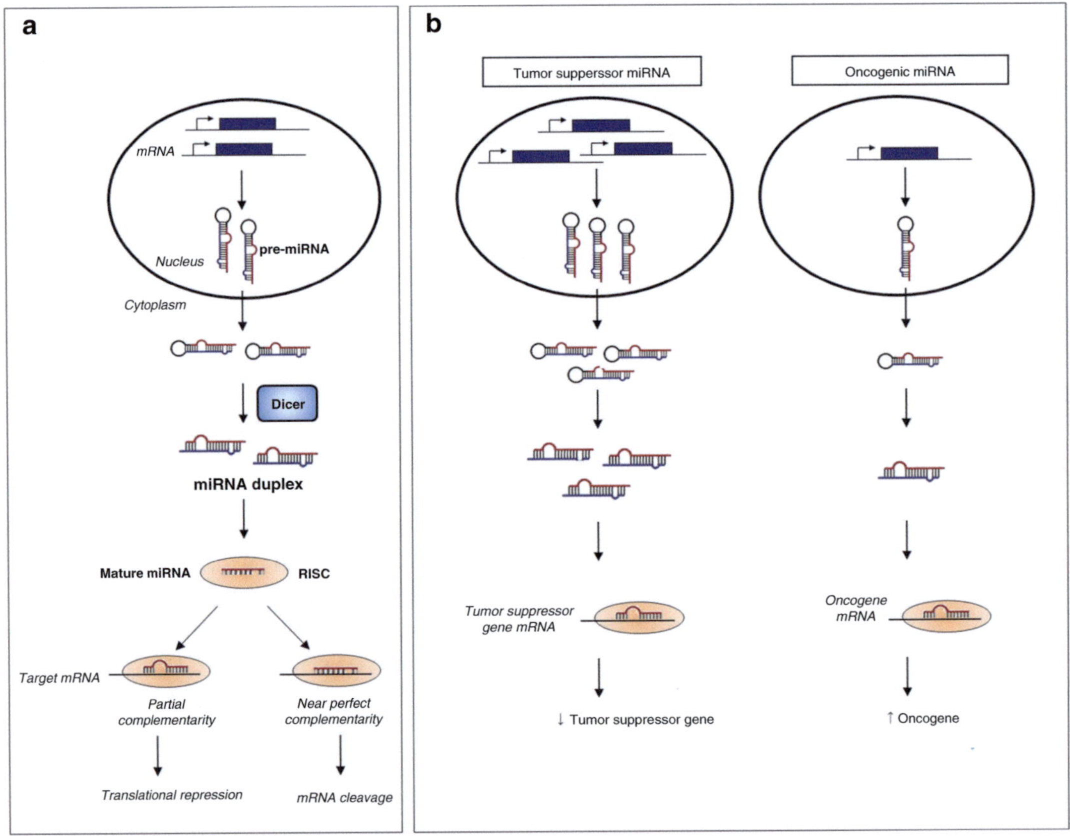

Fig. 1 (**a**) Schematic representation of miRNA processing pathway (see text for details). (**b**) micoRNA role in cancer. Overexpression of miRNAs could decrease tumor-suppressor gene levels (miRNA with a tumor suppressor function, *left*) or downregulation of miRNAs could result in increased expression of oncogenes (microRNA with an oncogenic function, *right*)

Based on their involvement in basic cellular functions, miRNAs may act as oncogenes (oncomirs) or tumor suppressors showing a great interest as therapeutics in cancer [5, 6] (Fig. 1b). One of the first examples is constituted by the let-7 family members (let-7a-i) that are widely reported as tumor suppressors [7–9]. In the case of the gene cluster miR15/16 it has been shown that the loss of the locus by a somatic deletion in chromosome 13q14.3 is involved in the B-cell chronic lymphocytic leukemia [10]. On the contrary, miR-155 and miRNA-21 are two of the first described oncogenic miRNAs that are overexpressed in many tumors [11, 12]. MiRNA-based therapies may have some important advantage over antibodies, small molecules and protein-based reagents, including the possibility to be easily synthesized and chemically modified [13]. However, their broad functionality underlines the need of a safe approach for their specific delivery to target tissues in order to control undesired off-target effects and toxicity. Here, we provide a protocol that we developed to address cell-specific delivery of therapeutic miRNAs by using nucleic acid aptamers as targeting moiety [14].

Different viral and non-viral approaches have been proposed to address this issue. Adenovirus- and adeno-associated virus-derived vectors can have high efficacy for miRNA delivery [15]. However, problems related to insertional mutagenesis and immunogenicity strongly restrict their applicability [16]. Alternatively, non-viral approaches including lipid and polymeric nanoparticles have been proposed and result to be more secure and simple to produce [17–19].

Aptamers are short structured single-stranded nucleic acids that can be developed against almost any target protein, including transmembrane receptors, by a combinatorial strategy named Systematic Evolution of Ligands by EXponential enrichment (SELEX) [20, 21]. Because of their high specificity and low toxicity, aptamers against cell surface molecules may reveal as the compounds of choice for the delivery of various secondary reagents. Indeed, in several cases has been shown that aptamers raised against cell surface receptors upon binding are rapidly internalized in a receptor-mediated manner [22]. Thus, aptamers may function as specific recognition ligands for target cells leading to the release of a secondary reagent only into this subset of cells.

Currently an increasing number of aptamers targeting cancer cell surface epitopes have been successfully and extensively used for the specific delivery of active drug substances including nanoparticles [23, 24], anticancer therapeutics [25], toxins [26], enzymes [27], radionuclides [28], viruses [29], and small interfering RNAs (siRNAs) [30–39].

A great effort have been devoted to the development of strategies for the direct conjugation of siRNA to aptamers, generating completely nucleic acid-based molecules either by extending the aptamer sequence with one strand of siRNA duplex and annealing with the other strand [30–37] or by combining the aptamer and the siRNA moieties *via* a stick bridge annealing [38, 39].

Basing on these rules, recently two chimeras that combine the anti Mucin 1 aptamer with let-7i miRNA [40] or miR-29b [41] have been reported to function in vitro increasing the response to paclitaxel [40] or inducing apoptosis [41] in ovarian cancer cells.

In our laboratory, we have characterized two RNA aptamers (named GL21.T and Gint4.T) that bind and inhibit the receptor tyrosine kinase Axl and platelet-derived growth factor receptor β (PDGFRβ) [42, 43], respectively. Both aptamers are rapidly internalized into target cells, thus providing innovative tools for tissue specific internalization.

By using the anti-Axl GL21.T aptamer, we have described the development of aptamer-miRNA conjugate for the delivery of the tumor suppressor let-7g miRNA [14]. The approach that we used is based on the annealing of two distinct strands of RNA, one strand consisting of the aptamer elongated at the 3′ end with the passenger strand of let-7g sequence and a second strand that consists of the complementary let-7g guide strand (Figs. 2 and 3). Notably in order to obtain a more effective Dicer substrate [44] we introduced internal partial complementarity and increase length extension by using the distal stem portion of the human let-7g pre-miRNA (26–27 bases).

Furthermore, we have obtained functional conjugates by using a different aptamer sequence (anti-PDGFRβ Gint4.T aptamer) [43] and a different conjugation strategy based on a stick-end annealing [38, 39]. In this approach, the aptamer and the passenger strand are elongated at their 3′ end with a 17-mer GC-rich “stick” complementary sequence (Fig. 4). The stick sequence (1) ensures a highly stable base pairing that facilitates the noncovalent binding between the miRNA and the aptamer; and (2) does not interfere with aptamer correct folding.

Here, we provide protocols for the approaches that we used. The strategy described has two important advantages. First, it is simply and uses only RNA-based reagents (an aptamer and a miRNA). RNAs are modified in all positions with 2′-fluoropyridines (2′-F-Py) that enhance their RNase resistance and reduce their immunogenicity [45]. Second, it permits to combine the tumor suppressive function of the specific miRNA used with that

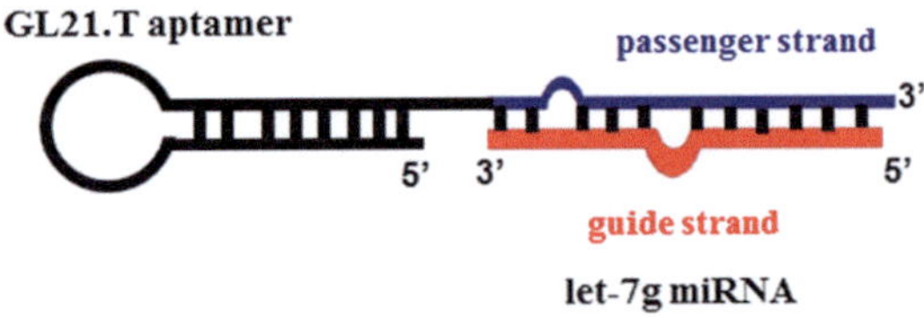

Fig. 2 Schematic representation of *GL21.T-let* conjugate

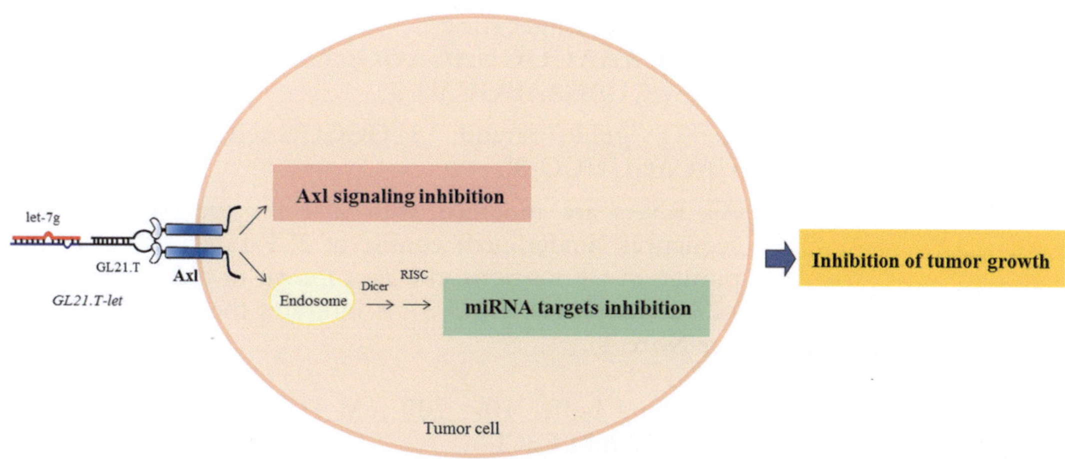

Fig. 3 Mechanism of *GL21.T-let* chimera multi-functional action. *GL21.T-let* combines aptamer-mediated Axl inhibition and let-7g miRNA-mediated target gene downregulation, resulting in a pronounced reduction of tumor growth

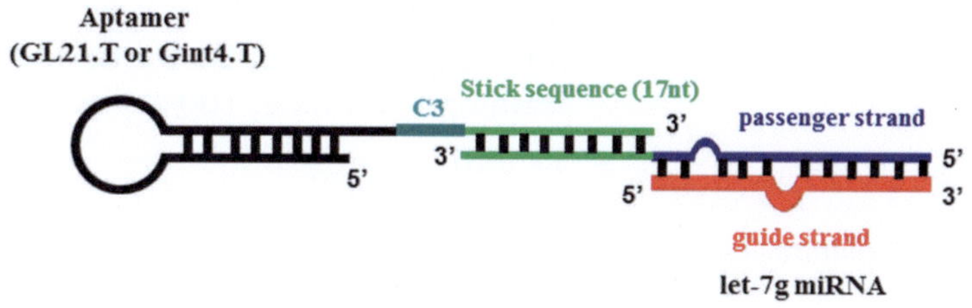

Fig. 4 Schematic representation of stick-based conjugates

of a functional aptamer that inhibits the tyrosine kinase activity of the target receptor (either Axl or PDGFRβ), thus further increasing the therapeutic potential of the resulting molecule. This is particularly relevant given the increasing hope placed in anticancer strategies based on combination therapies.

2 Materials

2.1 RNA Oligonucleotides

1. GL21.T-let passenger strand:
   5′-AUGAUCAAUCGCCUCAAUUCGACAGGAGGC UCACCAAACUGUACAGGCCACUGCCUUGC C**UU**-3′ (Bold UU are 3′-overhang).
2. 2.GL21.T-stick:
   5-AUGAUCAAUCGCCUCAAUUCGACAGGAG GCUCACX<u>GUACAUUCUAGAUAGCC</u>-3′.
3. Gint4.T-stick:
   5′-UGUCGUGGGGCAUCGAGUAAAUG CAAUUCGACAX<u>GUACAUUCUAGAUAGCC</u>-3′.

4. 4.Let-7g passenger-stick:
5′-CAAACUGUACAGGCCACUGCCUUGCCGG CUAUCUAGAAUGUAC-3′.
5. Let-7g guide strand: 5′-GGCUGAGGUAGUAGUUUG UACAGUUUG-3′.
6. All RNAs are modified with 2′-F-Py (*see* **Note 1**); stick sequences (underlined) consist of 2′-F-Py and 2′-O-methyl-purines at all positions, the *Free Energy* of the stick duplex is –27.2 (*see* **Notes 2** and **3**); X indicates the C3 carbon linker (*see* **Note 4**).

2.2 Aptamer-miRNA Conjugate

1. Annealing buffer 10×: 200 mM HEPES, pH 7.5, 1.5 mM NaCl, 20 mM $CaCl_2$.
2. 12 % non-denaturing polyacrylamide gel: 6.0 ml of 30% acrylamide/bis solution, 1.5 ml of 10× Tris–borate–ethylenediamine tetraacetic acid buffer (TBE) and water for a final volume of 15 ml.
3. 6× gel loading buffer: 4.62 g dithiothreitol (DTT), 5 % β-mercaptoethanol, 24 ml of 1M Tris–HCl, pH 6.8, 0.02 g bromophenol blue, 100 % glycerol to a final volume of 50 ml.
4. 1× TBE running buffer.

2.3 Cell Culture

1. Cell lines: U87MG glioblastoma (PDGFRβ and Axl-positive cells), NSCLC A549 (Axl-positive cells), and breast MCF-7 (Axl and PDGFRβ-negative cells) cells.
2. Growth medium: DMEM or RPMI supplemented with 10 % FBS.
3. Trypsin.
4. Ethylenediamine tetraacetic acid (EDTA).

2.4 Binding and Internalization Analysis

1. 10× dephosphorylation buffer: 500 mM Tris–HCl, pH 8.5, 1 mM EDTA.
2. Alkaline phosphatase (AP) inactivation buffer: 200 mM EGTA.
3. 10× phosphorylation buffer: 500 mM Tris–HCl, pH 8.2, 100 mM $MgCl_2$, 1 mM EDTA, 50 mM DTT, 1 mM spermidine.
4. Alkaline phosphatase.
5. T4 polynucleotide kinase.
6. [γ-^{32}P]-ATP (6000 Ci/mmol).
7. Recovering buffer: 1 % sodium dodecyl sulfate (SDS).
8. Proteinase K.

2.5 Cell Lysis and Western Blotting

1. Cell extraction buffer: 50 mM Tris–HCl, pH 8.0, 150 mM NaCl, 1 % Nonidet P-40, 2 mg/ml aprotin, 1 mg/ml pepstatin, 2 mg/ml leupeptin, and 1 mM sodium orthovanadate.
2. 4× Gel separating buffer: 1.5 M Tris–HCl, pH 8.7, 0.4 % SDS.
3. 4× Gel stacking buffer: 0.5 % Tris–HCl, pH pH 6.8, 0.4 % SDS.
4. 30 % acrylamide/bis solution (37.5:1).
5. 5× gel running buffer: 125 mM Tris, 960 mM glycine, 0.5 % SDS.
6. Laemmli buffer for sample loading: 2 % SDS, 5 % β-mercaptoethanol, 0.001 % BBF, 10 % glycerol.
7. Kaleidoscope molecular weight markers.
8. Polyvinylidenedifluoride (PVDF) membrane.
9. 3 mm chromatography paper.
10. Transfer buffer: 25 mM Tris, 190 mM glycine, 20 % methanol, 0.05 % SDS (*see* **Note 5**).
11. Antibody for miRNA validated target: anti-N-Ras antibody.
12. Antibody incubation buffer: 5 % nonfat dry milk in Tris-buffered saline with Tween (T-TBS).
13. T-TBS: dilute 10× stock TBS solution (250 mM Tris–HCl, pH 7.4, 1.37 M sodium chloride, 27 mM potassium chloride) and add 0.1 % Tween-20.
14. Enhanced chemiluminescent (ECL) reagent.
15. Bio-Max ML film.
16. Gas6 ligand.

2.6 Quantitative Real-Time PCR (RT-qPCR)

1. TRiZol reagent.
2. miScript Reverse Transcription Kit.
3. SYBR Green PCR Kit.
4. SiScript Primer Assay (let-7g and U6 RNA as housekeeping gene).

3 Methods

3.1 Aptamer-miRNA Conjugate Preparation

1. Dissolve each RNA oligo in sterile RNase-free water and determine concentration spectrophotometrically, assuming that one A260 unit is equal to 40 μg/ml of RNA (*see* **Note 6**).
2. Add 5 μM aptamer-passenger RNA strand to 5 μM of the appropriate guide RNA in annealing buffer to prepare *GL21.T-let* conjugates.
3. Heat the sample to 95°C for 15 min, incubate at 55 °C for 10 min and then at 37 °C for 20 min.

4. For stick-based conjugates, add 5 μM of the passenger-stick strand to 5 μM of the complementary guide partner in 1× annealing buffer.
5. Incubate the mixture at 95 °C for 15 min, 55 °C for 10 min and 37 °C for 20 min to form stick-miRNA duplex.
6. Refold the same amount of the stick aptamer (5 min 85 °C, 2 min snap-cooling on ice, warming up to 37 °C, *see* **Note 7**) and add to the stick-miRNA.
7. Incubate the mixture at 37 °C for 30 min.
8. Annealed chimeras can be stored at −20 °C (*see* **Note 8**).
9. Check the preparation of the conjugate by loading 15 pmol RNAs to a 12 % non-denaturing PAGE gel.
10. Run the gel in 1× TBE buffer (*see* **Note 9**).
11. Stain the gel with ethidium bromide and visualize with GEL. DOC XR gel camera. The annealing efficiency of the conjugate is assessed by measuring the shift in migration of the conjugate compare to the individual components.

3.2 Binding and Internalization Analysis

1. Dephosphorylate the RNA aptamers with alkaline phosphatase at 50 °C for 1 h.
2. End-label with aptamer with T4 polynucleotide kinase in the presence of [γ-^{32}P] ATP at 37 °C for 30 min.
3. Anneal the radiolabeled aptamer moiety to the corresponding guide strand or to the stick-miRNA duplex.
4. Seed 2×10^4 cells *per* well in a 24-well cell culture plate.
5. Incubate the cells with 100 nM individual RNAs for increasing incubation times (from 15 min up to 2 h) at 37 °C in the presence of 100 μg/ml polyinosine as a nonspecific competitor.
6. Remove unbound RNAs by washing with five times of 500 μl serum-free growth medium. Cells are left untreated (to measure total bound) or treated with 0.5 μg/μl proteinase K for 30 min at 37 °C.
7. Recover RNA in 300 μl of 1 % SDS.
8. Measure ^{32}P with a Beckman scintillation counter. The endocytosis rate is expressed as the amount of internalized RNA relative to total bound.

3.3 miRNA Activity Assays

1. The day before treatments, cells are seeded in 3.5 cm plates (1.4×10^5 cells/well, density of 60 % confluence) and left growing overnight.
2. Cells are treated with 400 nM aptamer or aptamer-miRNA conjugate (*see* **Note 10**) or are transfected with 100 nM commercial miRNA mimic as control.
3. Transfections are performed using Lipofectamine 2000 following the manufacturer's instructions.

4. Cells are grown for 2 days and then processed as detailed below for either immunoblotting or RT-qPCR.
5. For real-time PCR experiment, transfected or treated cells are recovered in 1 ml of TRiZol and RNAs are extracted by following the manufacturer's instructions. To assess the levels of the miRNA (let-7g in our example) or U6 RNA, as housekeeping gene, 1 μg total RNA is reverse transcribed by using miScript Reverse Transcription Kit and then amplified with the miScript-SYBR Green PCR Kit and specific miScript Primer Assay. Reactions are carried out according to the provider's instructions. The ΔΔCt method for relative quantization of gene expression is used.

   An example of the results produced by treating U87MG cells (PDGFRß receptor-positive cells) with a chimera containing the anti-PDGFRß internalizing aptamer [43] conjugated to let-7g miRNA by stick-end annealing (indicated as *Gint4.T-let-7g stick*) is shown in Fig. 5a. As expected, treating cells with the conjugate results in a significant increase of let-7g miRNA levels.
6. For immunoblot experiment, transfected or treated cells are washed with ice-cold D-PBS and lysed in lysis buffer. Protein concentration is determined by the Bradford assay using bovine serum albumin as standard. The cell lysates are subjected to SDS-PAGE. Gels are electroblotted into PVDF membranes and filters are probed with primary antibodies against the miRNA targets (for let-7g conjugates we used the antibodies against HMGA2 and N-Ras). Anti-αtubulin (DM 1A) antibody

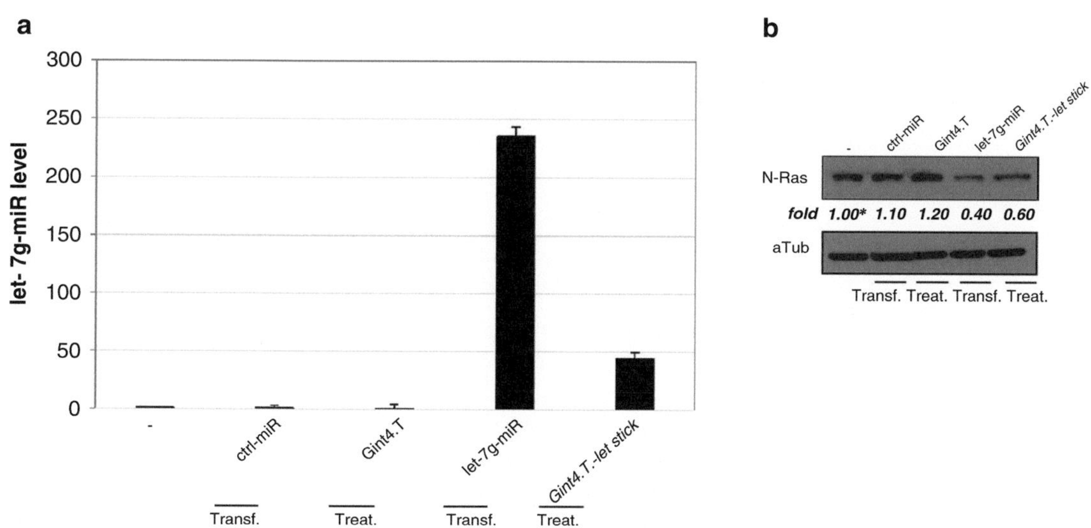

Fig. 5 U87MG cells were transfected (Transf.) with 100 nM let-7g-miR or control miR (Ctrl-miR) or were treated (Treat.) with 400 nM of Gint4.T, or *Gint4.T-let stick* chimera, as indicated. Following 48 h (**a**) let-7g miRNA levels were quantified by RT-qPCR. (**b**) Cell lysates were immunoblotted with anti-N-Ras and anti-αtubulin (αTub) antibodies. Values below the blots indicate signal levels relative to mock-treated cells, arbitrarily set to 1 ("-," labeled with *asterisk*)

is used to confirm equal loading. Proteins are visualized with peroxidase-conjugated secondary antibodies using the enhanced chemiluminescence system. The N-Ras protein levels detected following the treatment of U87MG cells with *Gint4.T-let stick* are shown in Fig. 5b. In accordance with the increase of let-7g miRNA levels, N-Ras expression is strongly reduced in cells treated with the chimera.

3.4 Aptamer Functional Activity Preservation

1. Starve A549 cells (1.5×10^5 cells per 3.5 cm plate) for 18 h, and then leave the cells untreated or treated with 200 nM aptamer or conjugate for 3 h.
2. Stimulate cell growth with 400 ng/ml Gas-6.
3. Lyse the cells as described in **step 6** of "miRNA Activity Assays".
4. Immunoblot with anti-phospho-Axl (tyr-702), anti-Axl and anti-αtubulin. This is a key experiment to demonstrate that conjugation of the miRNA to the aptamer does not abrogate aptamer function, thus confirming the multi-functionality of the generated construct.

4 Notes

1. 2′F-Py RNAs are used because of their increased resistance to degradation by nucleases.
2. The sequence of the stick portion depends on the specific aptamer to be used. It is important to check that (a) the stick duplex portion does not interfere with the correct aptamer folding and (b) the energy of the duplex produces the most favorable annealing in the conjugate.
3. The presence of 2′-O-methyl-purines further increases the stability of stick sequence base pairing.
4. The presence of a polycarbon linker between the aptamer and the stick sequence further increases stick flexibility, thus preventing the interference with the aptamer folding.
5. Transfer buffer can be used for up to three transfers within a week so long as the voltage is maintained constant for each successive run (the current will increase each time). Adequate cooling to keep the buffer lower than the room temperature is essential in order to prevent heat-induced damage to the apparatus and the experiment.
6. Once dissolved aliquot, store the solution at −20 °C. It is recommended to avoid repeated freezing-refreezing.
7. Stick-based approach has the advantage to permit the annealing of the miRNA moiety to the aptamer moiety. To ensure a correct aptamer folding, aptamer concentration must be low (not more than 20 μM).

8. Annealed chimeras can be stored at −20 °C. However, it is recommended to avoid many repeated freezing-refreezing cycles.
9. Run gel until the first dye front is close to the bottom.
10. Before each treatment, chimeras must be warming up to 37 °C and the aptamers must be subjected to a short denaturation-renaturation step (5 min 85 °C, 2 min snap-cooling on ice, warming up to 37 °C).

Acknowledgement

This work was supported by funds from: MIUR grant, MERIT RBNE08YFN3_001 (VdF), AIRC # 13345 (VdF); from the Italian Ministry of Economy and Finance to the CNR for the Project FaReBio di Qualità (VdF); Grant CNR "Medicina Personalizzata" (VdF); Compagnia San Paolo # 2011.1172 (VdF); CNR Flagship Project NanoMax (*DESIRED*) 2012–2014 (VdF).

References

1. Hanahan D, Weinberg RA (2011) Hallmarks of cancer: the next generation. Cell 144:646–674
2. Elbashir SM, Lendeckel W, Tuschl T (2001) RNA interference is mediated by 21- and 22-nucleotide RNAs. Genes Dev 15:188–200
3. Rand TA, Petersen S, Du F et al (2005) Argonaute2 cleaves the anti-guide strand of siRNA during RISC activation. Cell 123:621–629
4. Li Z, Rana TM (2014) Therapeutic targeting of microRNAs: current status and future challenges. Nat Rev Drug Discov 13:622–638
5. Di Leva G, Garofalo M, Croce CM (2013) MicroRNAs in cancer. Annu Rev Pathol 14:287–314
6. Garofalo M, Condorelli GL, Croce CM et al (2010) MicroRNAs as regulators of death receptors signaling. Cell Death Differ 17:200–208
7. Boyerinas B, Park SM, Hau A et al (2010) The role of let-7 in cell differentiation and cancer. Endocr Relat Cancer 17:F19–F36
8. Johnson SM, Grosshans H, Shingara J et al (2005) RAS is regulated by the let-7 microRNA family. Cell 120:635–647
9. Lee YS, Dutta A (2007) The tumor suppressor microRNA let-7 represses the HMGA2 oncogene. Genes Dev 21:1025–1030
10. Calin GA, Cimmino A, Fabbri M et al (2008) MiR-15a and miR-16-1 cluster functions in human leukemia. Proc Natl Acad Sci U S A 105:5166–5171
11. Bartels CL, Tsongalis GJ (2009) MicroRNAs: novel biomarkers for human cancer. Clin Chem 55:623–631
12. Rothschild SI (2014) microRNA therapies in cancer. Mol Cell Ther 2:7
13. Wu SY, Lopez-Berestein G, Calin GA et al (2014) RNAi therapies: drugging the undruggable. Sci Transl Med 6:240ps7
14. Esposito CL, Cerchia L, Catuogno S et al (2014) Multifunctional aptamer-miRNA conjugates for targeted cancer therapy. Mol Ther 22:1151–1163
15. Kota J, Chivukula RR, O'Donnell KA et al (2009) Therapeutic microRNA delivery suppresses tumorigenesis in a murine liver cancer-model. Cell 137:1005–1017
16. Kay MA (2011) State-of-the-art gene-based therapies: the road ahead. Nat Rev Genet 12:316–328
17. Zhou J, Shum KT, Burnett JC et al (2013) Nanoparticle-based delivery of RNAi therapeutics: progress and challenges. Pharmaceuticals (Basel) 6:85–107
18. Bouchie A (2013) First microRNA mimic enters clinic. Nat Biotechnol 31:577
19. Cedervall T, Lynch I, Lindman S et al (2007) Understanding the nanoparticle–protein corona using methods to quantify exchange

rates and affinities of proteins for nanoparticles. Proc Natl Acad Sci U S A 104:2050–2055

20. Tuek C, Gold L (1990) Systematic evolution of ligands by exponential enrichment RNA ligands to bacteriophage T4 polymerase. Science 249:505–510
21. Esposito CL, Catuogno S, de Franciscis V et al (2011) New insight into clinical development of nucleic acid aptamers. Discov Med 11:487–496
22. Cerchia L, de Franciscis V (2010) Targeting cancer cells with nucleic acid aptamers. Trends Biotechnol 28:517–525
23. Wang J, Sefah K, Altman MB et al (2013) Aptamer-conjugated nanorods for targeted photothermal therapy of prostate cancer stem cells. Chem Asian J 8:2417–2422
24. Farokhzad OC, Cheng J, Teply BA et al (2006) Targeted nanoparticle-aptamer bioconjugates for cancer chemotherapy in vivo. Proc Natl Acad Sci U S A 103:6315–6320
25. Bagalkot V, Farokhzad OC, Langer R et al (2006) Anaptamer doxorubicin physical conjugateas a novel targeted drug-delivery platform. Angew Chem Int Ed Engl 45:8149–8152
26. Chu TC, Marks JW 3rd, Lavery LA et al (2006) Aptamer:toxin conjugates that specifically target prostate tumor cells. Cancer Res 66:5989–5992
27. Chen CH, Dellamaggiore KR, Ouellette CP et al (2008) Aptamer-based endocytosis of a lysosomal enzyme. Proc Natl Acad Sci U S A 105:15908–15913
28. Hicke BJ, Stephens AW, Gould T et al (2006) Tumor targeting by an aptamer. J Nucl Med 47:668–678
29. Tong GJ, Hsiao SC, Carrico ZM et al (2009) Viral capsid DNA aptamer conjugates as multivalent cell targeting vehicles. J Am Chem Soc 131:11174–11178
30. McNamara JO II, Andrechek ER, Wang Y et al (2006) Cell type-specific delivery of siRNAs with aptamer-siRNA chimeras. Nat Biotechnol 24:1005–1015
31. Wullner U, Neef I, Eller A et al (2008) Cell-specific induction of apoptosis by rationally designed bivalent aptamer-siRNA transcripts silencing eukaryotic elongation factor 2. Curr Cancer Drug Targets 8:554–565
32. Dassie JP, Liu XY, Thomas GS et al (2009) Systemic administration of optimized aptamer-siRNA chimeras promotes regression of PSMA-expressing tumors. Nat Biotechnol 27:839–849
33. Neff CP, Zhou J, Remling L et al (2011) An aptamer-siRNA conjugate suppresses HIV-1 viral loads and protects from helper CD4(+) T cell decline in humanized mice. Sci Transl Med 3:66ra6
34. Ni X, Zhang Y, Ribas J et al (2011) Prostate-targeted radiosensitization via aptamer-shRNA chimeras in human tumor xenografts. J Clin Invest 121:2383–2390
35. Wheeler LA, Trifonova R, Vrbanac V et al (2011) Inhibition of HIV transmission in human cervicovaginal explants and humanized mice using CD4 aptamer-siRNA chimeras. J Clin Invest 121:2401–2424
36. Thiel KW, Hernandez LI, Dassie JP et al (2012) Delivery of chemo-sensitizing siRNAs to HER2+-breast cancer cells using RNA aptamers. Nucleic Acids Res 40:6319–6337
37. Zhu Q, Shibata T, Kabashima T et al (2012) Inhibition of HIV-1 protease expression in T cells owing to DNA aptamer-mediated specific delivery of siRNA. Eur J Med Chem 56:396–399
38. Zhou J, Swiderski P, Li H et al (2009) Selection, characterization and application of new RNA HIV gp 120 aptamers for facile delivery of Dicer substrate siRNAs into HIV infected cells. Nucleic Acids Res 37:3094–3109
39. Zhou J, Neff CP, Swiderski P et al (2013) Functional in vivo delivery of multiplexed anti-HIV-1 siRNAs via a chemically synthesized aptamer with a sticky bridge. Mol Ther 21:192–200
40. Liu N, Zhou C, Zhao J et al (2012) Reversal of paclitaxel resistance in epithelial ovarian carcinoma cells by a MUC1 aptamerlet-7i chimera. Cancer Invest 30:577–582
41. Dai F, Zhang Y, Zhu X et al (2012) Anticancer role of MUC1 aptamer-miR-29b chimera in epithelial ovarian carcinoma cells through regulation of PTEN methylation. Target Oncol 7:217–225
42. Cerchia L, Esposito CL, Camorani S et al (2012) Targeting Axl with an high-affinity inhibitory aptamer. Mol Ther 20:2291–303
43. Camorani S, Esposito CL, Rienzo A et al (2014) Inhibition of receptor signaling and of glioblastoma-derived tumor growth by a novel PDGFRβ aptamer. Mol Ther 22:828–841
44. Amarzguioui M, Rossi JJ (2008) Principles of Dicer substrate (D-siRNA) design and function. Methods Mol Biol 442:3–10
45. Robbins M, Judge A, MacLachlan I (2009) siRNA and innate immunity. Oligonucleotides 19:89–102

Chapter 17

Method for Confirming Cytoplasmic Delivery of RNA Aptamers

David D. Dickey, Gregory S. Thomas, Justin P. Dassie, and Paloma H. Giangrande

Abstract

RNA aptamers are single-stranded RNA oligos that represent a powerful emerging technology with potential for treating numerous diseases. More recently, cell-targeted RNA aptamers have been developed for delivering RNA interference (RNAi) modulators (siRNAs and miRNAs) to specific diseased cells (e.g., cancer cells or HIV infected cells) in vitro and in vivo. However, despite initial promising reports, the broad application of this aptamer delivery technology awaits the development of methods that can verify and confirm delivery of aptamers to the cytoplasm of target cells where the RNAi machinery resides. We recently developed a functional assay (*RIP assay*) to confirm cellular uptake and subsequent cytoplasmic release of an RNA aptamer which binds to a cell surface receptor expressed on prostate cancer cells (PSMA). To assess cytoplasmic delivery, the aptamer was chemically conjugated to saporin, a ribosome inactivating protein toxin that is toxic to cells only when delivered to the cytoplasm (where it inhibits the ribosome) by a cell-targeting ligand (e.g., aptamer). Here, we describe the chemistry used to conjugate the aptamer to saporin and discuss a gel-based method to verify conjugation efficiency. We also detail an in vitro functional assay to confirm that the aptamer retains function following conjugation to saporin and describe a cellular assay to measure aptamer-mediated saporin-induced cytotoxicity.

Key words Aptamers, Cytoplasmic delivery, Saporin, Cytotoxicity

1 Introduction

RNA aptamers are single-stranded nucleic acids whose binding properties are similar to those of antibody/antigen interactions (thus are often referred to as *nucleic acid antibodies*) [1, 2]. Aptamers have been developed for a variety of technical and therapeutic applications [1–3]. Therapeutic RNA aptamers are modified with fluoro or *O*-methyl groups at the 2′-position of their sugar moiety, which renders them resistant to degradation by serum nucleases and significantly reduces their immunogenicity. Thus, these aptamers can be used in preclinical as well as clinical applications [2]. Their development as therapeutics has primarily involved

Kato Shum and John Rossi (eds.), *SiRNA Delivery Methods: Methods and Protocols*, Methods in Molecular Biology, vol. 1364, DOI 10.1007/978-1-4939-3112-5_17,

their tendency to inhibit their targets upon binding [4–7], although aptamers have also been developed to activate cell-surface receptors [8]. In 2006 came the first description of aptamers for delivery of therapeutic siRNAs to target cells [9, 10]. Since then, multiple in vivo studies involving low-dose (<1 mg/kg), systemic administration of aptamer-targeted siRNA conjugates have reported efficacy, highlighting the promise of this approach for future clinical applications [11–16]. Despite these encouraging studies, which demonstrated aptamer-mediated delivery of functional RNAi to target cells in vivo, the broad application of the aptamer delivery technology awaits the identification of aptamer sequences capable of internalizing into target cells and delivering the siRNA cargo to the cytoplasm where the RNA-induced silencing complex (RISC) resides.

In a recent study, we described a functional assay (RIP assay) for assessing cell internalization and subsequent release of aptamers into the cytoplasm of target cells [17]. The RIP (ribosome-inactivating protein) assay is based on targeting a RIP (e.g., saporin) to the cytoplasm of target cells *via* conjugation to a ligand. Saporin is a so-called RIP due to its N-glycosidase activity which, leads to a single adenine base being removed from the ribosomal RNA of the large subunit of the ribosome. RIPs are some of the most toxic molecules (poisons) known. Notorious examples of these toxins include ricin and abrin [18]. These toxins contain a signal sequence that functions to insert the RIP into a cell, leading to cytoplasmic localization and subsequent enzymatic inactivation of ribosomes and inhibition of protein synthesis. This results in efficient cell death and eventually causes death of the victim. Unlike ricin and abrin, saporin lacks this signal sequence and thus, is unable to internalize into cells and is safe to handle. However, if given a method of entry into the cell (*via* conjugation to a ligand/aptamer), saporin becomes a very potent toxin, since its enzymatic activity is among the highest of all RIPs [18].

We previously employed this property of saporin to confirm cytoplasmic delivery of an RNA aptamer to prostate-specific membrane antigen (PSMA), a cell surface receptor expressed on prostate cancer cells [17]. Below, we describe the chemistry used to conjugate the PSMA RNA aptamer to the saporin toxin and propose an agarose gel-based method to quickly assess conjugation efficiency and purity. We also detail an in vitro functional assay (NAALADase assay) to confirm the function of the aptamer-saporin conjugate post-conjugation. In addition to being highly expressed on the surface of prostate cancer cells, the extracellular portion of PSMA is known to have multiple catalytic activities, including *N*-acetylated alpha-linked acidic dipeptidase (NAALADase), folate carboxypeptidase, and dipeptidyl peptidase IV functions [19, 20]. Importantly, PSMA's enzymatic activity has been linked to increased cellular migration and activation of oncogenic pathways [21]. RNA aptamers to PSMA have been shown to inhibit the enzymatic activity of the protein [22, 23] and thus, can be evaluated for function using this assay. Finally, we describe a standard cellular assay (MTS assay) to measure saporin-induced cytotoxicity. This assay is based on the principle that NAD(P)H-dependent cellular oxidoreductase enzymes within cells can be a measure of the number of viable cells present within a culture. These enzymes are capable of reducing the tetrazolium dye [3-(4,5-dimethylthiazol-2-yl)-5-(3-carboxymethoxyphenyl)-2-(4-sulfophenyl)-2H-tetrazolium; MTS] to its insoluble formazan, which has a purple color and can be quantitated using a plate reader [24]. Tetrazolium dye assays can also be used to measure cytotoxicity (loss of viable cells) or cytostatic activity (shift from proliferative to resting status) of potential medicinal agents and toxic materials (e.g., saporin) [25].

In conclusion, the methodologies and assays described herein, with the exception of the NALAADase assay, can be broadly applied to evaluating other aptamers with cell internalizing properties and are poised to expedite the identification of aptamer sequences capable of delivering therapeutics (e.g., RNAi modulators) to the cytoplasm of target cells. Functional assays (e.g., binding assays, inhibition assays) to assess aptamer function post-conjugation should be developed and optimized for each aptamer and used in place of the NAALADase assay.

2 Matertials

2.1 RNA-Saporin

1. Biotinylated RNA aptamers.
2. Streptavidin-ZAP.
3. FGF-Saporin.
4. 10× binding buffer: 200 mM HEPES, pH 7.4, 1.5 M NaCl, 20 mM calcium chloride, and 0.1 % BSA.
5. 10× Binding buffer: 10 mL of 1 M HEPES, pH 7.4, 15 mL of 5 M sodium chloride, and 0.5 g of bovine serum albumin. Bring to a final volume of 50 mL with UltraPure H_2O. Store at room temperature.
6. Molecular biology-grade agarose.
7. TAE buffer.
8. 6× gel loading dye.
9. 100 base pair DNA ladder.
10. 7 × 10 cm Minigel System.
11. PowerPac Basic Power Supply.
12. SYBR Gold Nucleic Acid Gel Stain.
13. Molecular Imager Gel Doc XR System.

2.2 Reagents for NAALADase Assay

1. 200 mM Tris–HCl, pH 7.5: Add 200 mL of UltraPure 1 M Tris–HCl buffer, pH 7.5 to 800 mL of UltraPure H_2O.
2. 10 mM $CoCl_2$: Dissolve 1189.6 mg of cobalt (II) chloride hexahydrate in 500 mL UltraPure H_2O.
3. Recombinant human PSMA/FOLH1/NAALADase I.
4. 100 μM NAAG.
5. [Glutamate-3,4-H^3]-NAAG.
6. AG 1-X8 Resin.
7. Flint glass Pasteur pipettes, 9 in.
8. Custom-made rack to hold 9 in. Pasteur pipettes in place and in alignment with rack scintillation vials (*see* Fig. 1.).
9. Borosilicate glass beads, 3 mm diameter.
10. 0.1 M Na_3HPO_4: Add 5.7 g of Na_3HPO_4 (anhydrous) to double-distilled H_2O. Bring to a final volume of 400 mL with double-distilled H_2O, and store at 4 °C until ready to use.
11. 1 M formic acid: Add 383 mL of double-distilled H_2O to 17 mL of 88 % formic acid.
12. Bio-Safe II scintillation fluid.
13. Polyethylene scintillation vials.
14. Beckman LS6500 scintillation counter.

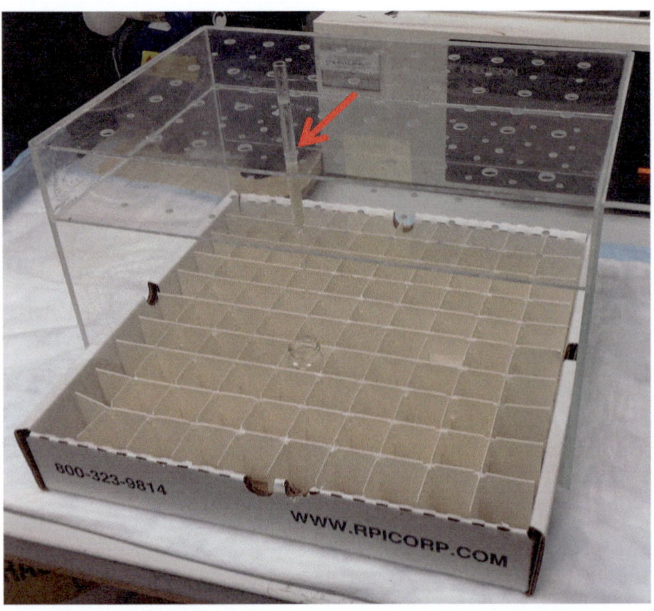

Fig. 1 NAALADase assay apparatus. Pictured is a custom-made plastic rack used to hold 9 in. long flint glass Pasteur pipettes in alignment with scintillation vials for measuring NAALADase activity. The *red arrow* points to a Pasteur pipette filled with 2 mL of AG 1-X8 resin

2.3 Reagents for MTS Assay

1. RPMI medium 1640.
2. 22Rv1.7 cells [11].
3. Fetal bovine serum: premium select.
4. 100× MEM nonessential amino acids solution.
5. Nunclon delta-treated 150 mm cell culture dishes.
6. 0.25 % Trypsin-EDTA.
7. CellTiter 96 AQueous One Solution Cell Proliferation Assay (MTS Assay).
8. 96-Well dishes.
9. Thermomax Microplate Reader.

3 Methods

3.1 RNA-Saporin Conjugation

1. Resuspend the biotinylated aptamer in ultrapure H_2O at a concentration of 100 μM, aliquot, and store at −80 °C.
2. Fold the biotinylated aptamer in 1× binding buffer (obtained by doing a 1:10 dilution of the 10× binding buffer) at a final concentration of 5 μM. The folding protocol is as follows: incubate aptamer solution at 65 °C for 10 min, followed by incubating at 37 °C for 10 min. The aptamer can be freeze-thawed several times after the folding protocol and still retain its activity.
3. Use a 1:4 molar ratio of streptavidin-ZAP and the folded biotinylated aptamer. Incubate at room temperature for 30 min in 1× binding buffer with a final concentration of 2.5 μM of the RNA.
4. The purity and efficiency of the conjugation of biotinylated aptamer to streptavidin-ZAP will need to be confirmed by running the aptamer-saporin conjugates on a 1 % agarose gel (*see* **Note 1**). For the 1 % agarose gel, dissolve 0.7 g of agarose into 70 mL of TAE buffer by heating in a microwave for approximately 2 min. Pour the agarose gel mixture into the Minigel system, and place the well combs into the gel, and allow the gel to cool and set for 30–45 min. After the gel has solidified, turn the gel so the wells are closest to the anode, and fill the apparatus with enough TAE to cover the gel. Mix 10 μL of the 2.5 μM biotinylated aptamer alone, streptavidin-ZAP alone, or biotinylated aptamer bound to saporin with 2 μL of the 6× gel loading dye. Carefully pipette the RNA/loading gel solution into the wells. Run the gel for 30 min at 100 V.
5. Stain the gel with 1× SYBR Gold for 30 min and perform UV imaging. The SYBR Gold stock solution is 10,000×, so dilute 5 μL of the 10,000× stock SYBR Gold solution into 50 mL of 1× TAE. Cover the gel with the 1× SYBR Gold solution and gently rock at room temperature in the dark for 30 min. Image the gel in a UV imager similar to the Molecular Imager Gel Doc XR System.

3.2 Activity of PSMA-Saporin Conjugate

1. A functional assay should be performed in order to confirm that the activity of the aptamer is not inhibited following conjugation to saporin (*see* **Note 2**). The functional assay we describe here measures PSMA enzymatic activity (NAALADase). The NAALADase assay is performed as follows (120 μL per sample replicate): A concentration of 0.333 μM is used for each aptamer, aptamer-saporin conjugate, and/or streptavidin-ZAP. 1× binding buffer alone is also included as a control.
2. Generate a Tris/$CoCl_2$ master mix by adding 500 μL of 200 mM Tris–HCl, pH 7.5 to 250 μL of 10 mM $CoCl_2$. Add 60 μL of the Tris–HCl/$CoCl_2$ master mix to each sample.
3. Upon receiving the rhPSMA, aliquot it into 2 μL aliquots under sterile conditions and store the aliquots at −80 °C. The rhPSMA aliquots should be diluted by adding 375 μL of ultrapure H_2O and 125 μL of 200 mM Tris–HCl, pH 7.5 to the 2 μL of the PSMA solution before using in the NAALADase assay. Add 10 μL of rhPSMA mixture to the RNA mixtures, and mix by pipetting. Incubate at 37 °C for 5 min, and start the timer after adding PSMA to the first sample.
4. Make a NAAG solution consisting of 992 μL of ultrapure H_2O, 5.5 μL of 100 μM NAAG, and 2.5 μL of ^{3}H-NAAG. Add 10 μL of the NAAG solution to each sample and incubate at 37 °C for 15 min. Mix the reaction after 7.5 min has past. Start the timer after adding the NAAG solution to the first sample.
5. Stop the reaction by adding 200 μL of ice cold 0.1 M Na_3HPO_4 to each sample.
6. Prepare columns by placing as many Pasteur pipettes as needed into the plastic rack (Fig. 1). Place one 3 mm borosilicate glass ball into each pipette.
7. Make a 1:1 mixture of AG 1-X8 resin and ultrapure H_2O to create a resin slurry. Add 2 mL of the resin slurry to each Pasteur pipette. Once all of the liquid has left the columns, add 2 mL of double-distilled H_2O to the columns, and allow this to flow through and discard.
8. Once all of the liquid has drained from the columns, place the scintillation vials under the columns. Add 190 μL of each reaction mixture to separate columns, and allow it to completely soak into the resin.
9. Add 2 mL of 1 M formic acid to each column. Allow the liquid to completely flow through the columns and collect 2 mL of eluent into each scintillation vial.
10. Add 10 mL of Bio-Safe II scintillation fluid to each scintillation vial and count in a liquid scintillation counter.

3.3 Measurement of Aptamer-Saporin Cytotoxicity

1. 22Rv1.7 cells are grown in RPMI medium 1640 with 10 % FBS and 1 % non-essential amino acids, and are propagated in 150 mm culture dishes (*see* **Notes 3** and **4**). The cells are split 1:3 when they are approximately 80 % confluent by trypsinizing with 0.25 % trypsin-EDTA for no more than 4 min for further propagation. After trypsinizing the cells, plate 1000 cells per well in a 96 well plate with 100 μL of media. Allow the cells to attach overnight.
2. Incubate the RNA-saporin conjugate, RNA alone, streptavidin-ZAP alone, or 1× binding buffer alone on the cells. Vary the time and treatment amounts in order to determine the EC_{50} and optimize treatment efficiency for each cell line. As a positive control for saporin-induced cytotoxicity, treat cells with FGF-saporin conjugate. The optimal amount of FGF-saporin needed to kill cells may have to be determined a priori. In our experience, the EC_{50} of FGF-saporin is anywhere from 0.15 to 1.4 nM depending on the cell line being tested.
3. Mix the cellular growth media in a 5:1 ratio with MTS (e.g., 100 μL of media and 20 μL of MTS per well in a 96-well plate). Remove media from cells and add 120 μL of the 5:1 media:MTS mixture to each well, followed by incubating for 1 h at 37 °C. Record the absorbance at 490 nm with a 96-well plate reader such as a Molecular Devices Thermomax microplate reader.

4 Notes

1. Conjugation problems: It is possible that conjugating the aptamer to saporin could ablate the activity of the aptamer. An aptamer-specific functional assay to verify activity of aptamer-saporin conjugate, such as the NAALADase describe for the PSMA aptamer, is recommended. Prior to conjugation to saporin-streptavidin (ZAP), it is advisable to determine that the biotinylated aptamer binds to its target and/or retains function. In the event that placing a biotin on the 5′ end of an aptamer disrupts its function, we advise positioning the biotin on the 3′ end. Additional linking chemistries could also be explored [26–29].
2. Variability in the expression of cell-surface receptors: The expression of cell-surface receptors being targeted with aptamers can vary from cell line to cell line and can also depend on passage number and growth conditions. We recommend verifying cell-surface protein expression levels for any given aptamer target receptor prior to performing the cell-based cytotoxic assay. Flow cytometry using antibodies specific to the receptor target should be performed.

3. Because modified nucleic acids are susceptible to degradation by mycoplasma nucleases [30], it is advisable to ensure that all cell cultures are free of mycoplasma contamination prior to performing the cell-based cytotoxic assays.
4. Cell-type variability: Different cell types have different sensitivities to saporin-induced toxicity. The streptavidin-ZAP kit contains saporin conjugated to FGF (saporin-FGF) as a positive control to confirm sensitivity to saporin-induced cell death.

References

1. Thiel KW, Giangrande PH (2009) Therapeutic applications of DNA and RNA aptamers. Oligonucleotides 19:209–222
2. Keefe AD, Pai S, Ellington A (2010) Aptamers as therapeutics. Nat Rev Drug Discov 9:537–550
3. Sundaram P, Kurniawan H, Byrne ME, Wower J (2013) Therapeutic RNA aptamers in clinical trials. Eur J Pharm Sci 48:259–271
4. Giangrande PH, Zhang J, Tanner A, Eckhart AD, Rempel RE, Andrechek ER, Layzer JM, Keys JR, Hagen PO, Nevins JR, Koch WJ, Sullenger BA (2007) Distinct roles of E2F proteins in vascular smooth muscle cell proliferation and intimal hyperplasia. Proc Natl Acad Sci U S A 104:12988–12993
5. Mi J, Zhang X, Rabbani ZN, Liu Y, Reddy SK, Su Z, Salahuddin FK, Viles K, Giangrande PH, Dewhirst MW, Sullenger BA, Kontos CD, Clary BM (2008) RNA aptamer-targeted inhibition of NF-kappa B suppresses non-small cell lung cancer resistance to doxorubicin. Mol Ther 16:66–73
6. Rusconi CP, Scardino E, Layzer J, Pitoc GA, Ortel TL, Monroe D, Sullenger BA (2002) RNA aptamers as reversible antagonists of coagulation factor IXa. Nature 419:90–94
7. Povsic TJ, Sullenger BA, Zelenkofske SL, Rusconi CP, Becker RC (2010) Translating nucleic acid aptamers to antithrombotic drugs in cardiovascular medicine. J Cardiovasc Transl Res 3:704–716
8. McNamara JO, Kolonias D, Pastor F, Mittler RS, Chen L, Giangrande PH, Sullenger B, Gilboa E (2008) Multivalent 4-1BB binding aptamers costimulate CD8+ T cells and inhibit tumor growth in mice. J Clin Invest 118:376–386
9. McNamara JO II, Andrechek ER, Wang Y, Viles KD, Rempel RE, Gilboa E, Sullenger BA, Giangrande PH (2006) Cell type-specific delivery of siRNAs with aptamer-siRNA chimeras. Nat Biotechnol 24:1005–1015
10. Chu TC, Twu KY, Ellington AD, Levy M (2006) Aptamer mediated siRNA delivery. Nucleic Acids Res 34, e73
11. Dassie JP, Liu XY, Thomas GS, Whitaker RM, Thiel KW, Stockdale KR, Meyerholz DK, McCaffrey AP, McNamara JO II, Giangrande PH (2009) Systemic administration of optimized aptamer-siRNA chimeras promotes regression of PSMA-expressing tumors. Nat Biotechnol 27:839–849
12. Pastor F, Kolonias D, Giangrande PH, Gilboa E (2010) Induction of tumour immunity by targeted inhibition of nonsense-mediated mRNA decay. Nature 465:227–230
13. Neff CP, Zhou J, Remling L, Kuruvilla J, Zhang J, Li H, Smith DD, Swiderski P, Rossi JJ, Akkina R (2011) An aptamer-siRNA chimera suppresses HIV-1 viral loads and protects from helper CD4(+) T cell decline in humanized mice. Sci Transl Med 3:66ra66
14. Zhou J, Shum KT, Burnett JC, Rossi JJ (2013) Nanoparticle-based delivery of RNAi therapeutics: progress and challenges. Pharmaceuticals (Basel) 6:85–107
15. Dassie JP, Giangrande PH (2013) Current progress on aptamer-targeted oligonucleotide therapeutics. Ther Deliv 4:1527–1546
16. Esposito CL, Cerchia L, Catuogno S, de Vita G, Dassie JP, Santamaria G, Swiderski P, Condorelli G, Giangrande PH, de Franciscis V (2014) Multi-functional aptamer-miRNA conjugates for targeted cancer therapy. Mol Ther 22(6):1151–1168
17. Hernandez LI, Flenker KS, Hernandez FJ, Klingelhutz AJ, McNamara JO II, Giangrande PH (2013) Methods for evaluating cell-specific, cell-internalizing RNA aptamers. Pharmaceuticals (Basel) 6:295–319
18. Walsh MJ, Dodd JE, Hautbergue GM (2013) Ribosome-inactivating proteins: potent poisons and molecular tools. Virulence 4: 774–784

19. Pinto JT, Qiao C, Xing J, Suffoletto BP, Schubert KB, Rivlin RS, Huryk RF, Bacich DJ, Heston WD (2000) Alterations of prostate biomarker expression and testosterone utilization in human LNCaP prostatic carcinoma cells by garlic-derived S-allylmercaptocysteine. Prostate 45:304–314
20. Carter RE, Feldman AR, Coyle JT (1996) Prostate-specific membrane antigen is a hydrolase with substrate and pharmacologic characteristics of a neuropeptidase. Proc Natl Acad Sci U S A 93:749–753
21. Ghosh A, Wang X, Klein E, Heston WD (2005) Novel role of prostate-specific membrane antigen in suppressing prostate cancer invasiveness. Cancer Res 65:727–731
22. Lupold SE, Hicke BJ, Lin Y, Coffey DS (2002) Identification and characterization of nuclease-stabilized RNA molecules that bind human prostate cancer cells via the prostate-specific membrane antigen. Cancer Res 62:4029–4033
23. Rockey WM, Hernandez FJ, Huang SY, Cao S, Howell CA, Thomas GS, Liu XY, Lapteva N, Spencer DM, McNamara JO, Zou X, Chen SJ, Giangrande PH (2011) Rational truncation of an RNA aptamer to prostate-specific membrane antigen using computational structural modeling. Nucleic Acid Ther 21: 299–314
24. Buttke TM, McCubrey JA, Owen TC (1993) Use of an aqueous soluble tetrazolium/formazan assay to measure viability and proliferation of lymphokine-dependent cell lines. J Immunol Methods 157:233–240
25. Wu Y, Tang M, Weng X, Yang L, Xu W, Yi W, Gao J, Bode AM, Dong Z, Cao Y (2014) A combination of paclitaxel and siRNA-mediated silencing of Stathmin inhibits growth and promotes apoptosis of nasopharyngeal carcinoma cells. Cell Oncol (Dordr) 37:53–67
26. Schulman LH, Pelka H, Reines SA (1981) Attachment of protein affinity-labeling reagents of variable length and amino acid specificity to E. coli tRNAfMet. Nucleic Acids Res 9:1203–1217
27. Guk K, Lim H, Kim B, Hong M, Khang G, Lee D (2013) Acid-cleavable ketal containing poly(beta-amino ester) for enhanced siRNA delivery. Int J Pharm 453:541–550
28. Bochkariov DE, Kogon AA (1992) Application of 3-[3-(3-(trifluoromethyl)diazirin-3-yl) phenyl]-2,3-dihydroxypropionic acid, carbene-generating, cleavable cross-linking reagent for photoaffinity labeling. Anal Biochem 204:90–95
29. Fink G, Fasold H, Rommel W, Brimacombe R (1980) Reagents suitable for the crosslinking of nucleic acids to proteins. Anal Biochem 108:394–401
30. Hernandez FJ, Stockdale KR, Huang L, Horswill AR, Behlke MA, McNamara JO II (2012) Degradation of nuclease-stabilized RNA oligonucleotides in mycoplasma-contaminated cell culture media. Nucleic Acid Ther 22:58–68

Chapter 18

Hapten-Binding Bispecific Antibodies for the Targeted Delivery of SiRNA and SiRNA-Containing Nanoparticles

Irmgard S. Thorey, Michael Grote, Klaus Mayer, and Ulrich Brinkmann

Abstract

Hapten-binding bispecific antibodies (bsAbs) are effective and versatile tools for targeting diverse payloads, including siRNAs, to specific cells and tissues. In this chapter, we provide examples for successful SiRNA delivery using this powerful targeting platform. We further provide protocols for designing and producing bsAbs, for combining bsAbs with SiRNA into functional complexes, and achieving specific mRNA knockdown in cells by using these functional complexes.

Key words Bispecific antibodies, Recombinant antibody expression, IgG, Fab, scFv, Hapten, Targeted delivery, SiRNA delivery, Lipid nanoparticles, Dynamic polyconjugates

1 Introduction

Since its discovery in 1998 by Fire, Mello, and colleagues [1], major improvements have been made in development of SiRNA-based drug candidates [2]. However, the major challenge in therapeutic application of SiRNA is still delivery to and into target organs, tissues, and cells. A multitude of SiRNA delivery systems, including formulations for targeted delivery, have been evaluated so far [3–11]. A delivery system for therapeutic use must (1) confer stability to SiRNA upon application; (2) enable targeting to and into the desired tissues and cell types; (3) enable release of the intact SiRNA from vesicles into the cytosol after uptake by the target cells, as a prerequisite for mRNA knockdown, and (4) facilitate all these requirements without toxic side effects.

Bispecific antibodies (bsAbs) that recognize cell surface antigens and haptens are effective tools for targeted drug delivery [12, 13]. We have recently developed a targeting platform consisting of IgG-derived bispecific antibodies (bsAbs) that simultaneously bind to both cell surface antigens as well as hapten-coupled payloads. This bsAb platform is highly flexible and allows rapid

Kato Shum and John Rossi (eds.), *SiRNA Delivery Methods: Methods and Protocols*, Methods in Molecular Biology, vol. 1364, DOI 10.1007/978-1-4939-3112-5_18, © Springer Science+Business Media New York 2016

adjustment of binding stoichiometry (e.g., 2:2; 2:1; 1:1; 1:2), as well as pharmacokinetic parameters (e.g., Fc-containing IgGs or Fabs without Fc, thus modulating binding capabilities to the neonatal Fc receptor). Using this delivery platform, we have shown highly effective and specific targeted delivery of small-molecule drugs and fluorophores in vitro and in vivo [14]. Hapten-binding bsAb formats can be easily combined with hapten-coupled DNA or RNA moieties as payloads. Thus, they are well suited for targeted delivery of nucleic acids, both DNA, (e.g., antisense oligonucleotides), and RNA (e.g., SiRNA or SiRNA formulations) [15].

Most approaches that combine protein binding modules (e.g., antibody-based molecules) with SiRNA formulations face the problem in which the manufacturing processes for proteins and SiRNAs (or SiRNA formulations) are different and hence not compatible [15]. Our delivery platform utilizing hapten-binding bsAbs provides a solution for this problem: First, hapten-binding bsAbs can be produced in bacterial or eukaryotic expression systems by established standard procedures. In parallel, hapten-coupled SiRNA or SiRNA formulations can be manufactured by chemical synthesis in organic solvents by established standard procedures. Both processes are separately performed, eliminating the need to invent one process which accommodates both protein expression (which requires aqueous buffer solutions) and chemical synthesis (preferentially performed in organic solvents). Finally, because of the separation of individual processes, protein and SiRNA production can be easily scaled up individually and independently. Another advantage of this modular system is the possibility to use various combinations of different payloads (e.g., SiRNAs against different target mRNAs) with different targeting entities and formats (e.g., bsAbs against different target antigens and varying binding stoichiometry). This allows the combination of a given bsAb with a multitude of payloads, or the combination of a defined payload with different hapten-binding bsAbs. Moreover, for each bsAb-payload combination, the binding of payload molecules to the bsAb at a predefined ratio is ensured. This ratio is defined by the number of binding sites in the chosen bsAb as well as the number of haptens that are coupled to the applied payload.

The following protocols describe established design, production, and purification procedures for generating recombinant IgG- or Fab-derived bsAbs (Subheadings 3.1, 3.2 and 3.3). These in turn can be combined with either SiRNA or a broad range of SiRNA formulations containing surface-exposed functional groups (e.g., amino or thiol groups), which can be chemically modified with hapten conjugates. We further describe protocols that ensure specific coupling of SiRNA or SiRNA-containing nanoparticles to the bsAb at defined positions and defined stoichiometry (Subheading 3.4). Finally, we show antibody-mediated SiRNA delivery to certain cell types (Subheading 3.5), followed by internalization and separation of SiRNA or SiRNA-containing vehicles from bsAbs. This facilitates efficient SiRNA transfer into the cytoplasm (Subheading 3.6), and enables knockdown of target mRNA in the desired cell types (Subheadings 3.7 and 3.8).

2 Materials

2.1 Bispecific Antibodies

The bsAbs that we use for SiRNA delivery are derivatives of IgGs or antibody Fab fragments that bind to cell surface antigens and to haptens, like digoxigenin (Dig) or modified biotin, as previously described [7, 8, 13]. For IgG formats, the cell surface-binding functionalities of these molecules are positioned in the two Fab arms of the IgG moiety, with two disulfide-stabilized anti-hapten single-chain Fv (scFv) modules recombinantly fused to the C-termini of either both heavy chains or both light chains, resulting in 2+2 formats (Fig. 1, top). The VH and VL domains of anti-hapten scFv domains [14] are held together by flexible linkers and by disulfide bonds between VH and VL to enhance stability (VHCys44 to VLCys100, [16–18]. For a different stoichiometry, a single anti-hapten specific scFv domain can be C-terminally linked to only one of the two CH3 domains of the antibody dimer, resulting in a 2+1 format. To enforce heterodimerization, the two Fc domains of this antibody format are reengineered at the dimerization interface within the CH3 region: one Fc domain carries a bulky amino acid substitution ("knob" mutation), the other Fc domain of the dimer harbors a corresponding "hole" mutation [19]. As an alternative that results in smaller molecules, 1+1 or 1+2 formats based on Fab fragments of cell-surface-specific antibodies can be designed, by adding one or two scFv domains C-terminally to the cell-surface targeting Fab (Fig. 1, bottom).

Cell-targeting entities, for example, include antibody modules that bind tumor-associated antigens such as Her2 [20], EGF receptor [21, 22], IGF1 receptor [23], CD22 [24], CD33, VEGF receptor 2 [15], the LeY carbohydrate antigen [25, 26], and others.

2.2 siRNAs and Hapten-Containing siRNA Derivatives

An important feature of our modular targeting approach is the linkage of the SiRNA to haptens or to a hapten-containing moiety without compromising either the SiRNA functionality or the interaction between the hapten and the hapten-binding domain of the bsAb. For direct coupling of hapten to the SiRNA of choice, we have focused on the 3′ end of the sense strand of double-stranded SiRNA derivatives. This position in SiRNA was previously found to tolerate added entities (such as cholesterol [27]) without affecting SiRNA activity. In accordance with this, we have observed the potency of 3′-hapten-conjugated SiRNA derivatives, i.e. their

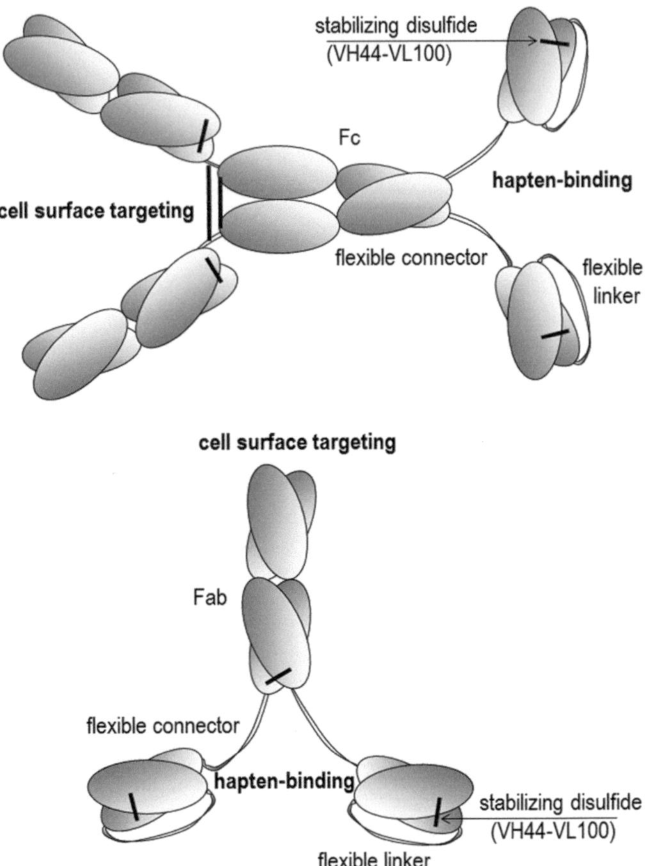

Fig. 1 Bispecific antibody (bsAb) formats for siRNA targeting. A 2 + 2 IgG format (*top*) which contains an IgG constant region (Fc); and a 1 + 2 Fab format (*bottom*) are shown. Hapten-binding disulfide stabilized single-chain Fv domains (scFv's) are fused at C-terminal to the natural IgG or Fab (which supplies the cell surface targeting function) by flexible (Gly4Ser)$_n$ modules (flexible connector). Likewise, VH and VL domains of hapten-binding scFvs are linked to each other by flexible (Gly4Ser)$_n$ modules (flexible linker). Hapten-binding scFv domains are stabilized by disulfide bonds between amino acids VH44 and VL100 (*arrows*)

ability to reduce target mRNA levels, to be the same as unmodified SiRNA [15]. As examples for hapten-coupled SiRNA that can be complexed to bsAbs, the structures of digoxigenin (Dig)-conjugated SiRNA molecules are shown in Fig. 2 (Dig-ds-SiRNA). As an additional modification, fluorescent compounds such as Cy5 can be attached to the 5′ end of the sense strand to enable visualization and tracking of SiRNA (Fig. 2, Dig-SiRNA-Cy5). Hapten coupling to the 5′ end of the sense strand also yields SiRNA with good activity in our experience, but is not superior to coupling to the 3′ end.

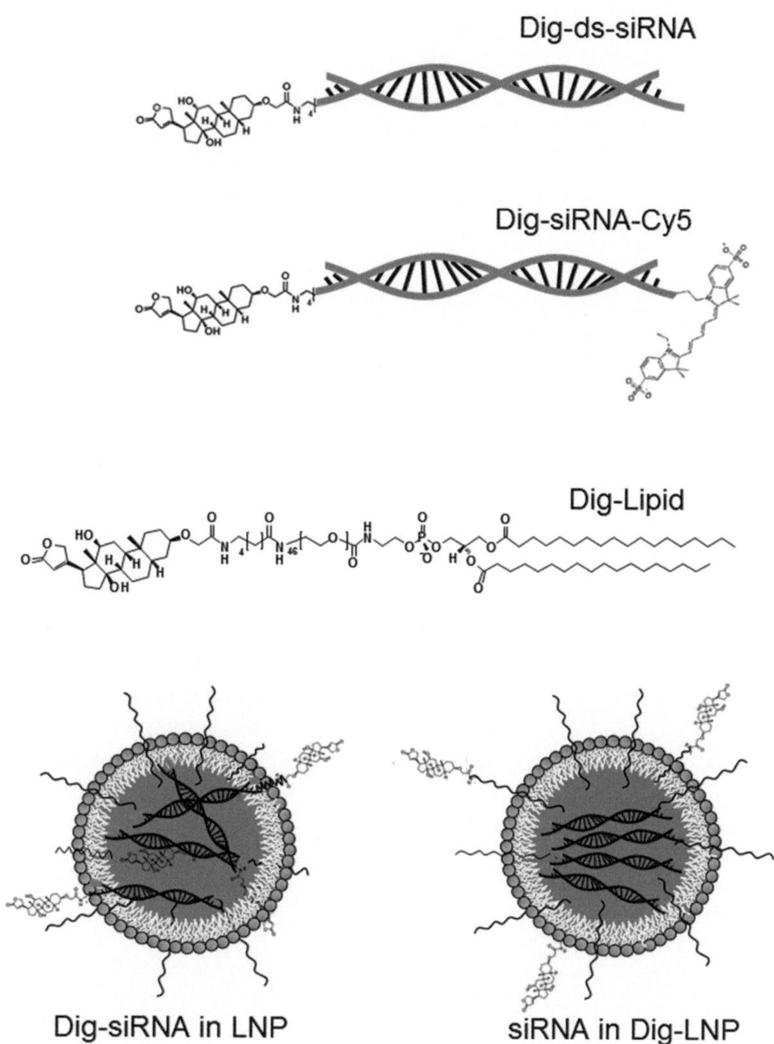

Fig. 2 Hapten-containing payloads for siRNA targeting

2.3 Hapten-Containing Nanoparticles

Hapten-coupled SiRNA can be directly complexed with bsAbs. These complexes are able to deliver SiRNA specifically to cells, which express the respective target antigen on their surface. However, specific accumulation of SiRNA in intracellular vesicles (upon internalization) by itself does not result in specific gene knock-down. The reason for this is that unmodified SiRNA accumulates in endosomal compartments but does not escape into the cytoplasm [15]. In order to enable delivery into the cytosol, the SiRNA can be packaged into nanoparticles [3–5, 7, 11, 28–34] which carry haptens on their surface. These haptens need to be accessible to bsAbs when incorporated into the structure of the (large) nanoparticle. Various kinds of nanoparticles for effective SiRNA delivery have been described, many of which contain

PEG. Therefore, one way of generating hapten-containing nanoparticles is the application of hapten-coupled PEG derivatives as components for nanoparticle generation (Fig. 2, Dig-Lipid). Incorporation of these reagents into formulations generates hapten-decorated nanoparticles (Fig. 2, SiRNA in Dig-LNP). Hapten-decorated nanoparticles can in some instances also be generated by formulating hapten-coupled SiRNAs into standard nanoparticles. Surprisingly, this results in nanoparticles which have hapten molecules exposed on their surface in an antibody-accessible way (schematic representation in Fig. 2, Dig-SiRNA in LNP). Fluorescently labeled SiRNA can be incorporated into hapten-decorated nanoparticles in the same manner without loss of the fluorescent signal. This can be used for visualization and tracking of hapten-containing bsAb-targeted nanoparticles.

Two established types of nanoparticles are dynamic polyconjugates (DPCs) and lipid-based nanoparticles (LNPs). We have successfully used DPCs and LNPs containing hapten-conjugated PEG lipids for the construction of functional SiRNA moieties for cellular targeting [15]. DPCs include scaffold-reagents, like poly butyl and amino vinyl ether (PBAVE), an endosomolytic polymer that is shielded from nonspecific cell interactions by reversible covalent modification with polyethylene glycol (PEG). Both SiRNA and hapten can be attached to the polymer by linkers which are either stable (e.g., for hapten linkage) or which enable pH-dependent payload release [32, 34, 35].

Hapten-decorated DPCs can be complexed with bsAb at a defined molar ratio (e.g., 1:1 or 2:1) to generate bsAb-targeted DPCs. Loading of bsAbs with SiRNA containing DPCs results in an increase in molecular weight and hydrodynamic radius which can be detected by SEC-MALLS. For instance, Dig-polymer-SiRNA DPCs without bsAb are a polydisperse solution with molecules of an estimated molecular weight between 300 and 720 kDa and a hydrodynamic radius from 7 to 10 nm. Addition of bsAb to form hapten-polymer-SiRNA DPC-bsAb complexes increases the molecular weight range to 500–1100 kDa and the hydrodynamic radius to 9–12.5 nm (Fig. 3) [15].

LNPs contain polyethylene glycol (PEG)-lipids whose lipophilic acyl chains anchor the hydrophilic PEG molecules in the particle. This ensures particle stability and structural integrity. The acyl chains of the PEG-lipids can be of various lengths. The LNPs which we have successfully used contain PEG-lipids either with a relatively long C18 anchor that consist of 18 methanediyl groups and is considered non-exchangeable, or with a shorter C16 anchor that consists of 16 methanediyl groups and is highly exchangeable [36, 37]. Hapten-conjugated SiRNA-containing LNPs alone or complexed with the bsAb in DPBS can be analyzed by dynamic light scattering (DLS) to determine their hydrodynamic radii and

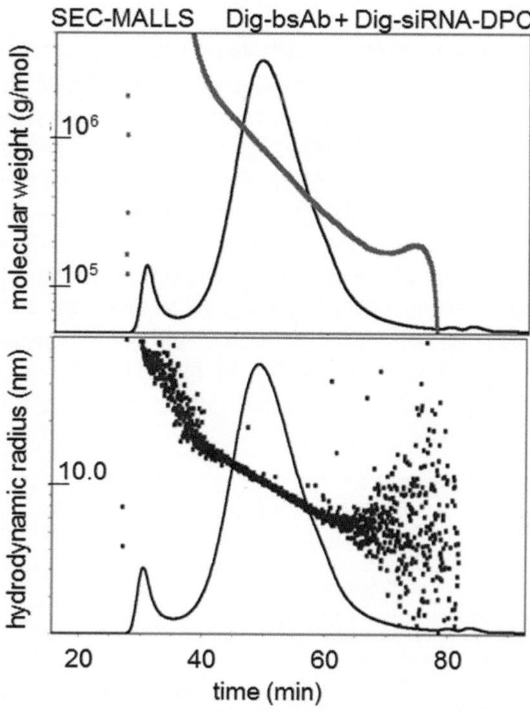

Fig. 3 SEC-MALLS of LeY-Dig bsAb complexed with Dig-siRNA-DPC nanoparticles. SEC was performed using a Superose 6 10/300 GL column, on a Dionex Ultimate 3000 HPLC system equipped with a three-angle laser light scattering (LS) detector (miniDAWN Treos), a differential refractive index (RI) detector (Optilab rEx), and a DLS detector (WyattQELS). The signal from the LS detector is indicated by the *black curves* in both panels. *Upper panel*: The molecular weight (calculated from the LS signal and the RI detector signal) is indicated by the *thick grey curve*. *Lower panel*: The hydrodynamic radius determined from the signal of the QELS detector is indicated by the *dot cloud*

polydispersity indices (Pdi). LNPs can be incubated together with bsAbs at room temperature (~25 °C; up to 3 h), and the time course of change in particle size and polydispersity can be determined by DLS [15, 38].

Hapten-decorated LNP formulations which were successfully used by us contained a total of 1.4 mol% PEG-lipids, of those were 0.4 or 0.04 mol% hapten-coupled (Dig-modified) C18 PEG-lipids. The remaining hapten-free PEG-lipids contained C16 lipid-anchors to enable effective LNP-deshielding and high SiRNA transfer/transfection potency [36–38]. Using such LNPs, we have achieved a knockdown efficiency of up to 90 % knockdown with IC50 of 1.7 nM (Table 1). Moreover, in contrast to results obtained with other SiRNA formulations [39], we did not observe any immunostimulatory effects with these LNPs [15].

Table 1
Specific and targeted mRNA Knockdown by LeY-Dig and CD33-Dig bsAbs complexed with Dig-siRNA nanoparticles. Top: DPCs containing Dig-siRNA specific for the Aha1 mRNA (target siRNA), or for the CD45 mRNA as negative control (cntrl siRNA), were prepared as payload. Payload alone, payload complexed with LeY-Dig bsAb (specific targeting) or payload complexed with CD33-Dig bsAb (non-specific bsAb) were applied for 30 min to MCF-7 cells (expressing LeY at high levels) or to Hep3B cells (expressing LeY at lower levels). Cells were washed with DPBS and further incubated for 24 hours. Cells were lysed and the ratio of mRNA levels of Aha1 to Cyclophilin A was assessed by quantitative RT-PCR assay.Bottom: LNPs containing Dig-siRNA specific for the Aha1 mRNA were prepared as payload. Payload alone, CD33-Dig bsAb alone, payload complexed with CD33-Dig bsAb (specific targeting) or payload complexed with CD22-Dig bsAb (non-specific bsAb) were applied for 30 min to CD33-expressing MOLM-13, Kasumi-1, or MV4-11 cells. Cells were washed with DPBS and further incubated for 24 hours. Cells were lysed and the ratio of mRNA levels of Aha1 to GAPDH was assessed by branched DNA assay.

Cell line (LeY$^+$, CD33$^-$)	Target siRNA w/o bsAb	cntrl siRNA + LeY bsAb	Target siRNA + LeY bsAb	Target siRNA + CD33 bsAb
MCF-7 (LeY^{++++})	<5 %	<5 %	>75 %	<5 %
Hep3B (LeY$^+$)	<5 %	<5 %	>15 %	<5 %

Cell line (CD33$^+$ CD22$^-$)	siRNA LNP w/o bsAb	CD33 bsAb w/o siRNA LNP	CD33 bsAb + siRNA LNP	CD22 bsAb + siRNA LNP
MOLM-13	<10 %	<5 %	>80 %	<10 %
Kasumi-1	<15 %	<5 %	>85 %	<5 %
MV4-11	<10 %	<5 %	>60 %	<5 %

2.4 Specific Reagents for bsAb Expression and Targeted Delivery

1. Non-adherent HEK 293 cells; FreeStyle™ 293-F Cells.
2. FreeStyle™ 293 Expression Medium.
3. 293-Free™ Transfection Reagent.
4. Protein A sepharose column; HiTrap Protein-A HP 5 ml.
5. Elution buffer; 0.1 M citric acid, pH 2.8.
6. Neutralization buffer; 1 M Tris–HCl, pH 8.5.
7. Gel filtration column; Superdex 200 HiLoad 120 ml 16/60.
8. Dulbecco's phosphate-buffered saline (DPBS); w/o Ca, w/o Mg.
9. Paraformaldehyde; 4 % (w/v) solution in DPBS.

3 Methods

3.1 Design of Bispecific Antibodies

Procedures for the production and purification of bsAbs have been described previously [40, 41]. One bsAb format that we have successfully used for SiRNA delivery consists of single-chain Fv (scFv) fragments fused C-terminally to natural IgG antibodies directed against cell surface antigens, resulting in a bsAb that is both bispecific and bivalent for each antigen (2 + 2 format). The scFv fragment can be fused either to the CH3 domain of the IgG (Fig. 1, top) or the constant region of the light chain. The IgG is separated from the VH of the scFv by a $(G_4S)_2$ linker; VH and VL of the scFv are separated by a $(G_4S)_3$ linker (Fig. 1), which in both cases was found to be the optimal length. The anti-hapten scFv module is stabilized by introduction of a VH44-VL100 disulfide bond to prevent domain dissociation and subsequent aggregation (Fig. 1) [16–18]. As an alternative IgG-based format, which leads to a different stoichiometry of binding specificities, a bsAb can be bivalent for one antigen (e.g., the cell-targeting specificity) and monovalent for the second (hapten-binding specificity) (2 + 1 format). This can be achieved by coupling one hapten-specific scFv domain to the CH3 region of a reengineered heavy chain which carries a "knob" mutation (e.g., T366Y). This bispecific heavy chain monomer is co-expressed with a monospecific heavy-chain monomer which carries a "hole" mutation (e.g., Y407T; [19]).

Bispecific formats can also be based on Fab antibody fragments. One or two scFv fragments can be fused at C-terminal to one or both chains of a Fab fragment, resulting in relatively small proteins with 1 + 1 or 1 + 2 stoichiometries, respectively (Fig. 1). Such formats may have better tissue penetration properties. However, they have different pharmacokinetic parameters (reduced serum half-lives), because they lack Fc domains [42, 43]. BsAbs can be expressed from expression plasmids containing sequences coding for a single heavy chain-based monomer (or in some cases one "knob" and one "hole" monomer) and a single light chain-based monomer, respectively. This avoids light chain or heavy chain mispairing problems which may arise with other bsAb formats.

3.2 Design of Expression Vectors for Bispecific Antibodies

For protein expression in mammalian cells, gene segments encoding light chains and heavy chains or domains (scFvs) of antibodies are synthesized with a 5′ end extension coding for a leader peptide. This targets proteins for secretion. Light-chain and heavy-chain modules are placed into expression cassettes that provide promoter sequences, e.g., the human cytomegalovirus (HCMV) immediate early enhancer and promoter, and the light and heavy chain constant gene segments. In addition to the antibody expression cassettes, the vectors contain: origins of replication, e.g., oriP of Epstein-Barr virus and pUC18 origin of replication, for eukaryotic and bacterial expression, respectively; a selectable marker to produce the DNA vector in bacteria; and a polyadenylation sequence in case of eukaryotic expression systems. The bsAbs described here can be produced in mammalian cells and purified from cell culture supernatants in the same manner as conventional IgGs.

3.3 Protocol for Transient Expression in Mammalian Cells and Purification of bsAbs

1. Seed non-adherent HEK 293 cells in fresh suitable cell culture medium at a density of $1\text{–}2 \times 10^6$ viable cells/ml.
2. On the same day, co-transfect equimolar ratios of heavy chain/scFv and light chain expression plasmids into the cells using 293-transfection reagents according to the manufacturer's instruction.
3. Harvest cell culture supernatants containing the bsAbs on day 7 after transfection, centrifuge at $14{,}000 \times g$ for 45 min at 4 °C, and filtrate through a 0.22 um filter. Cell culture supernatants can be stored at −20 °C until purification.
4. Apply sterile cell culture supernatant to a suitable Protein-A-Sepharose column, for example HiTrap Protein-A HP (5 ml).
5. Remove unbound proteins by washing, elute and recover bsAbs with 0.1 M citrate buffer, pH 2.8.
6. Neutralize immediately with 1 M Tris–HCl, pH 8.5 and concentrate if necessary.
7. Load bsAbs on a gel filtration column, for example Superdex 200 HiLoad 120 ml 16/60, collect fractions containing un-aggregated bsAbs, which have the expected MW (for example ~200,000 Da for the 2 + 2 IgG format) (*see* **Note 1**).

3.4 Protocol for the Preparation of siRNA-bsAb Complexes

Prior to cellular targeting, the exact coupling stoichiometry of the chosen targeting system should be verified experimentally. Therefore, bsAb and payload are combined in various molar ratios in DPBS, followed by an analysis of the resulting complexes. The protocol below and the results shown in Fig. 3 are based on use of a bsAb containing two cell-targeting and two hapten-binding specificities (2 + 2 format).

1. Add fluorescently labeled hapten-coupled SiRNA (for example, Dig-SiRNA-Cy5, shown in Fig. 2a) or hapten-decorated nanoparticles of choice (Fig. 2c), to hapten-binding bsAb of desired specificity (for example LeY-Dig-bsAb) (*see* **Note 2**).
2. Incubate at room temperature for 30 min.
3. Determine free and complexed SiRNA by size-exclusion chromatography; or free and complexed SiRNA-containing nanoparticles by SEC-MALLS or dynamic light scattering (Fig. 3).
4. At a ratio of two or less hapten-coupled SiRNA moieties per bsAb molecule, all fluorescence should be found in the high molecular weight fraction representing antibody-complexed SiRNA. Under these conditions, no free hapten-coupled SiRNA (or SiRNA containing nanoparticles) should be detected.
5. At ratios of more than two hapten-coupled SiRNA moieties per bsAb molecule, no further signal increase should be observed for the fractions that contain the bsAb-complex.

Instead, dose-dependent signal increases occur only in fractions corresponding to the lower molecular weight fraction representing free hapten-coupled SiRNA (nanoparticles).

3.5 Protocol for Delivery of siRNA-bsAb Complexes to Cells

1. For adherent cells: Seed cells into a flat-bottom 96-well plate to a final density of approximately 50–70 %. For non-adherent cells: Seed cells into a round-bottom 96-well plate to a final density of 3×10^5 cells per well (*see* **Notes 3** and **4**).
2. Add hapten-coupled SiRNA, or hapten-containing nanoparticles, to bsAb in cell culture medium at the molar ratio to occupy all hapten-binding sites (*see* **Note 5**).
3. In addition to cell surface antigen-specific SiRNA-bsAb complex, set up control reagents as follows: hapten-coupled SiRNA (nanoparticle) complexed with specific bsAb (which recognizes a surface antigen present on the cell type used), hapten-coupled SiRNA (nanoparticle) complexed with control bsAb (which recognizes an unrelated antigen), hapten-coupled control SiRNA (nanoparticle) (SiRNA targeting a gene not expressed in the cell type used) complexed with specific bsAb, hapten-coupled SiRNA (nanoparticle) without antibody and no reagent (cell culture medium only).
4. Incubate hapten-coupled SiRNA or hapten-decorated nanoparticle with bsAb at room temperature for 30 min.
5. Incubate cells with SiRNA-bsAb complexes or control reagents for 30 min at 37 °C (*see* **Note 6**).
6. Harvest cells at different time points after exposure to SiRNA or SiRNA complexes and analyze mRNA knockdown and specificity of knockdown.

3.6 Analysis of Specific Delivery: Protocol for FACS Analysis and Fluorescent Microscopy of siRNA-bsAb Complexes

Immediately following delivery of SiRNA-bsAb complexes to cells, targeted complexes become bound to the cell surface at significantly elevated levels in comparison to control bsAb or free payload. This can be verified by FACS analysis when using fluorescently labelled SiRNA derivatives. During the course of the next 24 h, payload becomes internalized along with the bsAb and accumulates in endosomal compartments. Within endosomes, SiRNA is separated from the bsAb. Intracellular trafficking and separation from the targeting vehicle can be monitored by fluorescent confocal microscopy when using fluorescently labeled SiRNA derivatives or nanoparticles as payloads (Fig. 4). Typically, SiRNA complexed with bsAb should accumulate on cell surfaces rapidly and become detectable in intracellular vesicular components within a couple of hours after application. During the course of 24 h, SiRNA should then separate from the bsAb.

1. For analysis of cell surface binding, incubate cells with fluorescently labeled SiRNA-bsAb complexes or control reagents (as described in the previous paragraph) for 1 h on ice (*see* **Note 7**).

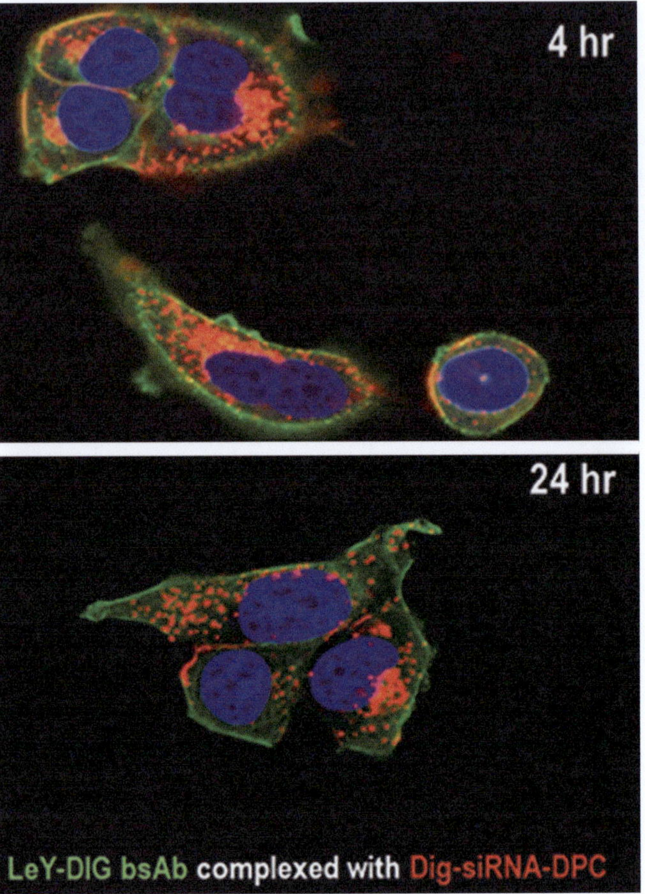

Fig. 4 Cellular targeting and internalization of LeY-Dig bsAb complexed with Dig-siRNA-DPC nanoparticles. LeY-Dig bsAb was complexed with Dig-Nu547-siRNA-DPCs; complexes were applied to MCF-7 cells for 30 min; cells were washed with DPBS and further incubated for 4 h (*top panel*) or 24 h (*bottom panel*). bsAb (*green signal*), Dig-Nu547-siRNA payload (Nu547; *red signal*), and nuclei (ToPro 3; *blue signal*) were visualized by laser scanning confocal microscopy. bsAb is initially co-localized with Dig-siRNA-DPC (resulting in *yellow-orange signal*) and is observed at the surface and within vesicular components (*top panel*). During the course of 24 h, payload accumulates within intracellular vesicles and separates from the antibody (*bottom panel*)

2. For analysis of internalization into cells (endosomes), incubate cells with fluorescently labeled SiRNA-bsAb complexes or control reagents for 30 min at 37 °C.
3. Remove unbound SiRNA-bsAb complexes by repeated washing and continue to incubate at 37 °C.
4. Track internalization and routing of the fluorescent SiRNA in time intervals for up to 24 h.
5. Fix cells with paraformaldehyde and analyze for SiRNA fluorescent label (*see* **Note 8**).

3.7 Analysis of siRNA-MediatedmRNA Knockdown: Quantitative PCR

The levels of target mRNAs in cells that were exposed to targeted delivery of SiRNA-bsAb complexes can be determined by any method of mRNA quantitation suitable for the gene of interest. We have successfully demonstrated knockdown of the Aha1 and CD31 mRNAs by using quantitative PCR as well as a branched DNA (bDNA) amplification assay (shown for Aha1 in Table 1). We therefore provide brief general protocols for these two assay systems.

1. Incubate cells with SiRNA-bsAb complexes or control reagents for targeted delivery as described above. Harvest cells after 24 h or determine the optimal time point of analysis experimentally.
2. Lyse cells, prepare mRNA and apply qPCR to quantitate mRNA levels of the target gene as well as of non-targeted reference genes. The ratio of target gene mRNA levels and those of reference genes (e.g., housekeeping genes such as GAPDH or Cyclophilin A) allows normalization between conditions and experiments.

3.8 Analysis of siRNA-Mediated mRNA Knockdown: Branched DNA Amplification Assay

The bDNA assay is particularly suited to determine the amount of a defined mRNA directly from cell lysates [44, 45].

1. Incubate cells with SiRNA-bsAb complexes or control reagents for targeted delivery as described above. Harvest cells and perform knock-down analyses after 24 h or determine the optimal time point of analysis.
2. Apply the protocol of the QuantiGene kit (Affymetrix) according to the manufacturer's instructions. In brief, transfer cell lysates to a capture plate in the presence of a gene-specific probe set, incubate at 53 °C overnight, wash, and subsequently incubate at 53 °C with an amplifier and an alkaline phosphatase-linked label probe (with washes between incubations).
3. After the final wash, add the luminescent alkaline phosphatase substrate dioxetane and incubate for 30 min at 53 °C. Detect luminescence with a suitable ELISA reader.

4 Notes

1. bsAbs are usually quite stable with a low propensity to aggregate. It is nevertheless advisable to store them in aliquots at −80 °C to avoid repeated freeze-thaw cycles.
2. For example, for 2 + 2 formats, apply SiRNA to bsAb at molar ratios of 1:10, 1:5, 1:4, 1:3, 1:2, 1:1, 1:0.5, 1:0, and 0:1.
3. The most effective concentration of complexes has to be determined experimentally. In our hands, using an IgG-based 2 + 2

bsAb targeting moiety, 25 nmol/l proved to be a good starting concentration.

4. Scale up appropriately if a larger sample size is desired.
5. Alternatively, determine the optimal bsAb to payload ratio experimentally.
6. Some background due to nonspecific attachment of hapten-coupled SiRNA or SiRNA-bsAb complexes to cells (e.g., with control samples) may occur in some systems and may prevent from adequate detection of bsAb and payload on the surface or within the cells. If this is observed, wash the cells in order to eliminate nonspecifically bound SiRNA-bsAb complexes and continue to incubate. However, this type of non-specific attachment is not expected to interfere with gene knockdown.
7. Cold incubation prevents from cellular internalization and enables surface staining by FACS or fluorescent microscopy (grow adherent cells on cover slips for fluorescent microscopy).
8. In order to visualize the bsAb, cells can be counterstained with a suitable secondary antibody specific for the human kappa light chain or the human IgG Fc domain (Fig. 4). Cells can be simultaneously stained to visualize F-actin (phalloidin) or nuclei (DAPI, shown in Fig. 4).

References

1. Fire A, Xu S, Montgomery MK et al (1998) Potent and specific genetic interference by double-stranded RNA in Caenorhabditis elegans. Nature 391:806–811
2. Kanasty R, Dorkin JR, Vegas A et al (2013) Delivery materials for siRNA therapeutics. Nat Mater 12:967–977
3. Akinc A, Querbes W, De S et al (2010) Targeted delivery of RNAi therapeutics with endogenous and exogenous ligand-based mechanisms. Mol Ther 18:1357–1364
4. Bhattarai SR, Muthuswamy E, Wani A et al (2010) Enhanced gene and siRNA delivery by polycation-modified mesoporous silica nanoparticles loaded with chloroquine. Pharm Res 27:2556–2568
5. Lee SK, Siefert A, Beloor J et al (2012) Cell-specific siRNA delivery by peptides and antibodies. Methods Enzymol 502:91–122
6. Leus NG, Talman EG, Ramana P et al (2014) Effective siRNA delivery to inflamed primary vascular endothelial cells by anti-E-selectin and anti-VCAM-1 PEGylated SAINT-based lipoplexes. Int J Pharm 459:40–50
7. Semple SC, Akinc A, Chen J et al (2010) Rational design of cationic lipids for siRNA delivery. Nat Biotechnol 28:172–176
8. Song E, Zhu P, Lee SK et al (2005) Antibody mediated in vivo delivery of small interfering RNAs via cell-surface receptors. Nat Biotechnol 23:709–717
9. Toloue MM, Ford LP (2011) Antibody targeted siRNA delivery. Methods Mol Biol 764:123–139
10. Yu B, Zhao X, Lee LJ et al (2009) Targeted delivery systems for oligonucleotide therapeutics. AAPS J 11:195–203
11. Zimmermann TS, Lee AC, Akinc A et al (2006) RNAi-mediated gene silencing in non-human primates. Nature 441:111–114
12. Beck A, Wurch T, Bailly C et al (2010) Strategies and challenges for the next generation of therapeutic antibodies. Nat Rev Immunol 10:345–352
13. Weidle UH, Tiefenthaler G, Weiss EH et al (2013) The intriguing options of multispecific antibody formats for treatment of cancer. Cancer Genomics Proteomics 10:1–18
14. Metz S, Haas AK, Daub K et al (2011) Bispecific digoxigenin-binding antibodies for targeted payload delivery. Proc Natl Acad Sci U S A 108:8194–8199
15. Schneider B, Grote M, John M et al (2012) Targeted siRNA Delivery and mRNA knock-

down mediated by bispecific digoxigenin-binding antibodies. Mol Ther Nucleic Acids 1:e46

16. Jung SH, Pastan I, Lee B (1994) Design of interchain disulfide bonds in the framework region of the Fv fragment of the monoclonal antibody B3. Proteins 19:35–47
17. Reiter Y, Brinkmann U, Jung SH et al (1995) Disulfide stabilization of antibody Fv: computer predictions and experimental evaluation. Protein Eng 8:1323–1331
18. Reiter Y, Brinkmann U, Lee B et al (1996) Engineering antibody Fv fragments for cancer detection and therapy: disulfide-stabilized Fv fragments. Nat Biotechnol 14:1239–1245
19. Ridgway JB, Presta LG, Carter P (1996) 'Knobs-into-holes' engineering of antibody CH3 domains for heavy chain heterodimerization. Protein Eng 9:617–621
20. Molina MA, Codony-Servat J, Albanell J et al (2001) Trastuzumab (herceptin), a humanized anti-Her2 receptor monoclonal antibody, inhibits basal and activated Her2 ectodomain cleavage in breast cancer cells. Cancer Res 61:4744–4749
21. Baselga J (2001) The EGFR as a target for anticancer therapy—focus on cetuximab. Eur J Cancer 37(Suppl 4):S16–S22
22. Kies MS, Harari PM (2002) Cetuximab (Imclone/Merck/Bristol-Myers Squibb). Curr Opin Investig Drugs 3:1092–1100
23. Chitnis MM, Yuen JS, Protheroe AS et al (2008) The type 1 insulin-like growth factor receptor pathway. Clin Cancer Res 14:6364–6370
24. Mansfield E, Amlot P, Pastan I et al (1997) Recombinant RFB4 immunotoxins exhibit potent cytotoxic activity for CD22-bearing cells and tumors. Blood 90:2020–2026
25. Brinkmann U, Pai LH, Fitzgerald DJ et al (1991) B3(Fv)-PE38KDEL, a single-chain immunotoxin that causes complete regression of a human carcinoma in mice. Proc Natl Acad Sci U S A 88:8616–8620
26. Pastan I, Lovelace ET, Gallo MG et al (1991) Characterization of monoclonal antibodies B1 and B3 that react with mucinous adenocarcinomas. Cancer Res 51:3781–3787
27. Soutschek J, Akinc A, Bramlage B et al (2004) Therapeutic silencing of an endogenous gene by systemic administration of modified siRNAs. Nature 432:173–178
28. Chan DP, Deleavey GF, Owen SC et al (2013) Click conjugated polymeric immuno-nanoparticles for targeted siRNA and antisense oligonucleotide delivery. Biomaterials 34:8408–8415
29. Jhaveri AM, Torchilin VP (2014) Multifunctional polymeric micelles for delivery of drugs and siRNA. Front Pharmacol 5:77. doi:10.3389/fphar.2014.00077
30. Malhotra M, Tomaro-Duchesneau C, Saha S et al (2013) Development and characterization of chitosan-PEG-TAT nanoparticles for the intracellular delivery of siRNA. Int J Nanomedicine 8:2041–2052
31. Miele E, Spinelli GP, Miele E et al (2012) Nanoparticle-based delivery of small interfering RNA: challenges for cancer therapy. Int J Nanomedicine 7:3637–3657
32. Rozema DB, Lewis DL, Wakefield DH et al (2007) Dynamic PolyConjugates for targeted in vivo delivery of siRNA to hepatocytes. Proc Natl Acad Sci U S A 104:12982–12987
33. Tiera MJ, Shi Q, Barbosa HF et al (2013) Polymeric systems as nanodevices for siRNA delivery. Curr Gene Ther 13:358–369
34. Wolff JA, Rozema DB (2008) Breaking the bonds: non-viral vectors become chemically dynamic. Mol Ther 16:8–15
35. Wong SC, Klein JJ, Hamilton HL et al (2012) Co-injection of a targeted, reversibly masked endosomolytic polymer dramatically improves the efficacy of cholesterol-conjugated small interfering RNAs in vivo. Nucleic Acid Ther 22:380–390
36. Akinc A, Goldberg M, Qin J et al (2009) Development of lipidoid-siRNA formulations for systemic delivery to the liver. Mol Ther 17:872–879
37. Sou K, Endo T, Takeoka S et al (2000) Poly(ethylene glycol)-modification of the phospholipid vesicles by using the spontaneous incorporation of poly(ethylene glycol)-lipid into the vesicles. Bioconjug Chem 11:372–379
38. Tao W, Davide JP, Cai M et al (2010) Noninvasive imaging of lipid nanoparticle-mediated systemic delivery of small-interfering RNA to the liver. Mol Ther 18:1657–1666
39. Robbins M, Judge A, MacLachlan I (2009) siRNA and innate immunity. Oligonucleotides 19:89–102
40. Grote M, Haas AK, Klein C et al (2012) Bispecific antibody derivatives based on full-length IgG formats. Methods Mol Biol 901:247–263
41. Haas AK, Mayer K, Brinkmann U (2012) Generation of fluorescent IgG fusion proteins in mammalian cells. Methods Mol Biol 901:265–276
42. Aigner A (2008) Cellular delivery in vivo of siRNA-based therapeutics. Curr Pharm Des 14:3603–3619

43. Leucuta SE (2013) Systemic and biophase bioavailability and pharmacokinetics of nanoparticulate drug delivery systems. Curr Drug Deliv 10:208–240
44. Burris TP, Pelton PD, Zhou L et al (1999) A novel method for analysis of nuclear receptor function at natural promoters: peroxisome proliferator-activated receptor gamma agonist actions on aP2 gene expression detected using branched DNA messenger RNA quantitation. Mol Endocrinol 13:410–417
45. Collins ML, Irvine B, Tyner D et al (1997) A branched DNA signal amplification assay for quantification of nucleic acid targets below 100 molecules/ml. Nucleic Acids Res 25: 2979–2984

Chapter 19

Stable Delivery of CCR5-Directed shRNA into Human Primary Peripheral Blood Mononuclear Cells and Hematopoietic Stem/Progenitor Cells via a Lentiviral Vector

Saki Shimizu, Swati Seth Yadav, and Dong Sung An

Abstract

RNAi is a powerful tool to achieve suppression of a specific gene expression and therefore it has tremendous potential for gene therapy applications. A number of vector systems have been developed to express short-hairpin RNAs (shRNAs) to produce siRNAs within mammalian T-cells, primary hematopoietic stem/progenitor cells (HSPC), human peripheral blood mononuclear cells, and in animal model systems. Among these, vectors based on lentivirus backbones have significantly transformed our ability to transfer shRNAs into nondividing cells, such as HSPC, resulting in high transduction efficiencies. However, delivery and long-term expression of shRNAs should be carefully optimized for efficient knock down of target gene without causing cytotoxicity in mammalian cells. Here, we describe our protocols for the development of shRNA against a major HIV co-receptor/chemokine receptor CCR5 and the use of lentiviral vectors for stable shRNA delivery and expression in primary human PBMC and HSPC.

Key words shRNA, Lentiviral vectors, Transfection, Transduction, 293-T cells, PBMC, CD34+ cells, CCR5

1 Introduction

Lentivirus vectors are powerful tool to efficiently transduce and express shRNAs in mammalian cells and to knock down specific target gene expression through RNA interference. Initially, shRNA was expressed from a transcriptionally strong U6 pol III promoter as an internal RNA polymerase promoter from a lentiviral vector. The U6 promoter was frequently used in order to effectively knock down a target mRNA. However, we and others recognized that continuous high level of shRNA over expression from the U6 promoter could cause cytotoxicity in vector transduced cells [1]. High level of shRNA expression might interfere the function of endogenous microRNA since maturation of shRNA to siRNA utilizes the

Kato Shum and John Rossi (eds.), *SiRNA Delivery Methods: Methods and Protocols*, Methods in Molecular Biology, vol. 1364, DOI 10.1007/978-1-4939-3112-5_19, © Springer Science+Business Media New York 2016

endogenous microRNA processing pathway and miRNAs plays a vital role in normal cellular functions. We learned from our experience of knocking down CCR5 expression in human peripheral blood mononuclear cells that a potent shRNA must be identified and expressed at lower level using transcriptionally weaker H1 RNA polymerase III promoter to stably knockdown CCR5 gene expression and to avoid cytotoxic effects [2]. Here, we describe our detailed protocol of a lentiviral vector construction for stable shRNA expression in human PBMC and CD34+ HSPC in vitro.

2 Materials

2.1 Construction of shRNA-Expressing Lentivirus Vector

1. pBS-hH1-3 plasmid DNA: The pBluescript II SK (−) plasmid DNA that contains a human H1 RNA polymerase III promoter, a RNA polymerase III promoter termination signal (5Ts), and restriction enzyme sites for a shRNA cloning (Fig. 1a, *see* **Note 1**). The plasmid DNA sequence data are available upon request.
2. FG12 HIV-1-based lentiviral vector plasmid DNA that contains restriction enzyme sites for an H1 promoter shRNA unit cloning. The plasmid DNA sequence data are available upon request (Fig. 2, *see* **Note 2**).
3. Restriction enzymes: Afe I enzyme, BamH I enzyme, Xba I enzyme, Xho I enzyme.

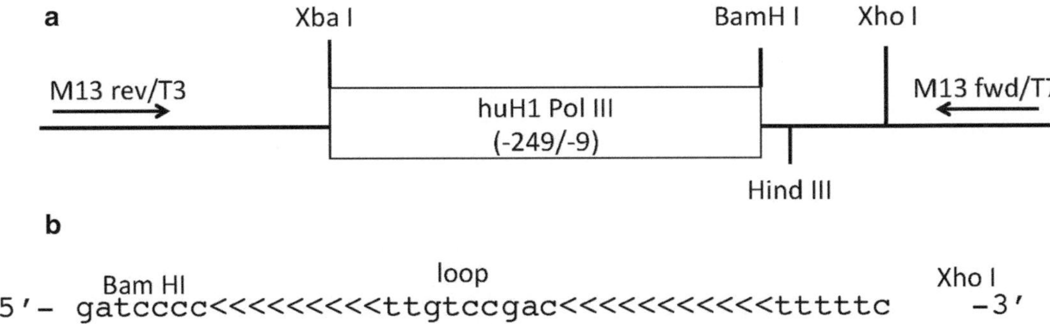

Fig. 1 (**a**) The map of pBS-hH1-3 plasmid DNA. A DNA fragment of human H1-RNA polymerase III promoter (−241/−9, amplified from HEK293T genomic DNA by PCR) was cloned between Xba I and Hind III sites of pBS-SKII(−) (Stratagene). BamH I and Xho I are designed for cloning oligo DNAs for a shRNA. XbaI and XhoI sites are designed for subcloning the shRNA expression cassette into FG12 lentivirus vector. huH1 Pol III: human H1 RNA polymerase III promoter. M13 rev/T3 and M13 fwd/T7: sequence primer-binding sites. (**b**) The design of siRNA oligo DNAs. A sense and antisense DNA including BamH1 restriction enzyme site at the 5′ prime, a 19–21mer antisense strand of the shRNA targeting sequence <<<<<<<<<, a loop, a corresponding reverse complementary sense sequence <<<<<<<<<, a RNA polymerase III promoter termination signal (ttttt) and a XhoI restriction enzyme site are required in the siRNA oligo DNAs

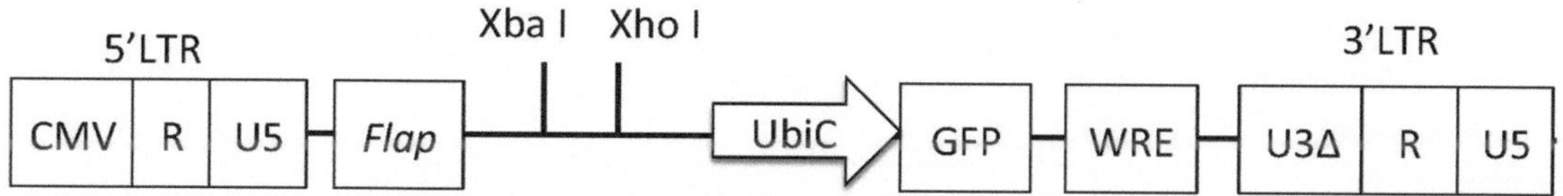

Fig. 2 The map of FG12 lentivirus vector. FG12 lentivirus vector is derived from FGUW lentivirus vector with modified multiple restriction sites. A shRNA expression cassettes cloned into pBS-hH1-3 plasmid DNA can be subcloned at XbaI and XhoI sites in FG12 vector. *LTR* long terminal repeat. *CMV* CMV promoter. *R* repeat sequence in HIV-1 LTR: Flap: UbiC: Human Ubiquitin C promoter. *GFP* green fluorescent protein. *WRE* woodchuck hepatitis virus posttranscriptional regulatory element. *U3Δ* nucleotide deletions in the U3 region for self-inactivation

4. Enzyme buffer 4.
5. TE buffer: 10 mM Tris-Cl, pH 7.5, 1 mM EDTA.
6. T4 DNA Ligase.
7. 10× T4 DNA ligase buffer.
8. DNA gel extraction kit.
9. DNA mini prep kit.
10. Electroporation cuvettes (2 mm gap).
11. XL-1 blue competent cells.
12. 2× YT medium.
13. 2× YT medium with 50 μg/mL ampicillin.
14. LB agar plate containing 50 μg /mL ampicillin.
15. Sterilized toothpicks (*see* **Note 3**).
16. Sequencing primers: M13 rev/ M13 fwd.

2.2 Transfection of 293-T Cells for Production of VSV-G Pseudotyped Lentiviral Vectors

1. 293-T cell line.
2. Culture medium: Iscove's modified Dulbecco's medium (IMDM) with 8 % Bovine serum, 2 % fetal calf serum, and 1 % glutamine/penicillin/streptomycin.
3. 10 mM Chloroquine.
4. 2 M Calcium chloride solution.
5. 2× HBS, pH 7.0: Dissolve 8 g NaCl, 0.2 g $Na_2HPO_4{\cdot}7H_2O$, 6.5 g HEPES in deionized water to a total volume of 500 ml.
6. phHCMV-G plasmid DNA for VSV-G expression.
7. pMDLg_pRRE plasmid DNA for HIV-1 gag and pol expression.
8. pRSV REV plasmid DNA for HIV-1 rev expression.
9. FG12 plasmid DNA for a HIV-1-based lentiviral vector.

2.3 Concentration of VSV-G Pseudotyped Lentiviral Vector, Resuspension, and Storage

1. Steriflip Filter Units; pore size 0.22 μm.
2. 1× Hanks' Balanced Salt Solution (HBSS).
3. 50 % Sucrose buffer: 125 g Sucrose dissolved in 245 ml HBSS supplemented with 10 mM EDTA.
4. 10 % Sucrose buffer: 50 % Sucrose buffer diluted five times with HBSS such that the final working concentration is 10 % sucrose, 2 mM EDTA.
5. Ultraclear centrifuge tubes.
6. SW32 Ti rotor with swinging bucket in Beckman Coulter Optima L-90k Ultracentrifuge.

2.4 Titration of shRNA Lentivirus Vector

1. 293-T cell line.
2. 2/8 IMDM: Iscove's modified Dulbecco's medium (IMDM) supplemented with 8 % bovine serum (BS) and 2 % fetal calf serum (FCS) and 1 % glutamine/penicillin/streptomycin (GPS).
3. 1 mg/ml Polybrene: Hexadimethrine bromide dissolved in double-distilled H_2O.
4. FACS buffer: 2 % FCS in 1× Phosphate buffer saline (PBS).
5. 10F RPMI: RPMI-1640 supplemented with 10 % FCS and 1 % GPS.
6. Trypsin-EDTA solution: 2.5 % Trypsin and 0.5 M ethylenediaminetetraacetic acid (EDTA).
7. Fix Buffer: 1 % Formaldehyde in 1× PBS.
8. Concentrated VSV-G pseudotyped lentiviral vector stocks.

2.5 Transduction of T-Cell Line with Lentiviral Vector Carrying shRNA

1. CCR5 expressing Molt-4/CCR5 T-cell line.
2. 1 mg/ml Polybrene (hexadimethrine bromide dissolved in double-distilled H_2O).
3. 10F RPMI: RPMI-1640 supplemented with 10 % FCS and 1 % glutamine/penicillin/streptomycin (GPS).
4. Titrated VSV-G pseudotyped lentiviral vector stocks.
5. 1× PBS.
6. FACS buffer: 2 % FCS in 1× PBS.
7. Fix buffer: 1 % Formaldehyde in 1× PBS.

2.6 Transduction of PBMC with Lentiviral Vector Carrying shRNA

1. Peripheral blood mononuclear cells (PBMC) can be isolated from whole blood from healthy donors. The PBMC can be isolated using Ficoll by centrifugation.
2. Anti-human CD8 mAb magnetic beads.
3. Magnetic box.
4. FACS buffer: 2 % FCS in 1× PBS.
5. 1 mg/ml Polybrene (hexadimethrine bromide dissolved in double-distilled H_2O).

6. 5 μg/ml PHA.
7. 200 units/ml Interleukin-2 human hIL-2 (IL-2).
8. 10F RPMI: RPMI-1640 supplemented with 10 % FCS and 1 % glutamine/penicillin/streptomycin (GPS).
9. 20F RPMI: RPMI-1640 supplemented with 20 % FCS and 1 % glutamine/penicillin/streptomycin (GPS).
10. Growth medium: 10 units/ml IL-2 in 20F RPMI.
11. Titrated lentiviral vector.
12. Fix buffer: 1 % Formaldehyde in 1× PBS.

2.7 Transduction of Human Fetal Liver-Derived CD34+ Cells with Lentiviral Vector Carrying shRNA

1. Human fetal liver for the source of CD34+ cells. The fetal liver can be provided from the UCLA CFAR Gene and Cellular Core laboratory or purchased through Advanced Bioscience Resources, Inc, Alameda, California.
2. CD34 microbead kit.
3. Cytokines: IL-6, IL-3, SCF.
4. 1 mg/ml Recombinant Fibronectin (RetroNectin) stock solution.
5. FACS Buffer: 2 % FCS in 1× PBS.
6. 10F RPMI: RPMI-1640 supplemented with 10 % FCS and 1 % glutamine/penicillin/streptomycin (GPS).
7. 20F RPMI: RPMI-1640 supplemented with 20 % FCS and 1 % glutamine/penicillin/streptomycin (GPS).
8. Infection medium: Yssel's Serum-Free T-cell Medium supplemented with 2 % bovine serum albumin (BSA), 2.5 μg/ml Fungizone and 0.45 mg/ml Piperacillin and Tazobactram (PT, also known as Zocyn).
9. Cytokine medium: 20F RPMI supplemented with 50 ng/ml IL-6, 50 ng/ml IL-3, 50 ng/ml SCF.
8. Titrated VSV-G pseudotyped lentiviral vector stocks.
10. Fix buffer: 1 % Formaldehyde in 1× PBS.

2.8 CCR5 Detection in Humanized Mouse Blood

1. 10× Red blood cell lysis buffer: 82.9 g of Ammonium Chloride, 10.0 g of potassium bicarbonate, 0.37 g of ethylenediamine tetraacetic acid (EDTA) disodium salt in 1L water (*see* **Note 4**).
2. 1× Red blood cell lysis buffer (*see* **Note 5**).
3. FACS buffer: 2 % FCS in 1× PBS.
4. Fix buffer: 1 % formaldehyde in 1× PBS.
5. Master Mix1: 0.5 μl of CD45-eFluore 450, 0.5 μl of CD3-APCH7, 0.5 μl of CD4-APC, 0.5 μl of CD8-PerCPCy5.5, 0.5 μl of CD19-V500, 0.5 μl of CCR5-PC7 in 100 μl of FACS buffer (*see* **Note 6**).

3 Methods

3.1 Construction of shRNA Expressing Lentivirus Vector

1. Synthesize a forward and a reverse oligonucleotide DNAs for a shRNA based on the design described in Fig. 1b.
2. Dilute the forward and reverse oligonucleotides in TE at a concentration of 10 pmol/μl each. Mix each 10 μl of forward and reverse oligo DNA into 0.5 mL tube and incubate at 95 °C for 10 min and slowly cool down to room temperature.
3. Digest 1 μg of pBS-hH1.3 with 1 U of BamH I and Xho I at 37 °C for 1 h in NEBuffer4. Run the product on 1 % agarose gel, cut the 3130 bp DNA fragment and purify using the gel extraction kit.
4. Incubate purified vector and 7 μl annealed oligos with T4 DNA ligase at 16 °C for 1 h in 10 μl of 1× T4 DNA polymerase buffer (*see* **Note 7**).
5. Transform competent cells with the product by electroporation. Thaw competent cells on ice. Chill an electroporation cuvette and ligation sample on ice. Add 1 μl of ligation sample into 55 μl of competent cells and transfer into electroporation cuvette gently, and apply one pulse.
6. Immediately add 250 μl of 2× YT medium (*see* **Note 8**).
7. Plate all of them into LB agar plate containing 50 μg/mL ampicillin using a spreader.
8. Incubate the plate at 37 °C overnight.
9. Pick 5–10 single colonies with toothpick and culture in 2 mL of 2× YT containing 50 μg/mL ampicillin medium overnight with constant shaking at 200 rpm (*see* **Note 9**).
10. Carry out Mini prep following the manufacturer's instructions.
11. Confirm sequence using M13 fwd/M13 rev primers.
12. Insert: Digest 1 μg of pBS-hH1-3 plasmid DNA with 1 U of Xho I and Xba I at 37 °C for 1 h in NEBuffer4 then, run the product on 1 % agarose gel, cut the 294 bp band and purify using gel extraction kit.
13. Vector: Digest 1 μg of FG12 vector with 1 U of Xho I and Xba I at 37 °C for 1 h in NEBuffer4. Run the product on 1 % agarose gel, isolate the 9865 bp band and purify using gel extraction kit.
14. Incubate purified vector (at least 50 ng) and insert with 1 μl of T4 DNA Ligase at 16 °C for 1 h in 10 μl of 1× T4 DNA polymerase buffer.
15. Transform competent cells with the product by electroporation. Thaw competent cells on ice. Chill an electroporation cuvette and ligation sample on ice. Add 1 μl of ligation sample into 55 μl of competent cells and transfer into electroporation cuvette gently, and apply one pulse.

16. Immediately add 250 μl of 2× YT medium into the cells.
17. Plate all of them into LB agar plate containing 50 μg/mL ampicillin using a spreader.
18. Incubate the plate at 37 °C overnight.
19. Pick 5–10 single colonies with toothpick and culture in 2 mL of 2× YT medium containing 50 μg/mL ampicillin overnight with constant shaking at 200 rpm.
20. Carry out Mini prep following the manufacturer's instructions.
21. Digest 5 μl of miniprep sample with 1 U of Afe I and Xba I enzyme in NEBuffer4 and confirm ~380 bp insert (*see* **Note 10**).

3.2 Transfection of 293-T Cells for Production of VSV-G Pseudotyped Lentiviral Vectors

1. Culture 15×10^6 293-T cells into T-175 flasks in 25 ml culture medium 1 day before.
2. On the day of transfection, change the medium into chloroquine medium (40 μl of 10 mM chloroquine + 10 ml culture medium).
3. Make a master mix of the viral DNA in a screw cap tube (viral DNA + 2 M calcium chloride + ddH_2O such that the total volume is 1100 μl.
4. Add 10 μg of vector DNA to the master mix. This is solution A.
5. Incubate solution A on ice for 5 min.
6. For solution B, add 1110 μl 2× HBS into a 50 ml conical tube.
7. Adding solution A to solution B: Using a p200 micropipette, Add solution A dropwise by the sides of the tube containing solution B. Shake tube to mix drops. The mixture will collect on the bottom of the tube. This is solution C.
8. Incubate solution C on ice for 20 min.
9. After 20 min, add solution C dropwise to the T-175 flask containing cells. Solution C should be added to the culture media in the flask and should not touch the cells.
10. Slowly swirl the media in the flask to mix the solution C.
11. Slowly lay the flask flat so that the media touched the cells.
12. Incubate for 8 h at 37 °C, 5 % CO_2.
13. After 8 h, slowly remove all the media in the flask and add 35 ml of culture media.
14. Incubate at 37 °C, 5 % CO_2.

3.3 Concentration of Vector, Resuspension, and Storage

1. Observe infected 293-T cells under the microscope for infection by checking for GFP expression.
2. Harvest supernatant in a 50 ml tube and filter the supernatant using the Steriflip filter unit.

3. Concentrate VSV-G-psuedotyped virions from the supernatant by ultracentrifugation. Transfer the filtrate (containing the vector) into an ultracentrifuge tube (Ultraclear 38.5 ml Beckman Coulter).
4. Slowly add 5 ml of 10 % sucrose buffer to the tubes by touching the pipette all the way to the bottom of the ultracentrifuge tube and releasing the sucrose buffer slowly. This will create a sucrose cushion and will help avoid the coprecipitation of unwanted particles, resulting in vector of significantly higher purity.
5. Use serum-free medium to balance the tubes to within 30 mg, and then load in an SW32 rotor.
6. Centrifuge at 67214 × *g* at 4 °C for 90 min. The brake must be inactivated to prevent disturbing the viral pellet during deceleration.
7. Discard the supernatant by inversion and leave the ultracentrifuge tube inverted for 90 s on absorbent paper towel.
8. Resuspend the viral pellet in 100 μl of HBSS and seal the ultracentrifuge tubes with parafilm. Store the tubes overnight at 4 °C.
9. Following overnight storage at 4 °C, carefully mix the vector by vigorous pipetting, and then store at −80 °C in small aliquots.

3.4 Titration of Lentivirus Vector Stocks

1. Plate 293-T cells at 0.5×10^5 cells/500 μl in 24-well plate 1 day before the infection.
2. Prepare titration medium containing 2/8 IMDM + 8 μl/ml polybrene.
3. Thaw an aliquot of the vector on ice, and prepare a serial dilution of vector with the titration media (*see* **Note 11**).
4. Dilution 1:1/300 diluted vector:1 μl of concentrated vector + 300 μl of titration media.
5. Dilution 2:1/3000 diluted vector:30 μl of 1/300 diluted vector + 270 μl of titration media.
6. Dilution 3:1/30,000 diluted vector:30 μl of 1/3000 diluted vector + 270 μl of titration media.
7. Remove media from plated 293-T cells and add 250 μl of the serially diluted vector to corresponding wells (*see* **Note 12**).
8. Incubate the infected cells for 2 h at 37 °C, 5 % CO_2, incubator.
9. After 2 h, remove the supernatant and add 1 ml of 2/8 IMDM to each well.

10. Three days post-infection, remove all the supernatant from infected 293-T cells and wash cells with 1× PBS.
11. Add 100 μl of trypsin-EDTA solution to each well and incubate for 2 min at 37 °C, 5 % CO_2 incubator.
12. Observe under the microscope to see if all the cells have detached from the well surface. If not then keep for an additional 2 min in the incubator.
13. Add 1 ml of 2/8 IMDM to wells to block trypsin-EDTA action and flush cells with the help of micropipette.
14. Count the cells and take 0.1×10^6 cells in 1.5 ml screw-cap tubes.
15. Centrifuge at 3500 × *g* for 1 min and aspirate supernatant.
16. Add 500 μl of fix buffer to each tube and transfer to FACS tubes.
17. Acquire samples by fluorescence-activated cell sorting (FACS) in order to measure % EGFP-positive cells.
18. Based on % EGFP-positive cells, calculate the titer of the vector according to the formula indicated below.
    (a) Select the dilution which shows closest to tenfold increase in % EGFP-positive cells as compared to previous dilution. For example: %EGFP+ cells with 1/30,000 dilution = 1.02; %EGFP+ cells with 1/3000 dilution = 10.8; %EGFP+ cells with 1/300 dilution = 63.7. Therefore, we will select 1/3000 as the dilution to calculate titer.
    (b) Formula for calculating titer units (TU/ml): [%EGFP+ cells/100] × [number of cells] × [dilution factor] × [1000/volume of vector (ml)], wherein the number of cells refers to the number of cells taken in screw caps tubes after harvesting the cells. For example: %EGFP+ cells = 10.8, number of cells = 0.1×10^6 cells, dilution factor = 3000, volume of vector = 250 μl. Therefore, $[10.8/100] \times [0.1 \times 10^6] \times [3000] \times [1000/250] = 1.296 \times 10^8$ TU/ml.
    (c) Calculation of amount of vector needed to reach a certain Multiplicity of Infection (MOI).

After titrating the vector, in order to perform immune cell transductions, it is important to calculate the amount of vector that will need to be added to the cell cultures to infect them at a certain MOI. The formula of calculation the amount of vector for a certain MOI is indicated below, wherein the total number of cells refers to the number of cells seeded in each well before infected them. Total plaque-forming units (PFU) = [Total number of cells] × [desired MOI], followed by volume of vector needed to reach desired MOI (μl) = [Total PFU] × [TU/ml].

3.5 Transduction of T-Cell Line with Lentiviral Vector Carrying shRNA

1. Aliquot 0.1×10^6 cells into sterile screw cap tubes.
2. Centrifuge the cells at $3,500 \times g$ for 1 min and carefully remove the supernatant.
3. Prepare 250 μl of polybrene/vector solution (248 μl of 10F RPMI + 2 μl of polybrene + calculated amount of vector). The final concentration of polybrene should be 8 μg/ml (*see* **Note 13**).
4. Add 250 μl of the polybrene/vector solution to each tube.
5. Loosen the caps of the tubes and incubate for 2 h at 37 °C, 5 % CO_2.
6. After 2 h, add 1 ml of 10F RPMI to the tubes.
7. Centrifuge the cells at $3,500 \times g$ for 1 min and carefully remove the supernatant.
8. Resuspend the cells in 1 ml 10F RPMI and plate cells in a 12-well plate.
9. Incubate at 37 °C, 5 % CO_2 for 3 days. Transgene expression can be assessed in 72 h.
10. After 72 h, collect and count the cells. Centrifuge the cells at $3,500 \times g$ for 1 min at room temperature.
11. Resuspend cells in 10F RPMI at 0.1×10^6 cells/well.
12. Centrifuge cells at $3,500 \times g$ for 1 min.
13. Add 300 μl Fix buffer and acquire by flow cytometer and check for % EGFP.

3.6 Transduction of PBMC with Lentiviral Vector Carrying shRNA

1. We deplete CD8+ cells from PBMC for investigation of lentiviral vector transduction and CCR5 knock down in CD4+ cells. For every 10×10^6 PBMC, add 70 μl of anti-human CD8 mAb magnetic beads into a 15 ml.
2. Add 7 ml of FACS buffer. Place the 15 ml tube in the magnetic box and wait for 3 min until the red magnet beads attach to the side of the tube.
3. Remove the supernatant by decanting. Repeat the washing (**step 2**).
4. Resuspend the beads in 400 μl cold FACS buffer for every 10×10^6 cells and keep on ice.
5. Aliquot 10×10^6 PBMC in a 15 ml tube and centrifuge at $453 \times g$ for 5 min. Aspirate the supernatant.
6. Mix the PBMC and washed anti-human CD8 mAb magnetic beads.
7. Incubate on ice for a total of 30 min, mixing every 10 min.
8. Add 7 ml of cold FACS buffer.
9. Place the 15 ml tube containing the cells and the beads in the magnet box and wait for 3 min until the red magnet beads attach to the side of the tube.

10. Decant supernatant (containing the cells of interest) into a fresh 15 ml conical tube.
11. Check percentage and purity of CD8+ depleted cells by staining 0.1×10^6 cells with antibodies against CD4, CD8, CD27, and CD45RA.
12. Spin down the CD8+ depleted cells and resuspend the cells in 20F RPMI. Count the cells.
13. Seed 2×10^6 CD8+ depleted cells per ml and activate the cells with PHA (Final concentration should be 2.5 μg/ml) in 20F RPMI for 3 days.
14. Post-activation, wash the CD8+ depleted cells with warm 10F RPMI and resuspend the cells in growth medium in a 96-well plate at 0.4×10^6 cells/100 μl/well.
15. Incubate cells for 4 h at 37 °C, 5 % CO_2.
16. Prepare 50 μl of polybrene/vector solution (48.8 μl of 10F RPMI + 1.2 μl of polybrene + calculated amount of vector). The final concentration of polybrene should be 8 μg/ml.
17. After 4 h, add the polybrene/vector solution to the cells and mix well by pipetting.
18. Incubate cells for 2 h at 37 °C, 5 % CO_2 incubator.
19. After 2 h, remove 140 μl of the culture medium carefully. Try not to disturb the cells.
20. Add 250 μl of growth medium and Incubate cells overnight at 37 °C, 5 % CO_2.
21. Incubate cells overnight at 37 °C, 5 % CO_2.
22. Collect all cells per well. Make sure that there are no cells left in the well by observing the wells under the microscope.
23. Wash the cells with 1 ml of pre-warmed 10F RPMI (optional).
24. Transfer cells to a 24-well plate and add 1 ml of growth medium into each well.
25. Incubate at 37 °C, 5 % CO_2. Transgene expression can be assessed within 48–72 h.
26. After 72 h, collect and count the cells. Centrifuge the cells at $3{,}500 \times g$ for 1 min at room temperature.
27. Resuspend cells in growth media at 1×10^6 cells/ml.
28. Stain remaining cell from each sample with 50 μl of FACS buffer + 1 ml of antibody against CD4, CD8, CD27 and CD45RA for 20 min at room temperature in the dark.
29. Add 1 ml 1× PBS and centrifuge at $3{,}500 \times g$ for 1 min.
30. Add 300 μl fix buffer and acquire cell events by flow cytometer and assess EGFP expression.

3.7 Transduction of CD34+ Cells with Lentiviral Vector Carrying shRNA

1. Prepare 20 μg/ml RetroNectin-PBS solution (working solution) for coating the 24-well plate.
2. Dispense 500 μl of RetroNectin-PBS solution into each well. Incubate for 2 h at room temperature in the dark.
3. Remove the RetroNectin-PBS solution. This can be reused if stored at 4 °C.
4. Dispense 500 μl of FACS buffer into each well for blocking. Incubate for 30 min at room temperature in the dark.
5. Aspirate FACS buffer and wash wells once with 500 μl 1× PBS. Do not remove PBS from wells.
6. Quick thaw frozen CD34+ cells into 5 ml pre-warmed infection medium.
7. Centrifuge the CD34+ cells at 314 × *g* for 5 min.
8. Resuspend cell pellet in 1 ml infection medium.
9. Count the cells. Bring concentration of cells to 1×10^6 cells/ml.
10. Remove PBS from the wells. CD34+ cells should be cultured at 0.25×10^6 cells per well; therefore add 250 μl of cells to each well.
11. Incubate cells for 1 h at 37 °C, 5 % CO_2, incubator.
12. Add 250 μl infection medium to each well such that there is 500 μl medium in each well.
13. A MOI of 0.3 is generally used to obtain transduction efficiencies of >50 %. CD34+ cells does not require the addition of polybrene and is not augmented by the presence of RetroNectin. Add calculated amount of vector to the designated well.
14. Incubate cells overnight at 37 °C, 5 % CO_2.
15. The next day, add 800–1000 μl of 10F RPMI to each well.
16. Collect all cells per well. Make sure that there are no cells left in the well by observing the well under the microscope.
17. Centrifuge tubes at 453 × *g* for 5 min at room temperature.
18. Resuspend cell pellet again in 800–1000 μL OF 10F RPMI. Repeat **step 17**.
19. Resuspend the cell pellet in 1 ml of cytokine medium. Transfer cells into a 24 well plate.
20. Incubate at 37 °C, 5 % CO_2. Transgene expression can be assessed within 48 h (*see* **Note 14**).
21. Collect and count the cells. Centrifuge the cells at 453 × *g* for 5 min at room temperature.
22. Resuspend cells in cytokine media at 0.1×10^6 cells/well.
23. Centrifuge remaining cell from each sample at 3,500 × *g* for 1 min.
24. Add 300 μl fix buffer and acquire cell events by flow cytometer and assess EGFP expression.

3.8 Detection of shRNA Expression by shRNA-Specific Real-Time RT-PCR

Real-time stem-loop RT-PCR method can be used to quantify the levels of the antisense strand of siRNA. For example, we successfully quantified antisense strand of siRNA directed to rhesus CCR5 [3].

1. Isolate 500 ng of total RNA from rhesus PBLs using TRIzol reagent (Invitrogen).
2. Subject the RNA to reverse transcription (RT) reaction (16 °C, 30 min; 42 °C, 30 min; 85 °C, 5 min).
3. Subject RT product to PCR (95 °C, 10 min, 1 cycle; 95 °C, 15 s; 58 °C, 1 min, 50 cycles). Primer and probe sequences used for rhesus CCR5siRNA were as follows: reverse transcription stem-loop primer, 5′-GTCGTATCCAGTGCAGGGTCCGA GGTATTCGCACTGGATACGACAAGAGCAA-3′; forward primer, 5′-GCGCGGTGTAAACTGAAC-3′; reverse primer, 5′-GTGCAGGGTCCGAGGT-3′; probe, 6-FAM-TGGATA CGACAAGAGCAA-MGB. A set of serially diluted synthetic 22-nt antisense strand of siRNA against rhCCR5 (GGU GUA AAC UGA ACU UGC UC; Sigma-Proligo) was used as standard for quantification [3].

4 Notes

1. Human H1-RNA polymerase III promoter DNA (−241/−9) is amplified from HEK293T genomic DNA and cloned between Xba I and Hind III site of pBS-SKII(−) (Fig. 1a).
2. FG12 lentiviral vector is constructed based on FUGW lentiviral vector [4, 5] with useful restriction enzyme sites (XbaI and XhoI) for shRNA cloning.
3. Sterilized pipette tips can be used instead of toothpick.
4. 10× Red lysis buffer is filtered with 45 μm filter and stored at 4 °C.
5. 1× red blood cell lysis buffer should be used at room temperature. Dilute 10× RBC buffer with ddH_2O and filter. Store at 4 °C.
6. We use clone 2D7 for anti-human CCR5 antibody for CCR5 cell surface detection by flow cytometry.
7. Make small aliquots (~10 μl) for T4 DNA ligase buffer to avoid repeated freeze and thaw cycles to preserve ATP.
8. S.O.C. can be used as a substitute to 2× YT medium.
9. Sometime lentiviral vector shows recombination during bacteria amplification; therefore bacteria should be cultured low speed (~100 rpm) and constant shaking.
10. For vector production, Midi prep sample is better than mini prep sample as it gives a high titer vector.

11. Though titers can be assessed using various techniques, titers based on transgene expression that can be monitored by flow cytometry are the simplest. It is best to evaluate titers at low dilutions or where the percentage of transduced cells is <30 %. This will help avoid issues of multiple integrations as at lower dilutions, the number of integrated copies is generally one.
12. 293-T cells can easily detach from the well surface. Media must always be carefully removed and cells should never be allowed to dry up.
13. One should avoid picking up very small volumes with a micropipette as there are high chances of pipetting errors. If the volume of the vector solution is too small, dilute the vector 1:10 in 1× PBS so that a sufficient amount can be picked up by micropipettes. For the diluted vector, add ten times the calculated volume of the vector to the cells.
14. Transduction efficiencies can vary significantly with cell types. Molt4CCR5 T cell line show 30–40 % GFP+ at 3 days after transduction at M.O.I. = 0.1. However, at the same M.O.I in primary cells (PBMC or FL derived CD34+ cells), transduction efficiency becomes almost half. It is hard to predict transduction efficiency in primary cells which have more variation to susceptibility in transduction. You should also consider donor to donor variation in primary cells lines. The timing at which transgene expression and transduction efficiency is critical to access vector transduction efficiency. Transgene expression from lentiviral vector requires more than 48 h after transduction.

References

1. An DS, Qin FX, Auyeung VC, Mao SH, Kung SK, Baltimore D, Chen IS (2006) Optimization and functional effects of stable short hairpin RNA expression in primary human lymphocytes via lentiviral vectors. Mol Ther 14:494–504
2. Shimizu S, Hong P, Arumugam B, Pokomo L, Boyer J, Koizumi N, Kittipongdaja P, Chen A, Bristol G, Galic Z, Zack JA, Yang O, Chen IS, Lee B, An DS (2010) A highly efficient short hairpin RNA potently down-regulates CCR5 expression in systemic lymphoid organs in the hu-BLT mouse model. Blood 115:1534–1544
3. An DS, Donahue RE, Kamata M, Poon B, Metzger M, Mao SH, Bonifacino A, Krouse AE, Darlix JL, Baltimore D, Qin FX, Chen IS (2007) Stable reduction of CCR5 by RNAi through hematopoietic stem cell transplant in non-human primates. Proc Natl Acad Sci U S A 104:13110–13115
4. Lois C, Hong EJ, Pease S, Brown EJ, Baltimore D (2002) Germline transmission and tissue-specific expression of transgenes delivered by lentiviral vectors. Science 295:868–872
5. Qin XF, An DS, Chen IS, Baltimore D (2003) Inhibiting HIV-1 infection in human T cells by lentiviral-mediated delivery of small interfering RNA against CCR5. Proc Natl Acad Sci U S A 100:183–188

Chapter 20

Hepatic Delivery of Artificial Micro RNAs Using Helper-Dependent Adenoviral Vectors

Carol Crowther, Betty Mowa, and Patrick Arbuthnot

Abstract

The potential of RNA interference (RNAi)-based gene therapy has been demonstrated in many studies. However, clinical application of this technology has been hampered by a paucity of efficient and safe methods of delivering the RNAi activators. Prolonged transgene expression and improved safety of helper-dependent adenoviral vectors (HD AdVs) makes them well suited to delivery of engineered artificial intermediates of the RNAi pathway. Also, AdVs' natural hepatotropism makes them potentially useful for liver-targeted gene delivery. HD AdVs may be used for efficient delivery of cassettes encoding short hairpin RNAs and artificial primary microRNAs to the mouse liver. Methods for the characterization of HD AdV-mediated delivery of hepatitis B virus-targeting RNAi activators are described here.

Key words RNA interference, microRNA, Adenoviral vectors, Northern blot, Q-PCR, Immunocytochemistry, X-gal staining

1 Introduction

Manipulation of the eukaryotic RNA interference (RNAi) pathway for therapeutic purposes has been extensively explored and demonstrates potential clinical applications [1–3]. The well-characterized endogenous RNAi pathway in humans uses microRNAs (miRs) to bind mRNA to cause target degradation or inhibition of translation (reviewed in [4]). Exploitation of this pathway involves use of either synthetic small interfering RNAs (siRNA), expressed short hairpin RNA (shRNA) or artificial miR (amiR) sequences that are designed to silence targets of interest. When introduced into eukaryotic cells, these sequences reprogram the endogenous RNAi pathway and achieve posttranscriptional inhibition of intended targets [1, 2, 5]. Typically, for in vivo application, synthetic RNAi effectors are delivered with non-viral vectors (NVVs), and expressed RNAi activators are delivered using viral vectors (VVs). VVs, such as recombinant adenoviral (Ad) vectors (AdVs), are highly efficient and are

Kato Shum and John Rossi (eds.), *SiRNA Delivery Methods: Methods and Protocols*, Methods in Molecular Biology, vol. 1364, DOI 10.1007/978-1-4939-3112-5_20,

therefore suited to many applications that include delivery of expressed hepatitis B virus (HBV) targeting RNAi activators [6, 7].

Ads are non-enveloped and the virions comprise icosahedral capsid particles with diameters of 70–90 nm. There are at least 57 known human Ad serotypes that belong to species A-G. Derivatives of human serotypes 5 (Ad5) and 2 (Ad2) from subgroup C are the best characterized, and have commonly been used as gene therapy vectors [8]. Serotype 5 Ad has a linear genome of 35 kbp, which is flanked by *cis*-acting inverted terminal repeats (ITRs) and contains a packaging signal (Ψ). There are two sets of genes: the early set (E1A, E1B, E2, E3, and E4) which is expressed before DNA replication and the late set (L1–L5) which is transcribed at high levels following initiation of DNA replication (reviewed in [9, 10]).

A major advantage of using AdVs for gene therapy to treat HBV infection is that the vectors are naturally hepatotropic [11]. Although the Coxsackievirus and adenovirus receptor (CAR) is an important primary attachment receptor for AdVs in vitro, interaction of AdVs with blood clotting factor X (FX) plays a significant role in hepatocyte transduction by these vectors. FX binds to the hexon protein of the AdV capsid and plays a role as a bridge receptor during hepatocyte transduction [12].

Efforts to improve adenoviruses as gene delivery vehicles have focused mainly on modifications of the viral capsid and vector genome. This has been carried out to increase stability, specificity, transgene capacity, reduce immunogenicity, and control transgene expression. Investigating the utility of deleting individual regions of the Ad genome led to the development of helper-dependent (HD) AdVs that have all the viral protein-coding sequences removed. Only the ITR sequences are retained in the HD AdVs. These sequences are necessary to enable replication and encapsidation of the vector genome [13–15]. The genome structure of the gutted Ads increases capacity for transgenes and the absence of viral protein-coding sequences reduces the vectors' immunostimulatory effects [16–18]. Although RNAi activator sequences are typically small, the increased transgene capacity of the HD AdVs may be used to accommodate larger combinatorial and gene editing cassettes.

To generate HD AdVs expressing an RNAi effector, a DNA insert bearing the expression cassette is initially inserted into the plasmid vector backbone (pHD Ad). Homologous recombination between a shuttle plasmid containing the expression cassette and the pHD Ad in *Escherichia coli* has been proposed as a more efficient alternative to direct ligation of cassettes into large pHD Ads [19]. The pHD Ad containing an RNAi cassette is then used to transfect a producer cell line. This is followed by infection with a helper virus (HV) to provide structural and functional components required to constitute the recombinant viral particles in trans (reviewed in [20]). Different HD AdV backbones are available, which provide the versatility required for different applications [19, 21, 22]. To reduce HV contamination in HD AdV preparations, ingenious strategies have been developed. Excision of the elements from HVs, based on use of the FLP/Frt and Cre/loxP systems [14, 23], has been used to enhance purity of HD AdVs. Typically, Cre recombinase-expressing packaging cells are infected with HVs that have their packaging elements flanked by recombinase recognition sites. Procedures for production and large-scale amplification of HD AdVs have been described in detail and their inclusion is beyond the scope of this chapter [24–26]. Here, the primary focus is on techniques that are required to characterize use of HD AdVs for hepatic delivery cassettes encoding amiRs and a *LacZ* reporter gene.

2 Materials

2.1 Detection of HD AdV Genomes and Reporter Gene Expression in the Liver

1. QIAamp DNA Mini Kit.
2. Dounce homogenizer.
3. Spectrophotometer, e.g., Nanodrop® Nd-100, Thermo Fisher Scientific, USA.
4. HD AdV-specific primers (*see* Subheading 3.2.1).
5. SYBR® Green *Taq* ReadyMix™.
6. Fixative: 1 % formaldehyde, 0.5 % glutaraldehyde in PBS.
7. X-gal staining solution: 4 mM potassium ferricyanide, 4 mM potassium ferrocyanide, 2 mM $MgCl_2$, 0.4 mg/ml β-galactosidase in dH_2O.
8. Roche LightCycler® v.2.

2.2 HD AdV Characterization

1. Aspartate transaminase (AST) and alanine transaminase (ALT) assay kits (Advia®1800 Chemistry System, Siemens, NY, USA).
2. Cytometric bead array (CBA) mouse inflammation kit.
3. MagNA Pure LC Total Nucleic Acid Isolation Kit.
4. MagNA Pure LC system (Roche Diagnostics, GmbH, Germany).
5. Spectrophotometer, e.g., Nanodrop® Nd-100.
6. SYBR® Green JumpStart™ *Taq* ReadyMix™.
7. HBV specific primers (*see* Subheading 3.4.1).
8. Roche LightCycler® v.2.
9. Acrometrix® HBV panel.
10. Dounce homogenizer.
11. MONOLISA® HBs Ag Assay kit.

2.3 Northern Blot Hybridization

1. Huh 7 and HEK293 cell lines. Cultured as monolayers in Dulbecco's modified Eagle medium (DMEM) supplemented with 10 % heat inactivated fetal bovine serum (FBS), 100 units/ml penicillin/streptomycin and 2 mM L-glutamine.
2. Trypsin/EDTA.
3. TRI Reagent®.
4. 0.5 % SDS: Dissolve 0.5 g SDS in 100 ml dH_2O and autoclave.
5. DEPC-water: Add 1 ml of 0.1 % Diethylpyrocarbonate (DEPC) to 1 L dH_2O, mix well and leave at room temperature for 1 h, autoclave at 121 °C for 1 h, and cool to room temperature prior to use.
6. TE buffer: 10 mM Tris–HCl, pH 8.0, 1 mM EDTA, pH 8.0, sterilize by autoclaving.
7. 500 ml 10× TBE buffer: Dissolve 27.5 g boric acid powder and 50 g Tris base in 400 ml dH_2O, add 20 ml 0.5 M EDTA, pH 8.0. Adjust pH to 8.0, make up volume to 500 ml, and autoclave.
8. 20× SSC buffer: 3 M NaCl, 0.3 M trisodium citrate, adjust pH to 7.0, and autoclave.
9. Ammonium persulfate (APS) 10 %: Dissolve 0.1 g APS in 1 ml sterile pure H_2O (prepare fresh).
10. 8 M Urea Polyacrylamide gel 15 % (60 ml): Mix 0.45 g bis-acrylamide, 8.55 g acrylamide, 28.8 g Urea, 6 ml 10× TBE, pH 8.0, and 20 ml H_2O. Dissolve in a beaker containing warm water, make the volume up to 60 ml with H_2O, cool to room temperature then add 250 μl of 10 % APS and 25 μl TEMED, pour the gel immediately, cover the top of gel apparatus with wet tissue and cling film, leave the gel overnight to set at room temperature.
11. Decade RNA molecular weight marker: Label the ladder with radioactive isotope according to the manufacturer's instructions.
12. ATP [γ-^{32}P] (3000 Ci/mmol, 10 mCi/ml).
13. T4 polynucleotide kinase (PNK).
14. Radioactively labeled amiR-specific probe (*see* Subheading 3.5).
15. Sephadex G-25 slurry: Add 5 g sephadex G-25 powder to 50 ml TE buffer, agitate overnight at room temperature, centrifuge at 2800 × *g* for 2 min, pour off supernatant and add 50 ml fresh TE, repeat 2–3 times, and then resuspend in 50 ml TE.
16. Hybond-N$^+$ positively charged nylon membrane.
17. Filter paper.
18. Semi-Dry Electroblotting apparatus.

19. UV cross-linker.
20. Rapid-hyb buffer.
21. Hybridization oven.
22. Phosphoimager.

3 Methods

3.1 Handling and Storage

AdVs are stable if stored correctly. Following mass production and purification, multiple aliquots should be prepared to minimize freeze-thaw cycles, and vectors should always be stored at −70 °C to −80 °C. When needed, thaw to room temperature then return to −70 °C to −80 °C as soon as possible. Avoid exposure to dry ice as the CO_2 decreases the pH of the storage solutions to destabilize the vector [27]. Vector administration, blood sampling, and killing of mice should be carried out according to ethically acceptable institutionally approved procedures.

3.2 Delivery of amiR- and LacZ-Expressing HD AdVs In Vivo

3.2.1 Quantitative PCR to Detect HD AdV Genomes in Hepatocytes

1. Systemically administer a bolus of 5×10^9 infectious viral particles to mice by tail vein injection.
2. To determine intra-hepatic HD AdV genome copy numbers, kill mice 7 days after vector administration, resect livers, and homogenize with equal volumes of saline.
3. Extract DNA from 200 μl of the homogenate using the QIAMP Mini extraction kit according to the manufacturer's instructions. Measure the DNA concentration using an appropriate spectrophotometer (e.g., Nanodrop® Nd-100, Thermo Fisher Scientific, USA).
4. Perform SYBR green-based quantitative PCR (Q-PCR) using a real time instrument. Following a hotstart at 95 °C to activate the Taq polymerase, carry out 50 cycles with the following parameters: annealing at 57 °C for 10 s, extension at 72 °C for 10 s, and denaturation at 95 °C for 10 s. Use 50 ng of DNA and primers specific for HD AdVs. An example of a suitable primer set that may be used to amplify HD AdV sequences is: pΔ28LacZ forward: 5′-GAA AAA ACA CAC TGG CTT GAA ACA-3′ and pΔ28LacZ reverse: 5′-TGC CAC CTC GTA TTT CAC CTC TA-3′.
5. Absolute copy numbers of HD AdV may be calculated from a standard curve generated using Q-PCR.

3.2.2 Detecting β-Galactosidase Activity in Liver Sections

1. To determine efficiency of HD AdV delivery to hepatocytes, harvest the livers at desired time interval following HD AdV administration.
2. Prepare cryosections according to standard protocols (*see* **Note 1**).

3. Fix the frozen sections to glass slides by incubation in fixative at room temperature for 5 min.
4. Remove the fixative and wash fixed sections twice for 5 min with PBS at room temperature.
5. Incubate the sections overnight at 37 °C in the X-gal staining solution.
6. Wash off the stain with H_2O and analyze the sections using a light microscope.

3.3 Assessing Toxicity of HD AdVs in Transduced Mice

3.3.1 Liver Function Assays

1. Collect blood samples at appropriate time points by using the appropriate procedure (e.g., retro-orbital puncture).
2. Allow blood to clot (leave for 2 h at 4 °C) before subjecting to centrifugation ($5900 \times g$ for 10 min).
3. Aspirate the serum and store at −20 °C until needed.
4. Dilute serum samples twofold with saline and measure AST and ALT levels using a kinetic assay such as the Advia® 1800 Chemistry System (Siemens, NY, USA) in a routine diagnostic chemistry laboratory.

3.3.2 Inflammatory Cytokine Analysis in Mice

Use the cytometric bead array (CBA) mouse inflammation kit to measure serum inflammatory cytokine concentrations. The assay measures IL-6, IL-10, MCP-1, IFN, TNF, and IL-12p70 protein levels in the mouse serum samples. Follow the protocol according to the manufacturer's instructions (*see* **Note 2**).

3.4 Efficacy of amiR Expressing HD AdVs Against HBV

The methods described in this chapter have been used successfully to develop amiR expressing HD AdVs, which target HBV. The protocols detailed below serve as an example of how HBV transgenic mice are used as a model to determine the effects of RNAi-activating AdVs on markers of HBV replication in vivo [28]. A dose of 5×10^9 infectious AdV particles is administered as a bolus via the tail vein, and blood samples are collected by retro-orbital puncture at relevant time points.

3.4.1 Quantitative-PCR Detection of Circulating HBV Particle Equivalents

An absolute Q-PCR method is used to measure serum HBV DNA as a determinant of circulating viral particle equivalents.

1. Isolate total DNA from 50 μl serum samples using the MagNA Pure LC Total Nucleic Acid Isolation Kit and MagNA Pure LC system (Roche Diagnostics, GmbH, Germany).
2. Perform Q-PCR on the DNA using SYBR® Green JumpStart™ *Taq* ReadyMix™ as described in Subheading 3.2.1. Amplify the surface antigen region of HBV DNA using the appropriate primer set, e.g., HBVs forward: 5′-TGC ACC TGT ATT CCC ATC-3′, and HBVs reverse: 5′-CTG AAA GCC AAA CAG

TGG-3′ using the Roche LightCycler v.2 (Roche Diagnostics, GmbH, Germany). Confirm specificity of the amplicons by melting curve analysis. Determine absolute value of viral particle equivalents by using a standard curve that is generated from a commercial HBV standard (Acrometrix® HBV panel, Life Technologies, USA).

3.4.2 HBsAg ELISA

Quantify mouse serum HBsAg levels using the MONOLISA® HBs Ag Assay kit. Dilute the mouse serum 50-fold with saline and use 100 μl of diluent to perform assay according to the manufacturer's instructions.

3.5 Characterization of amiR Expression in Liver-Derived Cells

The steps involved in processing of RNA for northern blot hybridization analysis are depicted schematically in Fig. 1 and is described in detail below.

1. For analysis of amiR expression in cultured cells: infect liver-derived Huh7 cells (~80 % confluent) at a multiplicity of infection (MOI) of 100 HD AdVs per cell. Include plates of untransduced control cells. Incubate cells at 37 °C for 48 h, aspirate the medium, and lyse cells with TRI Reagent®. Extract total RNA from the cell lysate according to the manufacturer's instructions. For assessment of amiR expression in mice: infect the animals by administering 5×10^9 infectious HD AdVs via tail vein injection. Harvest the left lobe of the liver 1 week after infection and homogenize the fresh liver in 1 ml TRI Reagent® using a hand-held Dounce homogenizer. Extract total RNA from the homogenized liver according to the manufacturer's instructions.
2. Resuspend the pellet in 0.5 % SDS.
3. Pour 15 % polyacrylamide gel on the day prior to carrying out the electrophoresis.
4. Also on the day before running the gel, label DNA probes and Decade RNA ladder with ATP [γ ^{32}P] using polynucleotide kinase. For probe labeling, take 2 μl from a 10 μM oligonucleotide probe stock, add 2 μl T4 Polynucleotide kinase (PNK) buffer, 10 U (1 μl) PNK, followed by 1–5 μl radioactive isotope (add more as radioisotope starts to decay). Make up to a final volume of 20 μl with DEPC-treated water and incubate at 37 °C for approximately 20 min. Prepare sephadex columns by inserting approximately 5 mm of compact filter fiber to the bottom of a 1-ml syringe, add 1 ml of sephadex to the syringe then centrifuge (700 × *g*) for 2 min in a 15-ml polypropylene tube. Add 30 μl of DEPC-water to the labeled probe, then pass through a sephadex column by centrifugation (700 × *g*) for 2 min and collect the probe. Label the Decade RNA ladder according to the manufacturer's instructions (*see* **Note 3**).

- Extract total RNA from the HD AdV-infected cells, e.g. murine hepatocytes

- Separate RNA on a 15 % polyacrylamide gel
- Include labeled molecular weight marker

- Transfer RNA onto a nitrocellulose membrane

- Fix RNA on the membrane
- Hybridize with a labelled probe
- Wash and expose to IP or X-ray film
- Develop the image

- Strip the membrane
- Reprobe for housekeeping gene (e.g. U6)

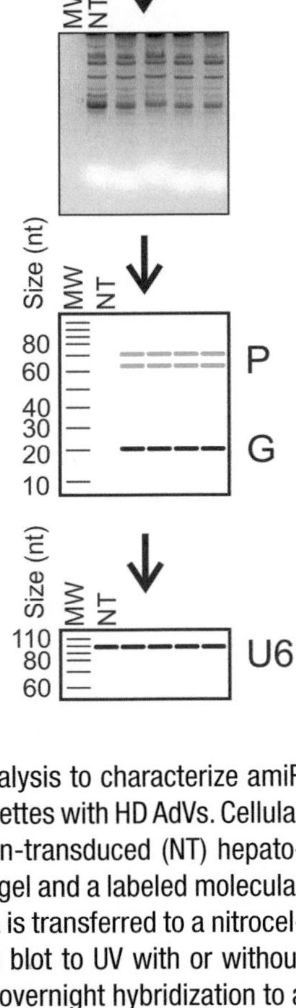

Fig. 1 Procedures involved in using northern blot analysis to characterize amiR expression following delivery of RNAi expression cassettes with HD AdVs. Cellular RNA is extracted from the transduced or control non-transduced (NT) hepatocytes. The samples are resolved on a polyacrylamide gel and a labeled molecular weight marker ladder (MW) is run alongside. The RNA is transferred to a nitrocellulose membrane and then fixed by exposure of the blot to UV with or without baking at 80 °C for 1 h. The membrane is probed by overnight hybridization to a radioactively labeled oligonucleotide. Thereafter the membrane is washed and exposed to an imaging plate (IP) or X-ray film for approximately 7 days. The IP is scanned or the X-ray film developed and the results analyzed. Molecular weight markers are used to determine the sizes of the mature processed guide of approximately 21 nt (G) and longer precursor (P) sequences. Bands representing the G and P sequences should be evident in RNA extracted from transduced but not from control NT cells

5. Fill electrophoresis buffer tanks with 500 ml of 0.5× TBE.
6. Before loading, pre-run the gel at 150 V for 30 min.
7. Add an equal volume of loading dye to 20–30 μg of RNA.
8. Denature the RNA by incubating at 80 °C for 5 min, place on ice, and then centrifuge briefly to collect the condensate.
9. Denature the radioactively labeled Decade RNA molecular weight marker at 95 °C for 5 min, place on ice, and then centrifuge briefly to collect the condensate.
10. Before loading, use a syringe to flush each well with running buffer.
11. Load samples and Decade RNA ladder carefully by pipetting to the very bottom of each well.
12. Run the gel at 150 V until the dye front has migrated a quarter of the way into the gel (~30 min) and then increase the voltage to 200 V and run until resolved.
13. Stain the gel by agitating in 100 ml running buffer with ethidium bromide at a final concentration of 4 μg/ml for 5 min and visualize using a transilluminator.
14. Trim the gel and measure the dimensions.
15. Cut six pieces of thick filter paper that are slightly larger (approximately 1 cm in each dimension) than the size of the gel.
16. Cut a similarly sized piece of positively charged nylon membrane.
17. Soak the filter paper and membrane in running buffer.
18. Make a sandwich comprising three pieces of filter paper, one piece of nylon membrane, the gel, and an additional three pieces of filter paper, then place in a Semi-Dry Electroblotting Unit Z34,050-2. Orientate the sandwich so that the membrane is on the anode (+) side of the apparatus.
19. Press out excess liquid and transfer at 3.3 mA/cm^2 of membrane for 1 h at 4 °C.
20. Remove the membrane from the blotter and cross-link with 200,000 μJ/cm^2 of energy using a UV cross-linker and/or bake at 80 °C for an hour.
21. Heat 10 ml/100 cm^2 Rapid-hyb buffer to 42 °C in a hybridization bottle.
22. Pre-hybridize the membranes in Rapid-hyb buffer at 42 °C for at least 15 min.
23. Denature the probe by heating at 80 °C for 5 min, add to the prehybridization solution bottle at a final concentration of 10 ng/ml, and incubate overnight with rotation.

24. Discard the hybridization solution using appropriate procedures for handling isotopes.
25. Wash the membrane once with 50 ml 5×SSC with 0.1 % SDS at room temperature for 20 min.
26. Wash the membrane twice with 50 ml 1×SSC, 0.1 % SDS at 42 °C for 15 min.
27. Seal the membrane in a plastic bag and place together with imaging plate (IP) in a cassette and incubate in the dark at room temperature for 7 days. If using X-ray film, carry out the exposure in an appropriate cassette at −70 °C.
28. In a dark room, place IP in the phosphoimager and scan the image. Alternatively, develop the X-ray film by placing it in developer, fixer, and then water for 2 min each.
29. To confirm equal loading and transfer of RNA, reprobing of the membrane with a probe specific for a housekeeping gene (e.g., U6 snRNA, 5′-TAG TAT ATG TGC TGC CGA AGC GAG CA-3′) should be undertaken. Strip the membrane by incubating with rotation at 80 °C in 50 ml 1 % SDS for 30 min (*see* **Note 4**).

4 Notes

1. Freshly resected livers may be wrapped in gauze saturated in saline and stored for up to 2 days at 4 °C before cryosections are prepared.
2. The samples may be diluted twofold if there is a limited volume of serum available.
3. A mixture of radioactively labeled DNA probes may be used as an alternative to RNA decade ladder.
4. Stripped membranes should be exposed to film for 2–3 days to confirm complete removal of the labeled probe.

Acknowledgments

The authors' laboratory receives financial assistance from the South African National Research Foundation (NRF, GUNs 81768, 81692, 68339, 85981 & 77954), Medical Research Council and Poliomyelitis Research Foundation.

References

1. Ely A, Naidoo T, Arbuthnot P (2009) Efficient silencing of gene expression with modular trimeric Pol II expression cassettes comprising microRNA shuttles. Nucleic Acids Res 37, e91
2. Marimani MD, Ely A, Buff MC, Bernhardt S, Engels JW, Arbuthnot P (2013) Inhibition of hepatitis B virus replication in cultured cells and in vivo using 2′-O-guanidinopropyl modified siRNAs. Bioorg Med Chem 21:6145–6155
3. Knoepfel SA, Centlivre M, Liu YP, Boutimah F, Berkhout B (2012) Selection of RNAi-based inhibitors for anti-HIV gene therapy. World J Virol 1:79–90
4. Azimzadeh Jamalkandi S, Azadian E, Masoudi-Nejad A (2014) Human RNAi pathway: crosstalk with organelles and cells. Funct Integr Genomics 14:31–46
5. Carmona S, Ely A, Crowther C, Moolla N, Salazar FH, Marion PL, Ferry N, Weinberg MS, Arbuthnot P (2006) Effective inhibition of HBV replication in vivo by anti-HBx short hairpin RNAs. Mol Ther 13:411–421
6. Crowther C, Ely A, Hornby J, Mufamadi S, Salazar F, Marion P, Arbuthnot P (2008) Efficient inhibition of hepatitis B virus replication in vivo, using polyethylene glycol-modified adenovirus vectors. Hum Gene Ther 19:1325–1331
7. Mowa MB, Crowther C, Ely A, Arbuthnot P (2012) Efficient silencing of hepatitis B virus by helper-dependent adenovirus vector-mediated delivery of artificial antiviral primary micro RNAs. MicroRNA 1:19–25
8. Russell WC (2000) Update on adenovirus and its vectors. J Gen Virol 81:2573–2604
9. Smith JG, Wiethoff CM, Stewart PL, Nemerow GR (2010) Adenovirus. Curr Top Microbiol Immunol 343:195–224
10. Russell WC (2009) Adenoviruses: update on structure and function. J Gen Virol 90:1–20
11. Huard J, Lochmuller H, Acsadi G, Jani A, Massie B, Karpati G (1995) The route of administration is a major determinant of the transduction efficiency of rat tissues by adenoviral recombinants. Gene Ther 2:107–115
12. Waddington SN, McVey JH, Bhella D, Parker AL, Barker K, Atoda H, Pink R, Buckley SMK, Greig JA, Denby L, Custers J, Morita T, Francischetti IMB, Monteiro RQ, Barouch DH, van Rooijen N, Napoli C, Havenga MJE, Nicklin SA, Baker AH (2008) Adenovirus Serotype 5 hexon mediates liver gene transfer. Cell 132:397–409
13. Sandig V, Youil R, Bett AJ, Franlin LL, Oshima M, Maione D, Wang F, Metzker ML, Savino R, Caskey CT (2000) Optimization of the helper-dependent adenovirus system for production and potency in vivo. Proc Natl Acad Sci U S A 97:1002–1007
14. Hardy S, Kitamura M, Harris-Stansil T, Dai Y, Phipps ML (1997) Construction of adenovirus vectors through Cre-lox recombination. J Virol 71:1842–1849
15. Alemany R, Dai Y, Lou YC, Sethi E, Prokopenko E, Josephs SF, Zhang WW (1997) Complementation of helper-dependent adenoviral vectors: size effects and titer fluctuations. J Virol Methods 68:147–159
16. Majhen D, Ambriovic-Ristov A (2006) Adenoviral vectors—how to use them in cancer gene therapy? Virus Res 119:121–133
17. Benihoud K, Yeh P, Perricaudet M (1999) Adenovirus vectors for gene delivery. Curr Opin Biotechnol 10:440–447
18. Brunetti-Pierri N, Stapleton GE, Law M, Breinholt J, Palmer DJ, Zuo Y, Grove NC, Finegold MJ, Rice K, Beaudet al, Mullins CE, Ng P (2009) Efficient, long-term hepatic gene transfer using clinically relevant HDAd doses by balloon occlusion catheter delivery in nonhuman primates. Mol Ther 17:327–333
19. Toietta G, Pastore L, Cerullo V, Finegold M, Beaudet al, Lee B (2002) Generation of helper-dependent adenoviral vectors by homologous recombination. Mol Ther 5:204–210
20. Mowa MB, Crowther C, Arbuthnot P (2010) Therapeutic potential of adenoviral vectors for delivery of expressed RNAi activators. Expert Opin Drug Deliv 7:1373–1385
21. Palmer DN, Ng P (2008) Methods for the production of helper dependent adenoviral vectors. In: LeDoux J (ed) Gene therapy protocols, vol 433, Methods in molecular biology. Springer, New York, pp 33–54
22. Shi CX, Graham FL, Hitt MM (2006) A convenient plasmid system for construction of helper-dependent adenoviral vectors and its application for analysis of the breast-cancer-specific mammaglobin promoter. J Gene Med 8:442–451
23. Ng P, Beauchamp C, Evelegh C, Parks R, Graham FL (2001) Development of a FLP/frt system for generating helper-dependent adenoviral vectors. Mol Ther 3:809–815
24. Palmer D, Ng P (2003) Improved system for helper-dependent adenoviral vector production. Mol Ther 8:846–852

25. Ng P, Parks RJ, Graham FL (2002) Preparation of helper-dependent adenoviral vectors. Methods Mol Med 69:371–388
26. Palmer DJ, Ng P (2004) Physical and infectious titers of helper-dependent adenoviral vectors: a method of direct comparison to the adenovirus reference material. Mol Ther 10:792–798
27. Nyberg-Hoffman C, Aguilar-Cordova E (1999) Instability of adenoviral vectors during transport and its implication for clinical studies. Nat Med 5:955–957
28. Marion PL, Salazar FH, Liittschwager K, Bordier BB, Seegers C, Winters MA, Cooper AD, Cullen JM (2003) A transgenic mouse lineage useful for testing antivirals targeting hepatitis B virus. In: Schinazi R, Sommadossi J-R and Rice C.M. (eds) Frontiers in viral hepatitis. Elsevier Science, Amsterdam, pp 197–202

Chapter 21

Intravascular AAV9 Administration for Delivering RNA Silencing Constructs to the CNS and Periphery

Brett D. Dufour and Jodi L. McBride

Abstract

Viral vector delivery of RNA silencing constructs, when administered into vasculature, typically results in poor central nervous system (CNS) transduction due to the inability of the vector to cross the blood–brain barrier (BBB). However, adeno-associated virus serotype 9 (AAV9) has the ability to cross the BBB and robustly transduce brain parenchyma and peripheral tissues at biologically meaningful levels when injected intravenously. Recent work by our lab has shown that this method can be used to deliver RNA silencing constructs, resulting in significant reductions in gene expression in multiple brain regions and in peripheral tissues. Here, we outline a method for delivery of AAV9 vectors expressing RNA interference (RNAi) constructs that lead to robust simultaneous transduction of mouse peripheral tissues and the CNS following a single injection into the jugular vein. Additionally, we outline methods for necropsy and immunofluorescence to detect AAV9 transduction patterns in the rodent CNS following a vascular delivery.

Key words AAV9, systemic, Jugular vein, Vascular, Gene therapy, RNAi

1 Introduction

RNA interference (RNAi) is a powerful biological tool to query basic gene function or to silence diseased genes in therapeutic applications. Viral vector delivery of RNAi constructs is most frequently achieved with focal injections into particular peripheral organs or into specific subregions of the CNS using stereotaxic methods. For widespread and simultaneous delivery to peripheral tissues, a vascular delivery approach can be used and transduction is typically limited only by the tropism of the viral vector serotype that is selected. Many AAV serotypes confer promiscuous binding to peripheral tissues (serotypes 1, 7, 8, 9), while others are more selective in the tissue types that they transduce (serotypes 2, 3, 4, 5, 6) following vascular injection [1]. However, widespread transduction from a single vector injection has not been a feasible strategy for the brain, as most viral vectors cannot cross the blood–brain barrier (BBB) in appreciable levels. For diseases involving

Kato Shum and John Rossi (eds.), *SiRNA Delivery Methods: Methods and Protocols*, Methods in Molecular Biology, vol. 1364, DOI 10.1007/978-1-4939-3112-5_21, © Springer Science+Business Media New York 2016

widespread CNS neuropathology (e.g., Alzheimer's disease, Huntington's disease, amyotrophic lateral sclerosis, lysosomal storage disorders, Rett Syndrome), the inability to achieve widespread transduction has been a significant roadblock in therapeutic development.

Historically, in both preclinical and clinical studies assessing vector-mediated gene delivery, one or a few focal injections have been used to fill particular brain structures. While these studies successfully demonstrate the utility of viral vectors to confer long-term gene expression, or gene suppression in some cases, the widespread pathology observed in many CNS diseases makes a global delivery strategy an appealing approach. Faust and colleagues demonstrated that AAV serotype 9 (AAV9) crosses the BBB and transduces both neurons and astrocytes in the mouse CNS following a singular tail-vein injection [2]. Recent studies have shown that AAV9 can be used as a delivery tool to replace or modify diseased genes and successfully improve disease phenotypes in a variety of mouse models of human disease, including amyotrophic lateral sclerosis [3], mucopolysaccharidosis III [4], and Rett syndrome [5, 6]. Vascular delivery of AAV9-RNAi has been employed to reduce gene expression in mouse models of cardiac myopathy [7, 8] and our laboratory was the first to demonstrate that a single jugular vein delivery of AAV9-RNAi crossed the BBB and significantly reduced expression of a disease-causing gene in multiple brain regions. Moreover, this global delivery strategy prevented both neuropathological and physiological manifestations of disease [9]. As expected, this delivery method also resulted in significant reduction in disease-causing gene expression in multiple peripheral tissues [9].

Here, we describe, in detail, the methodology that our laboratory uses to target the mouse CNS and periphery using systemic AAV9-RNAi and to assess transduction patterns in the injected animals: (1) Intra-jugular vein injection to administer vector; (2) mouse necropsy, perfusion, and tissue collection; and (3) immunofluorescent processing of brain sections to evaluate the expression and transduction patterns of AAV9 in mice.

2 Materials

2.1 Intra-jugular Vein Injection

1. PPE including scrubs, hair bonnets, water-resistant gown, surgical masks, and gloves.
2. Infusate (viral vector prep).
3. Ketamine (100 mg/ml).
4. Xylazine (100 mg/ml).
5. Ketamine/Xylazine Mix: 1 ml of 100 mg/ml ketamine, 0.1 ml of 100 mg/ml xylazine, 8.9 ml of 0.9 % sterile saline.

6. Hamilton glass luer tip (LT) syringes (25–500 μl or appropriate volume for infusion).
7. Hamilton Kel-F Hub removable surgical needles (30G, 20 mm length, beveled tip style 4).
8. 1 cc syringes with 27G needles.
9. Hair clippers.
10. Surgical lamp.
11. Sterile surgical chucks pads.
12. Sterile gauze.
13. Sterile surgical cotton swabs.
14. Surgical instruments (sharp-tipped surgical scissors, blunt-tipped dissecting scissors, forceps).
15. Bead sterilizer.
16. Laboratory tape.
17. Betadine wipes.
18. Alcohol wipes.
19. Carprofen.
20. Heating pad.
21. Wound clip applicator.
22. Wound clip removal tool.
23. 7 mm wound clips.

2.2 Necropsy, Tissue Collection, and Cutting for Histology

1. PPE: Gown or lab coat, hair bonnet, nitrile gloves, facemask.
2. 10 cc syringes.
3. 1 cc syringe with 27G needle.
4. BD Vacutainer Safety-Lok blood collection set (used for perfusion).
5. Dissection board, with needles to pin mouse.
6. Surgical instruments (surgical scissors, fine-tipped forceps, hemostats, small spatula, heavy duty scissors).
7. Spray bottle with 70 % Ethanol.
8. Sharps containers.
9. Waste disposal bags.
10. 0.9 % sterile saline.
11. 32 % Paraformaldehyde (liquid stock).
12. 0.1 M Phosphate buffer solution: 5.50 g sodium phosphate dibasic and 1.56 g sodium phosphate monobasic in 500 ml of dH_2O.
13. 4 % paraformaldehyde: 1:8 dilution of 32 % liquid paraformaldehyde stock to 0.1 M Phosphate buffer—i.e., 10 ml 32 % paraformaldehyde and 70 ml 0.1 M phosphate buffer.

14. Ice bucket with wet ice.
15. Labeled tissue culture plates (6–24 well, size depends on tissue harvested).
16. For molecular analyses, the following materials will also be used: mouse brain matrix, sterile razor blades, micro scissors or tissue punchers for regional brain dissection, small fine-tipped forceps, RNAse away, sterile petri dishes, 1.5 ml microcentrifuge tubes and/or DNAse/RNAse free foil packs, dry ice, or liquid nitrogen.

2.3 Immunofluorescence Staining

1. PPE: White lab coat, nitrile gloves.
2. Sliding microtome and microtome blade.
3. Paint brushes for tissue cutting.
4. Dry ice.
5. Trizma® pre-set crystals, pH 7.4.
6. Triton®X-100.
7. Serum (goat or donkey).
8. Dilution media: 7.46 g Trizma, 8.77 g sodium chloride, 0.5 ml Triton®X-100 in 1 L dH_2O.
9. TBS: 7.46 g Trizma and 8.77 g sodium chloride in 1 L dH_2O.
10. PBS: 5.47 g sodium phosphate dibasic, 1.60 sodium phosphate monobasic, 9.26 g sodium chloride in 1 L dH_2O.
11. Cryoprotectant solution: 300 g Sucrose, 300 ml ethylene glycol, 0.2 g Sodium Azide in 500 ml PBS.
12. Netted staining dishes (with glass dishes to contain fluid).
13. Orbital shaker.
14. Appropriate primary and secondary antibodies.
15. 24-well tissue culture plates.
16. Hook tools or small paint brushes for mounting tissue.
17. Clear shallow dish for mounting stained tissue.
18. Vectashield mounting medium.
19. Microscope slides.
20. Cover slips.
21. Clear nail polish.
22. Microscope with fluorescent capabilities.

3 Methods

3.1 Intra-jugular Vein Injection

1. Set up the surgical area. Tape down sterile chucks pads for the surgical space. Sterilize the surgical instruments (forceps, surgical scissors, blunt-tipped dissecting scissors) using a bead sterilizer. Illuminate the area with a light source.

2. Put on the appropriate personal protective equipment (PPE), including a hair bonnet, face mask, sterile gown or lab coat, and gloves.
3. Restrain the mouse for anesthetic injection. First, place the mouse on the cage lid, gently press the body down against the lid, and scruff the skin on the back to create a gentle but firm grasp on the mouse.
4. Hold the mouse up, with its ventral surface placed upward, and inject with ketamine/xylazine mix (*see* **Note 1**) at approximately 0.01 cc per gram bodyweight (i.e., 0.20 cc for a 20 g mouse) into the intraperitoneal space (IP).
5. Place the mouse back into the home cage and wait 5 min to check the mouse's anesthetic plane. To assess anesthetic plane, pinch the mouse's tail and foot pad and look for an involuntary reflex. If the mouse does not show a reflex following tail and foot pinch, it is now in a surgical plane of anesthesia. If there is a reflex, give the animal a supplemental dose of ketamine/xylazine, approximately 0.05–0.10 cc, and wait another 5 min and reassess anesthetic plane. Repeat the pinch assessment and supplement with more ketamine/xylazine (stepwise, 0.05–0.10 cc at a time) until an anesthetic plane is reached.
6. Place the mouse in a recumbent position (lying on its back) on a pad (separate from the surgical pad to be used shortly), and shave the upper chest and neck of the mouse.
7. Move the mouse to the clean surgical chucks pad, and again place in a recumbent position. Tape the limbs of the mouse down to hold it in place for surgery.
8. Sterilize the shaved surgical site using betadine wipes. Wipe the betadine off using ethanol wipes—this will help to visualize the jugular vein through the skin.
9. To expose the jugular vein for injection, first pinch the skin on the upper chest (above the pectoral muscle) and make a small incision in the skin over the mouse's right jugular vein (your left) using the sterile surgical scissors (Fig. 1a). Pull the skin up from the mouse with the forceps and keep tension on it, then carefully slide the scissors into the small opening made from the initial incision and cut through the skin upward toward the head (Fig. 1b). The final incision should be approximately 2.5 cm in length, and run parallel to the mouse's midline, about 0.5 cm to the left of the midline (mouse's right). If you can visualize the jugular vein through the skin prior to incision, the incision should run directly over the vein.
10. Expose the jugular vein by carefully cleaning off any connective or fat tissue and that may be in the way.

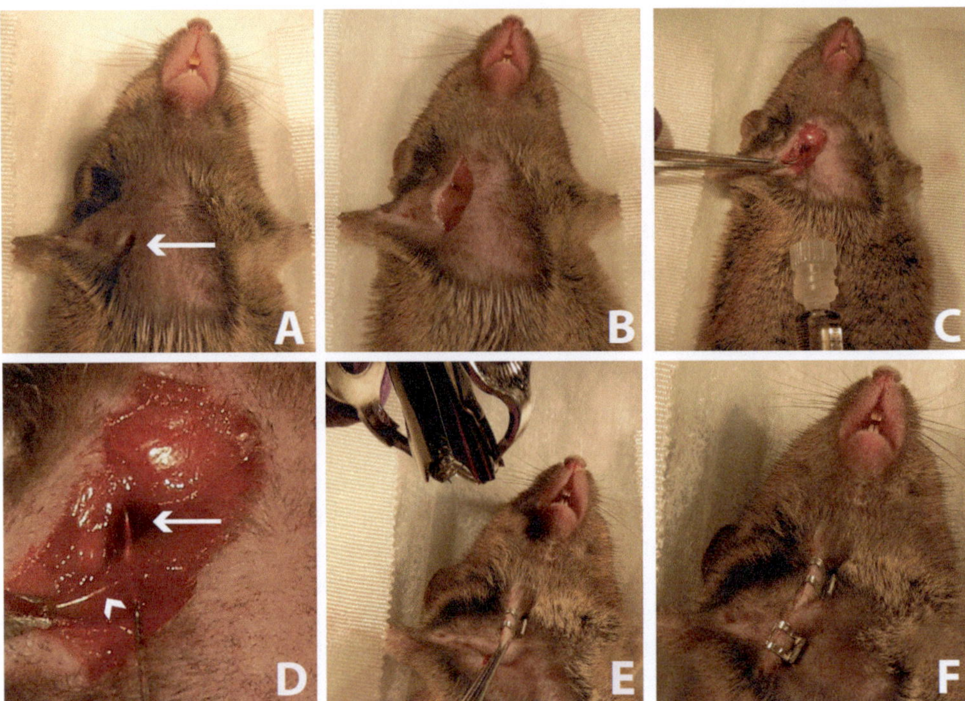

Fig. 1 Intra-jugular injection of AAV9-RNAi. (**a**) An incision is made in the upper chest and neck of the mouse. *Arrow* indicates the incision. (**b**) The jugular vein and pectoral muscle are exposed. (**c**) The pectoral muscle is stabilized using forceps and a 30G Hamilton syringe is inserted through the pectoral muscle and into the jugular vein lumen. AAV9 vector is infused into the vein. (**d**) Close-up image of jugular vein injection with needle tip inside of the vein lumen (*arrow*= jugular vein; *arrowhead*= pectoral muscle). (**e**) The skin is pinched and pulled away from the body and closed with wound clip. (**f**) The incision is completely closed with wound clips

11. Prime the Hamilton needle/syringe with virus by filling and emptying the syringe 3 times (*see* **Note 2**).
12. Fill the Hamilton syringe with the appropriate amount for virus for injection (*see* **Note 3**).
13. Hold onto the pectoral muscle with forceps, and push needle tip through pectoral muscle and into the lumen of the vein (Fig. 1c, d).
14. Visually confirm that the needle tip is in the lumen of the jugular vein, then slowly infuse virus (approximately 100 μl/min) (*see* **Notes 4** and **5**).
15. Slowly pull the needle out from the jugular vein and pectoral muscle. Keep pressure on the pectoral muscle with the forceps while pulling the needle out. Gently press a cotton tipped swab on the injection site to reduce backflow of blood or virus.
16. Gently pull the skin back together.
17. Pinch the incision closed with forceps and pull the closed skin away from the jugular vein with the forceps. Close skin incision

with 7 mm wound clips (Fig. 1e, f). Make sure that the skin is pulled away from the body when the wound clip is placed, to avoid damaging the muscle or veins under the skin.

18. Inject the mouse with analgesic—5 mg/kg carprofen once daily, for 2 days. The first injection should be immediately after surgery with the second at 24 h post-surgery.
19. Place the recovering mouse in a clean cage, with half of the cage placed on a heating pad, on low heat, until the mouse awakens.
20. Return mouse to the home cage only after it has awoken to avoid manipulation from cage mates. Wound clips can be removed 1 week after surgery.

3.2 Necropsy, Tissue Collection, and Sectioning for Histology

1. Prepare a space for the necropsy—set up with sterile pads, dissection board, light source, and clean surgical instruments. You will also need ice-cold sterile 0.9 % saline and 4 % paraformaldehyde (keep both on ice, until ready for injection).
2. Prepare tissue culture plates (6- to 24-well, use the plate with the appropriate size wells for the tissues you are planning to collect) by labeling and filling them with 4 % paraformaldehyde.
3. Fill two 10 cc syringes—one with 10 cc of 0.9 % saline and the other with 4 % paraformaldehyde. Keep both syringes on ice until ready for use. Connect each syringe to BD Vacutainer Safety-Lok blood collection set (this set consists of a needle, with a piece of plastic used to grip the needle, and thin tubing that connects to the syringe).
4. Modifications to this procedure can be made for collecting tissues for molecular analysis (i.e., qPCR or western blot to assess gene expression). *See* **Note 6** for modifications.
5. Anesthetize the mouse in its home cage by injecting IP with the ketamine/xylazine mix at approximately 0.01 cc per gram bodyweight (i.e., 0.20 cc for a 20 g mouse). The mouse should be in a surgical/deep plane of anesthesia before proceeding with the necropsy—there should be no response to toe pinch. If there is still a response, then supplement with additional ketamine/xylazine (0.05–0.10 cc at a time), and wait for suppression of response to toe pinch.
6. Once a deep plane of anesthesia is achieved, secure the animal to the dissection board in a recumbent position by placing 27G needles through all four paws (Fig. 2a). Spray the mouse's body with 70 % ethanol, to prevent hair from spreading over the surgical openings and tissues during necropsy.
7. Using forceps, grab the skin on the lower abdomen at midline and make an incision using surgical scissors to open the abdominal cavity (Fig. 2b).

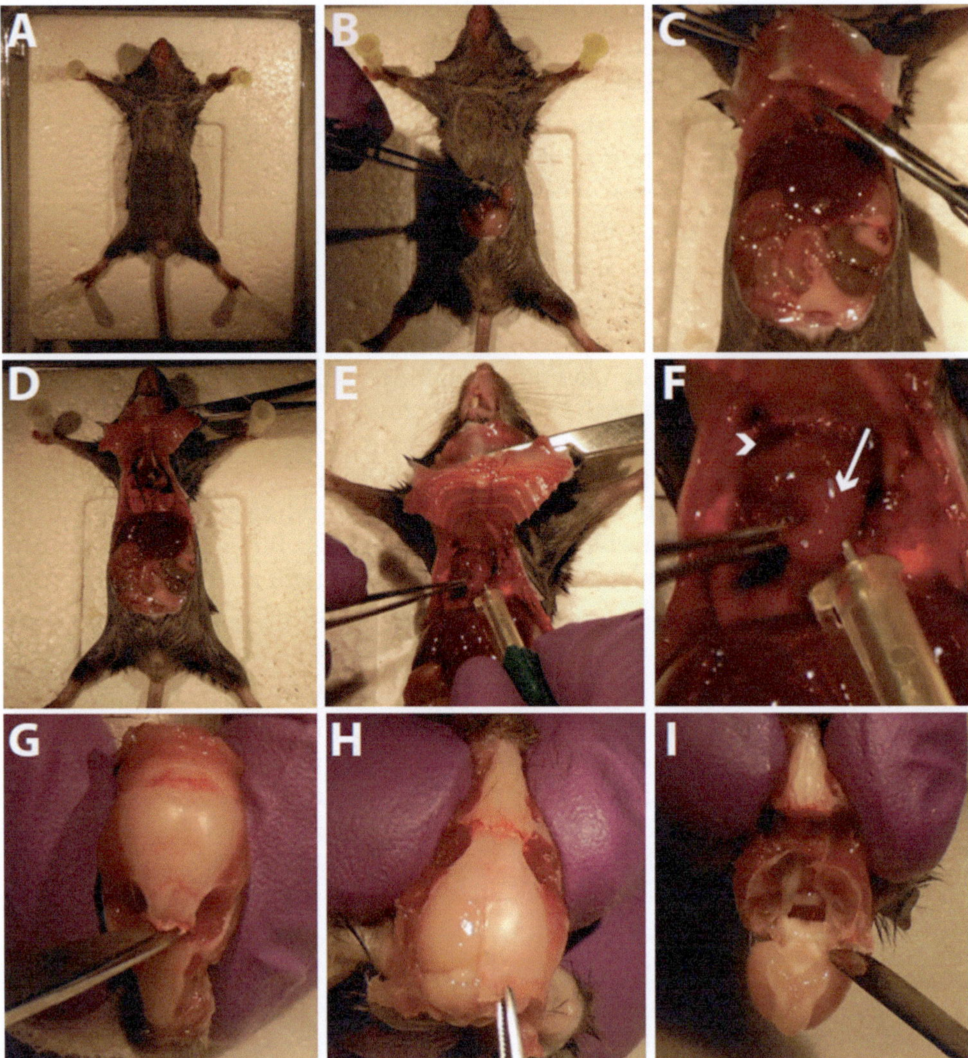

Fig. 2 Mouse necropsy and tissue collection. (**a**) The mouse is pinned down to the surgical board in all four paws. (**b**) An incision is made into the ventral abdomen. (**c**) The incision is extended up to the diaphragm, which is carefully cut open. (**d**) The incision is extended on both sides to open the chest cavity. Hemostats are used to hold the cavity open. (**e**) A needle is placed into the mouse's left ventricle for perfusion of saline and paraformaldehyde. (**f**) Close-up image of heart at perfusion. *Arrowhead* indicates the mouse's right atrium, which is cut at the start of the perfusion. *Arrow* indicates the left ventricle, where the needle is placed for perfusion. (**g**) After removing the mouse's head and exposing the skull, a small incision is made through the skull between the eye sockets. (**h**) A cut is made up the midline of the skull, starting at the cervical wound, and the cut skull is peeled off using forceps. (**i**) After full removal of the dorsal skull, the brain is gently removed using a spatula

8. Carefully cut up each side of the mouse's body until you reach the diaphragm (near the point when you reach the ribs with the lateral cuts).
9. Carefully cut through the diaphragm using surgical scissors (Fig. 2c). Cut along the outer ventral rim of the diaphragm to help avoid cutting organs (particularly the heart and lungs) with the scissors.

10. Continue cutting up the lateral sides of the mouse, through the ribs, to fully expose the heart (Fig. 2d).
11. Use hemostats to grip the lower sternum, clamp and drape hemostats over the rostral end of the mouse. Hemostats can be held in place by resting them on the needle holding the mouse's forepaw to the dissection board. This keeps the chest out of the way so that the heart is easily accessible during perfusion.
12. Gently grip the mouse's right atrium (*see* Fig. 2f) with forceps and cut it using surgical scissors. This creates an exit for the mouse's blood that will be pushed out during the perfusion.
13. Perfuse the mouse with 10 cc of 0.9 % saline. To do this, gently insert the needle tip approximately 2–3 mm into the mouse's left ventricle (Fig. 2e, f). Make sure not to breach the wall between the left and right ventricle. Slowly infuse the saline over 30 s (*see* **Note 7**). When done, remove the needle from the right ventricle.
14. Perfuse the mouse with 4 % paraformaldehyde. First, place the needle into the same opening in the left ventricle. Slowly perfuse with 10 cc of paraformaldehyde over 30 s. If the needle placement is correct, the animal will move slightly as the paraformaldehyde fixes proteins in skeletal muscle.
15. To harvest the brain, first cut off the head of the mouse using a large pair of scissors. Next, cut the skin on the dorsal surface of the head from the cervical cut down to the nose. Pull the skin down on the sides. Cover the loose skin with a Kimwipe, and use the skin to hold the head firmly in place while you open the skull and remove the brain.
16. To open the skull for brain removal, first use surgical scissors to cut the bone between the two eyes (Fig. 2g). Next, carefully make a 5 mm cut up the posterior skull (region covering the back of the cerebellum). Make sure not to insert the scissor blade into the tissue—it should slide in directly touching the inner surface of the skull. Gently pull off the pieces of skull adjacent to the cervical cut using forceps. Next make another cut up the midline of the skull using the same technique, approximately 1 cm. Again, remove the skull covering each hemisphere of brain using forceps (Fig. 2h). Finally, make one final cut upward, to connect the midline cut to the cut between the eye sockets. Again, remove the skull covering the final rostral portion of the brain.
17. Slide a surgical spatula under the brain at the rostral end and continue to slide under the ventral surface of the brain. Tip the skull/brain upside down (at a 45–90° angle), and use the surgical spatula or dissecting scissors to gently cut the large trigeminal and other cranial nerves that hold the brain in place

(Fig. 2i). Gently remove the brain and place it into a tissue culture plate or other collection vessel filled with 4 % paraformaldehyde overnight (approximately 12–16 h) to postfix.

18. Harvest all peripheral tissues of interest. Antal et al. [10] provide a good overview for mouse organ dissection. Store tissues in a tissue culture plate filled with 4 % paraformaldehyde overnight (approximately 12–16 h) to postfix.
19. Dispose of mouse carcass.
20. After an overnight postfix, move all harvested tissues into a new tissue culture plate filled with 30 % sucrose. Once the 30 % sucrose fully permeates the tissues and they sink to the bottom of the wells, they are ready for sectioning.
21. Cut brains using a frozen, sliding microtome at a thickness of 40 μm and collect tissues in 24-well plates filled with cryoprotectant solution.
22. Cut peripheral tissues using a cryostat, at an appropriate temperature and thickness for the tissue being cut.

3.3 Immunofluorescence Staining of Brain to Assess Transduction

Day 1

1. Wash free-floating tissue sections in dilution media in netted staining dishes (5 × 5 min) (*see* **Notes 8** and **9**).
2. Block tissue in 5 % of the appropriate serum (goat, donkey, etc.) in netted staining dishes (5 ml serum per 100 ml dilution media).
3. Incubate tissue in primary antibody solution (primary antibody solution: 3 ml serum and 400 μl triton-X per 100 ml PBS) in 24-well tissue culture plates, either shaking at room temp overnight or at 4° for 48 h. Each primary antibody should be used at its own concentration.

Day 2

1. Wash tissue in dilution media in netted staining dishes (5 × 5 min).
2. Incubate tissue in the appropriate fluorophore-conjugated secondary antibody solution in 24-well tissue culture plates. Secondary antibody solution: 3 ml normal serum per 100 ml dilution media. Each secondary antibody should be added to this solution at its own required concentration, which is typically 1:500. Incubation time will depend on each primary and secondary antibody, which can range from 15 min to 4 h (*see* **Notes 10** and **11**).
3. Wash tissue in TBS in netted staining dishes (3 × 5 min).
4. Incubate tissue in Hoechst 333258 pentahydrate solution (diluted to a concentration of 1:10,000 in water) for 1 min.

Hoechst stains DNA fluorescent blue which is used to visualize nuclei.

5. Wash tissue again in TBS in netted staining dishes (3 × 5 min)
6. Store tissue at 4° in PBS either in the netted staining dishes or in a new storage container, covered in foil to protect from light, until able to mount onto slides.
7. Mount tissue onto either subbed (gelatin coated) or charged microscope slides. Allow for tissue to partially dry (approximately 5 min) so that they are well adhered to the slide, yet not completely dried out.
8. Apply approximately 75 μl of wet mounting medium to the slide (*see* **Note 12**), and place coverslip. Allow to dry for 30–60 min. Seal the outside of the slide (around the coverslip) using clear nail polish, to prevent the wet-mount and tissue from drying out.
9. Store slides at 4° in an opaque container.

4 Notes

1. Some mouse strains (general background and/or particular transgenic lines) can be sensitive to anesthesia. Exercise caution when injecting with anesthesia for the first time. The dosing can be altered slightly to accommodate the required dose for the particular strain utilized in a study.
2. As there is a significant amount of dead space in regular syringes and needles, we use a glass Hamilton luer tip syringe, which dramatically reduces the amount of virus lost during each injection (roughly 50 μl vs. 5 μl). A small gauge needle is also desirable for jugular injection as to cause a minimal amount of tissue damage and to reduce the amount of blood and virus that backs out of the opening. 700 series Hamilton syringes, which fit Hamilton 30G Kel-F hub needles, work particularly well.
3. From our own studies, we have discovered that the amount of AAV9 used is critical for successful silencing using RNAi. The degree of silencing will depend on various factors including the promoter used, the potency of the silencing sequence, the total number of cells transduced, total viral genomes injected, and bodyweight of the animals injected. We found that when we injected approximately 7.5e10 viral genomes per gram bodyweight (i.e., 7.5e11 tvg for a 10 g mouse), we achieved significant 12–33 % reductions of our target gene within various regions of the CNS. Theoretically, increased dosing (increased viral genomes per gram body weight) should increase the number of cells transduced and the degree of silencing

achieved. In a previous study in which we injected mice in the range of 2e10–3e10 vg/g bodyweight, we did not achieve significant reductions in our target gene in the CNS using RNAi. However, at both doses, significant reductions of our target gene in the periphery were achieved. For further information, *see* Dufour et al. [9].

4. To further confirm correct placement in the jugular vein, you can pull back on the plunger of the Hamilton and backfill the syringe with a small amount of blood from the vein. When infusing, the viral solution visibly mixes directly with the blood, providing further confirmation of correct placement. If the needle tip is not accurately placed in the jugular vein lumen, virus will typically collect under the connective tissue that covers the jugular and create a bubble there. If this occurs, remove the needle from the pectoral muscle and jugular vein and replace the needle adjacent to the first site.

5. Fu et al. [4] showed that vascular pretreatment with 25 % mannitol, prior to tail-vein injection of AAV9, increased transduction of the CNS. In our studies, we found that pretreatment with 25 % mannitol, when compared to saline pretreatment, did not increase AAV9 transduction following intra-jugular administration [9].

6. The protocol listed here is easily modifiable for collecting tissues for molecular analyses instead of histology. The main modification is that paraformaldehyde is not used with tissues that are used for molecular analyses. **Step 14** (perfuse with 10 cc of 4 % paraformaldehyde) should be omitted entirely. To account for reduced solution clearing the blood from the circulatory system, **step 13** should be modified so that 15 cc of saline is used for the perfusion instead of 10 cc. For **steps 17** and **18**, do not postfix tissues in paraformaldehyde for molecular analyses—instead, tissues should be harvested/dissected, placed in RNAse/DNAse free microcentrifuge tubes, frozen on dry ice, and stored at −80 °C until use. For region-specific analyses of the CNS—immediately after removal, keep the brain cold in a saline-filled petri dish sitting on wet ice, then cut the brain into 1 mm thick coronal slabs using an ice-cold mouse brain matrix and sterile razor blades. Regions of interest can be dissected using micro scissors and small forceps, placed into microcentrifuge tubes, frozen on dry ice, and stored at −80 °C until use. If both molecular and histological analyses are desired from the same animal, collect and store specimens for molecular analysis first; next, postfix any tissues for histological analysis in 4 % paraformaldehyde for 24 h, then sink and store in 30 % sucrose until ready to section.

7. If the saline is pushing through the vasculature appropriately, the mouse's blood should drain from the right atria, visible

blood vessels should clear, and the liver should lighten in color from a deep red-brown to a light pink-tan color. If the needle breaches the wall between the ventricles, then the lungs fill with saline and saline will be pushing out through the mouth and/or nose of the mouse. When this occurs, the blood does not clear from the circulatory system well, which will create artifacts for histological analysis. If this occurs, you can push more saline and/or paraformaldehyde through (approximately an extra 5 ml) and do so at increased pressure/rate. Sometimes pulling back on the needle can also help.

8. In our experience, reporter gene expression delivered with AAV9 is visualized more completely when the tissue is stained using immunofluorescence methods as opposed to simply visualizing the native reporter gene expression (i.e., native GFP expression) or staining using standard DAB immunohistochemistry. For this reason, immunofluorescence staining methodology is included here.
9. For staining sections that are not free-floating (i.e., paraffin embedded or cryostat cut sections that are immediately placed on slides), the immunofluorescence protocol outlined here will work by making minor modifications. The sections should be stained directly on the slide. First, bring slides/sections to room temperature, by leaving them on a lab bench for 20 min. Next, use a pap pen to outline the tissue on the slide, which creates a wax barrier to hold solutions onto the slide for staining. Follow the same incubation times and sequence of reagents included in the free-floating protocol outlined in this chapter.
10. With slight modifications to this protocol, double-label immunofluorescence can also be used to identify specific cell types that are transduced by AAV9. First, the tissue should be simultaneously incubated in both primary antibodies of interest, in Day 1—**step 3**, ensuring that the primary antibodies are made in different animals. The other modification is for Day 2—**step 2**: the tissue should be incubated in the appropriate fluorophore-conjugated secondary antibodies in sequence (i.e., incubate in secondary antibody #1, followed by 3×5 min washes in dilution media, then incubate in secondary antibody #2). After this modification, the rest of the procedure can proceed as outlined, continuing again with Day 2—**step 3**.
11. When staining for GFP expression, we were able to best detect our signal when using our secondary antibody (Alexafluor 488 Goat anti Rabbit) at a concentration of 1:500 and incubating the tissue in this antibody solution for approximately 3–4 h (Fig. 3). When we increased the concentration of secondary antibody, there was too much background, and when we decreased the concentration we lost the signal. By extending the incubation time from our 1 h standard, we were able

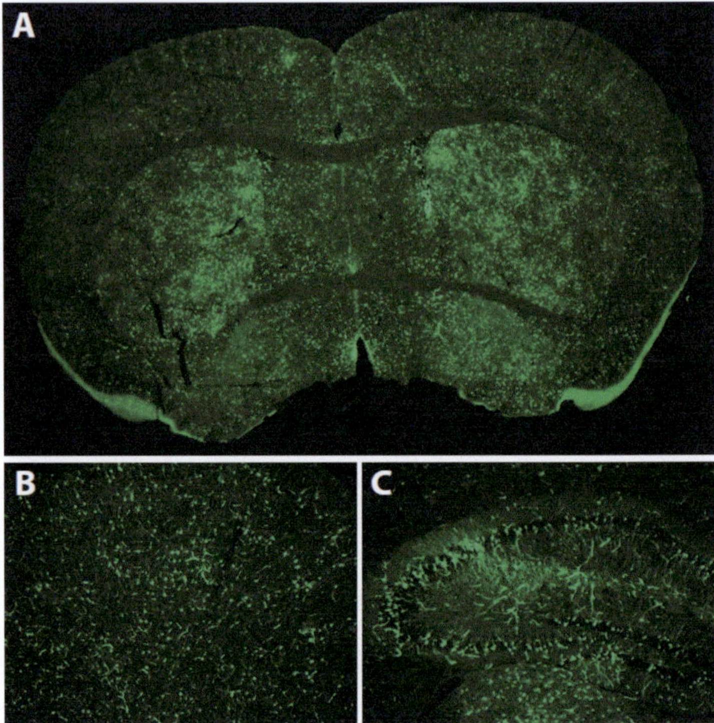

Fig. 3 Immunofluorescence detection of expressed transgene. (**a**) Immunofluorescence-stained coronal brain section of an AAV9-GFP injected mouse, showing widespread GFP expression. Examples of regional brain expression are shown: (**b**) GFP expression in frontal cortex and (**c**) hippocampus

to obtain a really good signal-to-noise ratio. This secondary was used in conjunction with a rabbit anti-GFP antibody used at a 1:1000 concentration in Day 1—**step 3**.

12. Various companies offer mounting media that slows the degradation of the fluorophore used for immunofluorescence (e.g., Vectashield mounting medium, Vector laboratories), which is helpful for maintaining the life of the tissue as well as reducing photobleaching during examination under the microscope.

Acknowledgments

This research was supported by a research grant from the Hereditary Disease Foundation (J.L.M.), ONPRC Core Grants RR000163 (J.L.M.) and P51OD011092 (J.L.M.), and a T32 Neuroscience training grant NS7466-14 (B.D.D.). Confocal microscopy was supported by grants S10RR024585 and P30-NS061800.

References

1. Zincarelli C, Soltys S, Rengo G, Rabinowitz JE (2008) Analysis of AAV serotypes 1–9 mediated gene expression and tropism in mice after systemic injection. Mol Ther 16:1073–1080
2. Foust KD, Nurre E, Montgomery CL, Hernandez A, Chan CM, Kaspar BK (2009) Intravascular AAV9 preferentially targets neonatal neurons and adult astrocytes. Nat Biotechnol 27:59–65
3. Foust KD, Salazar DL, Likhite S, Ferraiuolo L, Ditsworth D, Ilieva H, Meyer K, Schmelzer L, Braun L, Cleveland DW, Kaspar BK (2013) Therapeutic AAV9-mediated suppression of mutant SOD1 slows disease progression and extends survival in models of inherited ALS. Mol Ther 21:2148–2159
4. Fu H, Dirosario J, Killedar S, Zaraspe K, McCarty DM (2011) Correction of neurological disease of mucopolysaccharidosis IIIB in adult mice by rAAV9 trans-blood–brain barrier gene delivery. Mol Ther 19:1025–1033
5. Gadalla KK, Bailey ME, Spike RC, Ross PD, Woodard KT, Kalburgi SN, Bachaboina L, Deng JV, West AE, Samulski RJ, Gray SJ, Cobb SR (2013) Improved survival and reduced phenotypic severity following AAV9/MECP2 gene transfer to neonatal and juvenile male Mecp2 knockout mice. Mol Ther 21:18–30
6. Garg SK, Lioy DT, Cheval H, McGann JC, Bissonnette JM, Murtha MJ, Foust KD, Kaspar BK, Bird A, Mandel G (2013) Systemic delivery of MeCP2 rescues behavioral and cellular deficits in female mouse models of Rett syndrome. J Neurosci 33:13612–13620
7. Fechner H, Sipo I, Westermann D, Pinkert S, Wang X, Suckau L, Kurreck J, Zeichhardt H, Muller O, Vetter R, Erdmann V, Tschope C, Poller W (2008) Cardiac-targeted RNA interference mediated by an AAV9 vector improves cardiac function in coxsackievirus B3 cardiomyopathy. J Mol Med (Berl) 86:987–997
8. Suckau L, Fechner H, Chemaly E, Krohn S, Hadri L, Kockskamper J, Westermann D, Bisping E, Ly H, Wang X, Kawase Y, Chen J, Liang L, Sipo I, Vetter R, Weger S, Kurreck J, Erdmann V, Tschope C, Pieske B, Lebeche D, Schultheiss HP, Hajjar RJ, Poller WC (2009) Long-term cardiac-targeted RNA interference for the treatment of heart failure restores cardiac function and reduces pathological hypertrophy. Circulation 119:1241–1252
9. Dufour BD, Smith CA, Clark RL, Walker TR, McBride JL (2014) Intrajugular vein delivery of AAV9-RNAi prevents neuropathological changes and weight loss in Huntington's disease mice. Mol Ther 22:797–810
10. Antal C, Teletin M, Wendling O, Dgheem M, Auwerx J, Mark M (2007) Tissue collection for systematic phenotyping in the mouse. Curr Protoc Mol Biol Chapter 29, Unit 29A.24

Chapter 22

Efficient Gene Suppression in Dorsal Root Ganglia and Spinal Cord Using Adeno-Associated Virus Vectors Encoding Short-Hairpin RNA

Mitsuhiro Enomoto, Takashi Hirai, Hidetoshi Kaburagi, and Takanori Yokota

Abstract

RNA interference is a powerful tool used to induce loss-of-function phenotypes through post-transcriptional gene silencing. Small interfering RNA (siRNA) molecules have been used to target the central nervous system (CNS) and are expected to have clinical utility against refractory neurodegenerative diseases. However, siRNA is characterized by low transduction efficiency, insufficient inhibition of gene expression, and short duration of therapeutic effects, and is thus not ideal for treatment of neural tissues and diseases. To address these problems, viral delivery of short-hairpin RNA (shRNA) expression cassettes that support more efficient and long-lasting transduction into target tissues is expected to be a promising delivery tool. Various types of gene therapy vectors have been developed, such as adenovirus, adeno-associated virus (AAV), herpes simplex virus and lentivirus; however, AAV is particularly advantageous because of its relative lack of immunogenicity and lack of chromosomal integration. In human clinical trials, recombinant AAV vectors are relatively safe and well-tolerated. In particular, serotype 9 of AAV (AAV9) vectors show the highest tropism for neural tissue and can cross the blood–brain barrier, and we have shown that intrathecal delivery of AAV9 yields relatively high gene transduction into dorsal root ganglia or spinal cord. This chapter describes how to successfully use AAV vectors encoding shRNA in vivo, particularly for RNA interference in the central and peripheral nervous system.

Key words RNA interference, Short-hairpin RNA, Adeno-associated virus, Intrathecal administration, Spinal cord, Dorsal root ganglia

1 Introduction

RNA interference (RNAi), a well-documented technique that induces loss-of-function phenotypes through posttranscriptional gene silencing, can be used to develop treatments for refractory degenerative diseases or cancers [1, 2]. The RNAi pathway is initiated by the endoribonuclease Dicer, which cleaves long, double-stranded RNAs into short (21- to 23-nucleotide) interfering RNA

Kato Shum and John Rossi (eds.), *SiRNA Delivery Methods: Methods and Protocols*, Methods in Molecular Biology, vol. 1364, DOI 10.1007/978-1-4939-3112-5_22, © Springer Science+Business Media New York 2016

molecules (siRNAs) that mediate sequence-specific gene silencing [3, 4]. However, it is still difficult to deliver siRNAs into the central nervous system (CNS) because the blood–brain barrier (BBB) limits their transfection efficiency, resulting in insufficient inhibition of gene expression and limited duration of therapeutic effects. Viral delivery of short-hairpin RNA (shRNA) expression cassettes may enable more efficient and long-lasting transduction into the target CNS to achieve effective siRNA-based gene suppression. Non-pathogenic recombinant adeno-associated virus (AAV) vectors encoding therapeutic transgenes have been used in clinical studies, and vectors based on AAV can be injected intravenously for delivery of target genes to multiple organs, such as the liver, skeletal muscle and cardiac muscle. However, to avoid systemic side effects, selective administration routes should be considered. For example, our group previously showed that intrathecal administration of an AAV9-based vector did not damage tissues histologically or alter the expression of non-targeted endogenous mRNAs in the spinal cord and posterior root ganglion (DRG) [5]. Thus, we describe herein a method to successfully design and deliver an AAV vector encoding shRNA to neuronal tissue.

To efficiently silence genes in vivo by RNAi, it is first necessary to understand the elements that influence the function and specificity of siRNA to select appropriate siRNA sequences [6]. These elements include (1) sequence space restrictions that define the boundaries of siRNA targeting, (2) structural and sequence features, (3) mechanisms underlying non-specific gene modulation, and (4) additional features specific to the intended use. To select potent siRNA sequences, it is important to initially determine the appropriate candidate sequence in the target mRNA; therefore, we provide steps for choosing functional siRNA sequences [7] in Subheading 3.1.

Once an siRNA sequence is selected, knockdown efficiency should be tested in a transfectable cell line with stable gene expression to obtain reproducible research results (*see* **Note 1**). In particular, the cells used should express a specific endogenous target gene or transfect cells with a target gene of interest. The protocol in Subheading 3.2 thus describes an example using HEK293T cells expressing a target gene from a specific plasmid using firefly and Renilla luciferase activity as reporters for successful translation [8]. In particular, the psiCHECK™-2 plasmid provides a quantitative and rapid approach for optimization of RNAi.

After the selected siRNA sequence is validated in vitro, the AAV vector for in vivo siRNA expression is developed as described in Subheadings 3.3 and 3.4. When developing an AAV vector for in vivo siRNA expression, it is important to first select an AAV serotype that has tropism for the tissue of interest. In general, AAV serotypes 1, 5, 7, 8, and 9 yield widespread transgene expression in the adult rodent brain [9]; however, AAV9 shows the highest tropism for neural tissue and can cross the BBB [10]. Furthermore, upon intrathecal administration of recombinant AAV vectors (rAAV1, 6, 8 or 9) encoding green fluorescent protein, rAAV9 showed the greatest biodistribution and transduction of the spinal cord and DRG in mice [11, 12], leading us to select AAV9 for CNS administration in the protocol presented here. To generate the AAV vector, we amplified the selected siRNA together with a hairpin-forming sequence by PCR to form an shRNA sequence and then inserted this sequence into the pU6icassette vector, which contains a human U6 promoter and two BspMI restriction sites, to combine the AAV vector with the shRNA sequence; this procedure is described in Subheading 3.3.

In addition to selecting the appropriate serotype, it is also necessary to consider administration routes, such as oral ingestion, intravenous injection, or transdermal application. The DRG consists of highly vascular tissue lacking a blood-nerve barrier; thus, the permeability of cerebrospinal fluid into the DRG is relatively high [13, 14]. We therefore chose to inject the AAV9 vector intrathecally to deliver a transgene to the spinal cord and DRG, and we found that AAV9 can be used to stably express transgenes in both tissues. However, intrathecal injection is a technically demanding procedure, as microsurgery techniques are required for laminectomy and insertion of a tube into subarachnoid space without neural tissue damage. We have thus described the steps for the surgery and intrathecal AAV administration in mice in Subheading 3.4.

After intrathecal administration of the AAV vector, gene-silencing effects in the spinal cord and DRG are evaluated at the mRNA and protein level by qPCR and western blotting/immunohistochemistry, respectively, as described in Subheading 3.5. Potential side effects after injection are evaluated through assessment of the general health including body weight and laboratory data, and motor/sensory functions of the hindlimbs of the mice. In addition, expression of endogenous microRNAs (miRNA), such as Let-7 or miR-124 is measured by qPCR because overexpression of shRNA may decrease the physiological processing of endogenous miRNA [15, 16]. Steps for side effect evaluation are described in Subheading 3.6.

Thus, the protocol described here shows readers how to design an siRNA sequence to target a gene of interest, validate the sequence in vitro, incorporate the sequence into an shRNA-expressing AAV vector, deliver the vector in vivo to the spinal cord and DRG, and then evaluate both targeted gene suppression and potential side effects.

2 Materials

2.1 Reagents and Supplies to Determine Knockdown Efficacy of siRNA In Vitro

1. Plasmid DNA carrying the gene of interest; psiCHECK™-2.
2. Cells expressing a specific endogenous target gene or HEK293T cells transfected with a target gene of interest.
3. Lipofectamine LTX Reagent with PLUS™ Reagent.
4. Opti-MEM reduced-serum medium.
5. Fetal bovine serum.
6. Illuminometer.
7. Dual-Luciferase® Reporter Assay System.
8. AAV vectors (can be prepared by any method; *see* also Subheading 3.3).

2.2 Reagents and Supplies for Intrathecal Delivery

1. Eight-week-old female ICR mice weighing 25–35 g.
2. Friedman-Pearson rongeur for laminectomy.
3. Dumont #5.
4. Adson.
5. Lexer-Baby scissor.
6. 27-gauge needle.
7. PE-10 tube.
8. 10-μl Hamilton syringe.
9. Surgical microscope.
10. Magnetic stand with holder to accommodate Hamilton syringe.

2.3 Reagents and Supplies for Evaluation of In Vivo Delivery of AAV

2.3.1 qPCR

1. mRNA can be extracted from target tissue using any standard technique.
2. QuantiTect Reverse Transcription Kit.
3. Forward and reverse primers for interest genes.
4. PrimeScript RT master mix.
5. Light Cycle 480 Real-Time PCR instrument.

2.3.2 Western Blotting

1. Target tissues extracted as described below in Subheading 3.5.2.
2. Cold homogenization buffer containing 0.1 % sodium dodecylsulfate (SDS).
3. 1 % sodium deoxycholate.
4. 1 % Triton X-100.
5. 1 mM phenylmethylsulfonyl fluoride (PMSF).
6. Protease inhibitor cocktail.
7. Laemmli sample buffer.
8. 15 % SDS–PAGE gel.

9. Polyvinylidene difluoride membrane.
10. Rabbit polyclonal antibody targeting the protein of interest.
11. Mouse anti-GAPDH monoclonal antibody.
12. Horseradish peroxidase-conjugated secondary antibodies, e.g., goat anti-rabbit HRP IgG and goat anti-mouse HRP IgG.
13. SuperSignal West Femto Maximum Sensitivity Substrate.

2.3.3 Immunohistochemistry

1. Frozen sections of target tissues obtained as described below in Subheading 3.5.
2. Phosphate-buffered saline (PBS).
3. Normal goat serum (NGS).
4. Appropriate anti-mouse antibody and/or rabbit-polyclonal antibody.
5. Appropriate secondary fluorescent antibody.
6. 4′,6-Diamidino-2-phenylindole, dihydrochloride (DAPI).

2.3.4 Behavioral Testing

1. Rotarod.
2. Semmes-Weinstein monofilament anesthesiometer (Aesthesio; Precision Tactile Sensory Evaluator).
3. Syringe (1-mL; Terumo) filled with acetone.
4. Hot plate.

3 Method

3.1 Designing siRNAs with High Functionality and Specificity

1. Use siRNA Wizard (http://www.invivogen.com/sirna-wizard) to identify candidate siRNA sequences based on the sequence of the gene of interest.
2. For quality control, ensure that the selected sequence has a guanine at the 5′ end and an adenine or thymidine at the 3′ end of the sense strand. Duplexes with an adenine at the 5′ end and an adenine or thymidine at the 3′ end are less favored but may still be used, whereas duplexes with a guanine or cytosine at the 3′ end are discarded (*see* **Note 2**).
3. Ensure that the candidate siRNA sequence does not have three or five consecutive guanine or cytosine residues.
4. Ensure that the sequence has low GC content (between 30 % and 55 %, *see* **Note 3**).
5. Use the BLAST program (http://blast.ncbi.nlm.nih.gov/Blast.cgi) to evaluate the homology of the selected sequence with non-targeted genes (significant homology with non-targeted genes can lead to off-target effects).
6. After identifying a suitable candidate sequence, add a dTdT overhang to the 3′ end of the sequence to facilitate processing into manufactured siRNA duplexes.

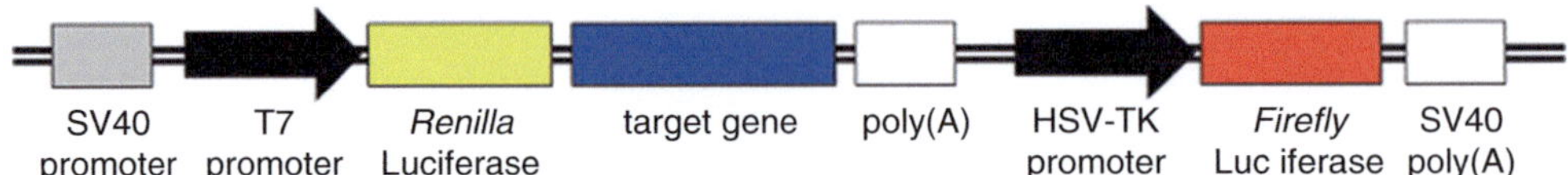

Fig. 1 Schematic representation of the plasmid vector used in vitro. This plasmid encodes firefly luciferase and a fusion of Renilla luciferase and a target gene (e.g., TRPV1)

7. Determine the silencing efficacy of these sequences should be determined by quantitative RT-PCR or an in vitro luciferase assay as described in Subheading 3.2. The design of the shRNA sequence is based on the validated siRNA but with inclusion of structural elements required for insertion into and expression from an AAV vector (*see* Subheading 3.3.1).

3.2 Knockdown Efficacy of siRNA In Vitro

1. Generate a plasmid that encodes a fusion of Renilla luciferase and the protein expressed by the target gene, in addition to a transfection control plasmid encoding firefly luciferase (Fig. 1). In this case, we used the psiCHECK-2 vector encoding TRPV1 as the target gene [8].
2. Suspend HEK293T cells in Opti-MEM medium with 10 % FBS. Apply 250 μl of the suspension per well (8×10^4 cells/well) on a 24-well plate.
3. After incubating the cells for 24 h (day in vitro 1; DIV1), use Lipofectamine Plus reagent to transfect the cells with the psi-CHECK-2 plasmid encoding TRPV1 (80 ng).
4. On DIV 3, add 10 or 25 nM siRNA to the cells. At least eight siRNAs with different sequences and siRNA containing a scrambled sequence are prepared.
5. On DIV4, rinse the cells with PBS two times gently and harvest using Passive Lysis Buffer, followed by freezing at −80 °C for 15 min.
6. Process the lysate to measure firefly and Renilla luciferase activity. Use a luminometer (GL-200: Microtec CO., LTD, Japan) and Dual-Luciferase® Reporter Assay System (Promega) (according to the manufacturer's instructions) to determine relative luminescence units.
7. Select the siRNA sequence exhibiting the highest suppression of Renilla luciferase activity, as this sequence has the strongest effect against the target gene transcript fused to the Renilla luciferase transcript. A tested siRNA that reduces Renilla luciferase activity by at least approximately 90 % relative to control Renilla luciferase activity, measured relative to firefly luciferase activity, should be selected.
8. Based on the selected siRNA sequence, design the shRNA sequence (*see* Subheading 3.3.1) for use in the AAV vector.

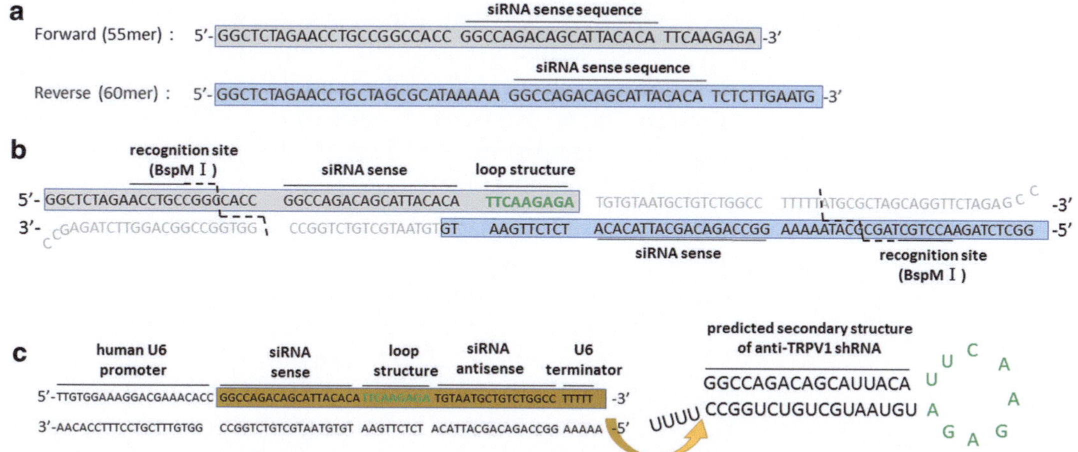

Fig. 2 Construction of anti-TRPV1 shRNA expression vector. (**a**) Forward and reverse primers used for the amplification of shRNA against target sequence contain a loop structure, terminator, BspMI site, and siRNA sense sequence. (**b**) The loop structures of the forward and reverse primers are complementary with each other. The DNA insert digested with BspMI was cloned into the pU6icassette plasmid. (**c**) The anti-TRPV1 shRNA cassette was inserted downstream of the polymerase III human U6 promoter in the AAV vector (*left*). The transcribed RNA forms a hairpin. The secondary structure of anti-TRPV1 shRNA is shown (*right*)

3.3 Generation of AAV Vector

3.3.1 Generation of AAV Encoding shRNA Targeting Gene

1. Design forward and reverse primers containing a loop structure, terminator, BspMI site and the siRNA sense sequence validated in the in vitro experiments above (Fig. 2a; *see* **Note 4**). For our anti-TRPV1 vector, the inserts of the constructs were generated by PCR with the following primers; 5′-ggctctagaacctgccggccacc-(TRPV1-siRNA sense)-ttcaagagaga-3′ and 5′-ggctctagaacctgctagcgcataaaaa-(TRPV1-siRNA sense)-tctcttgaatg-3′. Part of the sense sequence and the loop structure of these two primers are complementary to each other (Fig. 2b).
2. Amplify the shRNA construct using the two primers and a DNA polymerase with good proof-reading ability, such as Pyrobest.
3. After obtaining the DNA construct (*see* **Note 5**), digest the pU6icassette vector and DNA product with BspMI [17].
4. Ligate the vector and insert, and then transform competent DH5α cells according to the manufacturer's protocol. Plate cells on LB agar plates with tetracycline and then incubate overnight. Use sterile tips to pick individual colonies and inoculate bacterial cultures. Incubate cultures for 3 h in a shaker at 37 °C.
5. Centrifuge the cultures at 60–200 ×*g*. and decant and discard the supernatant. Use a QIAprep kit to extract the plasmid according to the manufacturer's protocol.

6. After checking whether the sequence was inserted in the pUC19U6 plasmid using specific primers targeting the regions upstream and downstream of the multi-cloning region in the plasmid, ligate the insert with the pAAV-MCS plasmid (Fig. 2c). Transform the ligated plasmid into competent DH5α cells.
7. Purify the resulting plasmid carrying the shRNA, and then co-transfect approximately 70 % confluent AAV-293 cells with the recombinant pAAV-MCS plasmid, the packaging plasmid (pAAV-RC vectors) and an adenovirus helper plasmid (pHelper) at a ratio of 2:1:1 using the established calcium phosphate method. Incubate transfected cells in Dulbecco's Modified Eagle Medium containing 10 % fetal bovine serum with 1 % penicillin/streptomycin at 37 °C [18].
8. Six hours after transfection, replace the medium with fresh culture medium, and culture the cells for 48 h at 37 °C.
9. Harvest and pellet the cells by centrifugation at 3000 ×*g* at 4 °C. Resuspend the pellet in phosphate-buffered saline, pH 8.5 (PBS). Freeze the sample at −70 °C using an ethanol-dry ice bath and then thaw the cells in a 37 °C water bath. This step should be performed 4 times for effective viral assembly.
10. Treat the suspension with Benzonase at 37 °C for 20 min to digest and remove the cellular genomic DNA and plasmids, followed by half-saturated ammonium sulfate, pH 8.5 precipitation on ice for 20 min. After centrifugation at 17,400 ×*g* for 30 min at 4 °C, discard the supernatant and keep the pellet.
11. Centrifuge the AAV vector-containing cells in PBS at 130,000 ×*g* for 15 h, collect the viral fraction from the bottom of the gradient.
12. Mix and reload the vector-containing fractions from the first iodixanol linear gradient centrifugation onto an iodixanol continuous gradient for further purification using the same conditions used for the first gradient.
13. Perform size-exclusion chromatography with an AKTA Explorer 100 HPLC system equipped with a 2-mL sample loop. Equilibrate a Superdex 200 HR 10/30 GL column (GE Healthcare) with MHA buffer (3.3 mM MES, 3.3 mM HEPES, 3.3 mM sodium acetate, 50 mM sodium chloride, pH 6.5). Load the vector-containing fraction onto the column at a flow rate of 0.5 mL/min, and collect the eluate as 0.5-mL fractions over the duration of 1 column volume (23 mL). The AAV peak fractions can be identified by absorbance at 280/260 nm. The human growth hormone (hGH) polyadenylation (poly(A)) signal inserted downstream of the shRNA sequence in the recombinant AAV vector can be quantified for viral titration by quantitative real time-PCR using the TaqMan system (Life Technologies). The

following primers and probes targeting the poly(A) signal of hGH were used: 5′-CAGGCTGGTCTCCAACTCCTC-3′ and 5′-GCAGTGGTTCACGCCTGTAA-3′ served as the primer set, and 5′-TACCCACCTTGGCCTC-3′ served as the probe.

3.3.2 AAVKnockdown of a Target Gene In Vitro

1. Suspend HEK293T cells in Opti-MEM with 10 % FBS. Apply 250 µl of the suspension per well (8×10^4 cells/well) on a 24-well plate.
2. On DIV1, add different concentrations (e.g., 1×10^{11}/mL, 3×10^{11}/mL, 1×10^{12}/mL) of the AAV vector produced in Subheading 2.3.2 to the cells.
3. On DIV3, transfect cells with the psiCHECK™-2 plasmid encoding a target gene.
4. On DIV4, harvest the transduced/transfected cells and then evaluate knockdown efficiency by luciferase assay (*see* Subheading 3.2). This step will confirm the functionality of the generated AAV vector.

3.4 Intrathecal Delivery of AAV

1. Anesthetize mice using intraperitoneal injection of chloral hydrate (0.5 mg per g body weight) or any anesthetic agents.
2. Place the mouse in the prone position on heating pad and perform partial laminectomy of the caudal portion of the second lumbar vertebra (L2) and the rostral portion of the third lumbar vertebra (L3).
3. Expose the dura mater and puncture it using a 27-gauge needle.
4. Connect a PE-10 tube to a 10-µl Hamilton syringe and caudally insert the tube into the subarachnoid space at the laminectomy site, i.e., between L2 and L3 (*see* **Note 6**).
5. Inject the AAV vector (10 µl) slowly over a 2-min period. A high-titer stock of AAV should be prepared (e.g., a titer of 6×10^{12} vector genomes per microliter was prepared in our study) and diluted with sterile PBS immediately before injection.
6. Remove the catheter, suture the incision, and maintain the mouse in the head-up-position and allowed it to recover on heating pad.

3.5 Evaluation of In Vivo Delivery

3.5.1 Evaluation of Suppression of mRNA in the Target Tissues by qPCR

1. Extract total RNA from tissues harvested at an appropriate time (e.g., 4 weeks) after AAV injection using Isogen. In our study, the RNA samples were obtained from the lumbar spinal cord (located at the first to second vertebral level) and the lumbar DRGs. Target tissues are quickly isolated on ice with microscopy, and then frozen with liquid nitrogen. An appropriate homogenizer should be prepared for a small amount of tissues (e.g., Micro Smash MS-100R TOMY SEIKO Co., LTD. Tokyo, Japan).

2. Perform real-time quantitative polymerase chain reaction using vector-specific primers.
3. Calculate the ratio of targeted mRNA expression to GAPDH mRNA expression to estimate the shRNA silencing efficiency.

3.5.2 Evaluation of Suppression of Protein in the Target Tissues by Western Blotting

1. Prepare lysates of the harvested tissue as described previously [5]. Briefly, the tissues are homogenized in cold homogenization buffer containing 0.1 % sodium dodecylsulfate (SDS), 1 % sodium deoxycholate, 1 % Triton X-100 and 1 mM phenylmethylsulfonyl fluoride together with a protein inhibitor cocktail.
2. Mix 5 μg of extracted protein from each sample with Laemmli sample buffer, denature the protein at 37 °C for 60 min, and separate the protein on a 15 % SDS–PAGE gel.
3. Transfer the separated proteins to a polyvinylidene difluoride membrane (BioRad) and then incubated the membrane with a specific rabbit polyclonal antibody and a mouse anti-GAPDH monoclonal antibody.
4. Rinse the membranes and incubate them with 0.1 % horseradish peroxidase-conjugated secondary antibody appropriate for the primary antibodies described above, e.g., goat anti-rabbit HRP IgG and goat anti-mouse HRP IgG.
5. Visualize protein–antibody interactions using SuperSignal West Femto Maximum Sensitivity Substrate.

3.5.3 Evaluation of Suppression of Protein in the Target Tissues by Immunohistochemistry

1. Intracardially perfuse animals with paraformaldehyde under terminal anesthesia. Immediately remove the target tissues (e.g., lumbar spinal cord and DRG) and post-fix them in 4 % PFA in PBS at 4 °C overnight.
2. After post-fixation, transfer the samples to PBS containing 30 % sucrose and dehydrated them for 3 days.
3. Embed the fixed and cryoprotected tissues in 4 % CMC (carboxymethylcellulose sodium salt) (*see* **Note** 7) and section them at 20 μm.
4. Incubate the sections for 30 min at room temperature in a blocking solution (5 % NGS) and then with a specific antibody for 24 h at 4 °C.
5. Wash the sections and incubated them for 1 h at room temperature with an appropriate secondary fluorescent antibody in 2 % NGS.
6. Wash the sections and counterstain the nuclei using 4′,6′-diamidino-2-phenylindole (DAPI).
7. Image the sections using a fluorescence microscope.

3.6 Evaluation of Side Effects After AAV Administration

3.6.1 Body Weight and Laboratory Data

1. Measure body weight weekly to assess the general health of mice after AAV injection; the mice should gain weight at the normal rate.
2. Quantify serum levels of alanine aminotransferase (ALT), aspartate aminotransferase (AST), alkaline phosphatase (ALP), lactate dehydrogenase (LDH), total bilirubin (T-BIL), albumin (ALB), blood urea nitrogen (BUN), and creatinine (Cre) to evaluate liver and kidney dysfunction.

3.6.2 Behavioral Phenotype

1. Perform rotarod (Ugo Basile Biological Research Apparatus, Varese, Italy) testing weekly. Subject mice to forced motor activity using the rotating rod, and measure riding time (seconds) to determine endurance and motor coordination.
2. Evaluate tactile thresholds weekly. Measure mechanical sensitivity by applying a series of calibrated Semmes-Weinstein monofilaments (0.02–8 μg) to the plantar aspect of the hindpaw, with each filament applied once to each mouse. Beginning with the 1-g filament, apply each filament perpendicularly to the hindpaw for 4–6 s. Record a brisk withdrawal of the hindpaw as a positive response, and a lack of withdrawal as a negative response. Repeat the filament testing 2 times; at least two responses to the filament out of the three trials indicates an overall positive response. If the mouse demonstrates an overall positive response, apply the filament with the next lower force as described above. If no overall positive response is observed (0/3 or 1/3 responses), apply the filament with the next higher force as described above. Once the response threshold is determined (i.e., from response to no response, or vice versa), the responses to the next five filament tests are recorded to determine the median withdrawal threshold.
3. Every week, perform acetone testing to determine cold sensitivity. Using a plastic tube connected to a 1-mL syringe and without touching the skin, apply 100 μL of acetone to the plantar surface of the hindpaw. Apply acetone 5 times to each paw at an interval of at least 30 s, and record the number of brisk foot withdrawals in response to the acetone application.
4. Every week, record responses to noxious heat using a hot plate to determine heat sensitivity. Place the mice in a transparent plastic chamber on a 55 °C metal hot plate and record the time until paw flinching, licking, or withdrawal occurs. Stop measurement at a maximum cutoff of 30 s to prevent tissue damage, and wait 5 min between consecutive stimulations of the same hindpaw. Perform the test 3 times for the plantar hindpaw, and average the withdrawal latencies.

3.6.3 Evaluation of microRNA Expression to Detect Off-Target Effects of shRNA-AAV

1. Extract miRNA from the target tissue (e.g., DRG or spinal cord) using Isogen II (Nippon Gene, Tokyo, Japan).
2. Reverse transcribe 100 ng of miRNA using the PrimeScript RT master mix.
3. Amplify miRNAs of interest (e.g., Let-7 or miR-124) using miRNA-specific primers in a Lightcycler 480 real-time PCR instrument, with U6 small nuclear RNA used as an endogenous control.
4. Calculate the ratio of Let-7 or miR-124 expression to U6 expression and compare the ratio between the shRNA-AAV and the control group to determine whether shRNA affected endogenous miRNA expression.

4 Notes

1. HEK293 cells, HeLa cells or Cos7 cells are recommended. To study neuronal differentiation, axonal growth or signaling pathways, use the mouse neural crest-derived cell line Neuro 2A [19].
2. Statistical analysis of published siRNA sequences reveals that, in contrast to non-functional duplexes, functional duplexes display lower internal stability at the 5′ antisense end relative to the rest of the sequence [20, 21]. The pyrimidines C and T should be avoided because expression of RNAs from RNA polymerase III promoters is only efficient when the first transcribed nucleotide is a purine. In cases where the siRNA sequence starts with a C or T, we recommend adding an A as the first nucleotide. This addition will not affect the activity of the siRNA; rather, the addition will generate a T at the end of the antisense siRNA strand that will be included in the termination signal, thereby maintaining complementarity with the target sequence.
3. Although experts recommend the use of siRNA with a low GC content, there are many examples of active siRNAs with high GC content [22, 23].
4. The loop structures of these two primers are complementary to one another.
5. It is not necessary to ensure that the construct of interest can be amplified using DNA sequencing because it is quite difficult to read the sequence that makes the loop structure at this step.
6. The tip of the PE catheter is finely drawn by hand with steam from hot water to reduce its diameter.
7. Curling of sections is prevented by embedding tissue in 4 % CMC rather than Tissue-Tek OCT.

Acknowledgements

This work was supported in part by a Ministry of Health Labour and Welfare Sciences research grant, a Grant-in-Aid for Scientific Research (C) from the Japan Society for the Promotion of Science, and grants from the Uehara Memorial Foundation and the General Insurance Association of Japan.

References

1. Davidson BL, McCray PB (2011) Current prospects for RNA interference-based therapies. Nat Rev Genet 12:329–340
2. Takeshita F, Ochiya T (2006) Therapeutic potential of RNA interference against cancer. Cancer Sci 97:689–696
3. Mikami M, Yang J (2005) Short hairpin RNA-mediated selective knockdown of NaV1.8 tetrodotoxin-resistant voltage-gated sodium channel in dorsal root ganglion neurons. Anesthesiology 103:828–836
4. Li G, Li D, Xie Q, Shi Y, Jiang S, Jin Y (2008) RNA interfering connective tissue growth factor prevents rat hepatic stellate cell activation and extracellular matrix production. J Gene Med 10:1039–1047
5. Hirai T, Enomoto M, Machida A, Yamamoto M, Kuwahara H, Tajiri M et al (2012) Intrathecal shRNA-AAV9 inhibits target protein expression in the spinal cord and dorsal root ganglia of adult mice. Hum Gene Ther Methods 23:119–127
6. Birmingham A, Anderson E, Sullivan K, Reynolds A, Boese Q, Leake D et al (2007) A protocol for designing siRNAs with high functionality and specificity. Nat Protoc 2:2068–2078
7. Reynolds A, Leake D, Boese Q, Scaringe S, Marshall WS, Khvorova A (2004) Rational siRNA design for RNA interference. Nat Biotechnol 22:326–330
8. Hirai T, Enomoto M, Kaburagi H, Sotome S, Yoshida-Tanaka K, Ukegawa M et al (2014) Intrathecal AAV serotype 9-mediated delivery of shRNA against TRPV1 attenuates thermal hyperalgesia in a mouse model of peripheral nerve injury. Mol Ther 22:409–419
9. Lawlor PA, Bland RJ, Mouravlev A, Young D, During MJ (2009) Efficient gene delivery and selective transduction of glial cells in the mammalian brain by AAV serotypes isolated from nonhuman primates. Mol Ther 17:1692–1702
10. Bourdenx M, Dutheil N, Bezard E, Dehay B (2014) Systemic gene delivery to the central nervous system using Adeno-associated virus. Front Mol Neurosci 7:50
11. Storek B, Reinhardt M, Wang C, Janssen WG, Harder NM, Banck MS et al (2008) Sensory neuron targeting by self-complementary AAV8 via lumbar puncture for chronic pain. Proc Natl Acad Sci U S A 105:1055–1060
12. Snyder BR, Gray SJ, Quach ET, Huang JW, Leung CH, Samulski RJ et al (2011) Comparison of adeno-associated viral vector serotypes for spinal cord and motor neuron gene delivery. Hum Gene Ther 22:1129–1135
13. Allen DT, Kiernan JA (1994) Permeation of proteins from the blood into peripheral nerves and ganglia. Neuroscience 59:755–764
14. Jimenez-Andrade JM, Herrera MB, Ghilardi JR, Vardanyan M, Melemedjian OK, Mantyh PW (2008) Vascularization of the dorsal root ganglia and peripheral nerve of the mouse: implications for chemical-induced peripheral sensory neuropathies. Mol Pain 4:10
15. Grimm D, Streetz KL, Jopling CL, Storm TA, Pandey K, Davis CR et al (2006) Fatality in mice due to oversaturation of cellular microRNA/short hairpin RNA pathways. Nature 441:537–541
16. Rossi JJ (2008) Expression strategies for short hairpin RNA interference triggers. Hum Gene Ther 19:313–317
17. Miyagishi M, Taira K (2004) RNAi expression vectors in mammalian cells. Methods Mol Biol 252:483–491
18. Hermens WT, ter Brake O, Dijkhuizen PA, Sonnemans MA, Grimm D, Kleinschmidt JA et al (1999) Purification of recombinant adeno-

associated virus by iodixanol gradient ultracentrifugation allows rapid and reproducible preparation of vector stocks for gene transfer in the nervous system. Hum Gene Ther 10:1885–1891

19. Olmsted JB, Carlson K, Klebe R, Ruddle F, Rosenbaum J (1970) Isolation of microtubule protein from cultured mouse neuroblastoma cells. Proc Natl Acad Sci U S A 65: 129–136
20. Khvorova A, Reynolds A, Jayasena SD (2003) Functional siRNAs and miRNAs exhibit strand bias. Cell 115:209–216
21. Schwarz DS, Hutvágner G, Du T, Xu Z, Aronin N, Zamore PD (2003) Asymmetry in the assembly of the RNAi enzyme complex. Cell 115:199–208
22. Hasuwa H, Kaseda K, Einarsdottir T, Okabe M (2002) Small interfering RNA and gene silencing in transgenic mice and rats. FEBS Lett 532:227–230
23. Bertrand JR, Pottier M, Vekris A, Opolon P, Maksimenko A, Malvy C (2002) Comparison of antisense oligonucleotides and siRNAs in cell culture and in vivo. Biochem Biophys Res Commun 296:1000–1004

Chapter 23

Synthetic SiRNA Delivery: Progress and Prospects

Thomas C. Roberts, Kariem Ezzat, Samir EL Andaloussi, and Marc S. Weinberg

Abstract

Small interfering RNA (siRNA) is a powerful tool for modulating gene expression by RNA interference (RNAi). Duplex RNA oligonucleotides induce cleavage of homologous target transcripts, thereby enabling posttranscriptional silencing of potentially any gene. As such, siRNAs may have utility as novel pharmaceuticals for a wide range of diseases. However, a lack of "drug-likeness," physiological barriers, and potential toxicities have meant that systemic delivery of SiRNAs in vivo remains a major challenge. Here we discuss various strategies that have been employed to solve the problem of SiRNA delivery. These include chemical modification of the SiRNA, direct conjugation to bioactive moieties, and nanoparticle formulations.

Key words SiRNA delivery, Cell-penetrating peptides, Lipid nanoparticles, Lipidoids, SiRNA conjugates

1 Introduction

Small interfering ribonucleic acids (SiRNAs) are a class of RNA oligonucleotides that act as effectors of RNA interference (RNAi) in higher organisms. SiRNAs induce highly specific post-transcriptional gene silencing (PTGS) by catalytically degrading homologous RNA target transcripts. Since the first reports in *C. elegans* [1], and subsequent demonstration in human cells [2], RNAi has attracted huge interest and SiRNAs are now routinely used as research tools for the in vitro study of gene function in the laboratory. Furthermore, SiRNAs are promising novel therapeutic agents for a wide range of pathological conditions including viral infection, cancers, inherited diseases, toxic RNA disorders, and many others. SiRNA-based therapeutics present a number of advantages over conventional pharmaceuticals: (1) the mechanism of action is well understood and is identical for every target, (2) identification of lead compounds is relatively fast, and (3) any RNA transcript can be targeted, including transcripts coding for undruggable protein products. However, oligonucleotide delivery remains

Kato Shum and John Rossi (eds.), *SiRNA Delivery Methods: Methods and Protocols*, Methods in Molecular Biology, vol. 1364, DOI 10.1007/978-1-4939-3112-5_23, © Springer Science+Business Media New York 2016

a significant obstacle to the development of SiRNA pharmaceuticals. Here we describe the challenges facing synthetic SiRNA therapeutics and discuss recent advances in the nonviral delivery of SiRNAs in vivo.

SiRNAs consist of a double-stranded RNA duplex of 19–21 base pairs with 2 nucleotide single-stranded overhangs at the 3′ end of each strand (often referred to as a 21mer or 19+2mer). This configuration is characteristic of the products of processing by the ribonuclease (RNase) III family enzyme Dicer, which is involved in the maturation of endogenous-SiRNAs and microRNAs (miRNAs) [3]. Following entry into the cell, SiRNAs are bound by the Argonaute protein AGO2 and one strand (the passenger or sense strand) of the duplex is discarded [4, 5]. The remaining SiRNA strand (the guide or antisense strand) then directs the Argonaute protein (and associated RNA-Induced Silencing Complex [RISC] factors) to complementary mRNA targets which are catalytically cleaved by the slicer activity of AGO2 [6].

A number of variations on this generic SiRNA design theme have also been shown to induce RNAi. Dicer-substrate SiRNAs are 27mer duplex RNAs with a 3′ dinucleotide overhang on one termini and a blunt end at the other [7]. Processing by Dicer is coupled to RISC loading, and as a result Dicer substrate SiRNAs have improved potency relative to an equivalent 21mer SiRNAs [7]. Recently, in vivo RNAi was even demonstrated using single-stranded SiRNAs (ss-SiRNAs) that are extensively chemically modified [8, 9], suggesting that the passenger strand is not required for RNAi activity. The major advantage of ss-SiRNAs is that they can be delivered to peripheral tissues without the need for lipid formulation. An alternative approach developed by RXi Pharmaceuticals, called self-delivering SiRNAs (sd-rxRNA), has also achieved in vivo delivery in the absence of a delivery vehicle. sd-rxRNAs consist of only a short region (<15 base pairs) of duplex sequence, contain extensive chemical modifications and have a single-stranded thiol tail consisting of multiple phosphorothioate linkages [10]. Further variants on the traditional SiRNA design include blunt-ended duplexes [11], asymmetric SiRNAs [12], small internally segmented interfering RNAs (sisiRNAs) [13], and synthetic short hairpin SiRNAs (sshRNAs) [14]. In summary, SiRNAs come in a variety of flavors and, while some show potency in the absence of delivery vector, high doses of oligonucleotides are typically required. As a result, multiple strategies have been employed in order to improve SiRNA activity.

The physicochemical properties of SiRNA molecules are sub-optimal in terms of pharmaceutical development. SiRNAs are relatively large (~13 kDa) polyanions and therefore exhibit low hydrophobicity. As a result, they do not readily cross the cell membrane and experience electrostatic repulsion from the negatively charged cell surface. Despite their large size, SiRNAs are still small enough (<10 nm) that they can pass through the fenestrations of the glomerulus in the kidney and are therefore rapidly cleared by renal filtration [15]. The limited capacity of SiRNAs to associate with plasma proteins further contributes to their low tissue permeability and high renal clearance. Furthermore, extracellular biofluids such as plasma contain high concentrations of RNases. Unmodified RNA duplexes are rapidly hydrolyzed (within minutes) following intravenous injection [16, 17], and therefore systemic injection of naked SiRNA typically results in negligible gene silencing [18].

The post-transcriptional gene silencing activity of SiRNAs is generally restricted to the cytoplasm [19] (although there are exceptions to this [20–22]). Consequently, SiRNAs must escape from the endosome in order to be incorporated in cytoplasmic RISC. Failure to do so results in the SiRNA being degraded in the lysosome or re-exported by exocytosis [23]—gene silencing activity being reduced in the case of both eventualities.

A limitation of the RNAi approach is that SiRNAs have the potential to be toxic in a variety of ways. Incorporation of the passenger strand into RISC may induce off-target slicing. SiRNA strand incorporation is asymmetric, such that the strand with the most thermodynamically unstable 5′ terminus is preferentially loaded into RISC [24, 25]. Consequently, mismatches can be introduced in the sequence of the passenger strand in order to bias incorporation of the guide strand (and thereby minimize off-target silencing). Both passenger and guide strands may trigger off-target silencing of transcripts with similar sequences to the on-target sequence by both slicing (in cases where there is a high degree of homology between SiRNA and cognate target), or by miRNA-like effects (which typically only require homology over a 7 nucleotide "seed" region [26–28]). As a result, selection of the target sequence must be considered carefully so as to minimize the probability that such off-target effects may occur. Furthermore, certain sequence motifs (e.g., UGUGU and GUCCUUCAA) are recognized by Toll-like receptors (i.e., TLR7 and TLR8) and therefore trigger the innate immune system [29–32].

Given that the cellular machinery that executes SiRNA-mediated gene silencing is the same as that used by the endogenous miRNAs [33]. It is possible that successful delivery of high concentrations of exogenous SiRNAs may saturate the miRNA processing pathway leading to wide-spread off-target changes in gene expression, reduced gene-silencing efficiency [34] and, in extreme cases, fatality [35]. Although the majority of studies have ignored this potential problem, at least one study has shown effective on-target silencing in vivo without perturbation of endogenous miRNA activity [36].

2 Chemical Modifications

Chemical modification has been extensively used in the development of RNAi therapeutics in order to confer drug-like properties on the SiRNA molecule. Chemical modification of SiRNAs can improve serum nuclease stability, reduce passenger strand off-target silencing, abrogate immunostimulatory effects, improve pharmacokinetics and biodistribution, or enhance potency. The introduction of chemical modifications at certain positions may disrupt the interaction between the SiRNA and Argonaute protein, thereby reducing or abolishing gene silencing activity. Additionally, in the case of SiRNA variants that require enzymatic cleavage (e.g., Dicer substrate SiRNAs) chemical modification may also influence oligonucleotide processing. As a result, the beneficial properties of introducing specific chemical modifications must be carefully balanced against any loss in activity. Conversely, chemical modifications which disrupt RISC incorporation can be included in the passenger strand in order to abrogate potentially harmful off-target effects [37].

Several studies have systematically investigated which nucleotide positions are amenable to modification [38, 39]. However, much information regarding which modification combinations and positions are effective remains proprietary, and therefore not publicly accessible. Furthermore, due to the number of available chemistries and the vast number of possible combinations there are no unified rational modified SiRNA design rules. Here we discuss common classes of SiRNA chemical modification (Fig. 1).

2.1 Backbone Modifications

Chemical modification of the SiRNA phosphodiester (PO) backbone typically involves substitution of a non-bridging oxygen atom with another atom or group. In the case of the phosphorothioate (PS) modification the oxygen is replaced with sulfur. PS-SiRNAs have been developed with several configurations, including fully PS-substituted guide strands, alternating PS and PO backbones in both strands, terminal PS linkages, and 3′ PS tails [10, 40]. PS-containing SiRNAs exhibit improved nuclease resistance, enhanced binding to plasma proteins [41] and enhanced uptake [42]. A disadvantage of PS modifications is that each modification reduces the oligonucleotide *Tm* by 0.5–0.8 °C. Importantly, extensive PS-modification can cause nonspecific protein binding, cytotoxicity, and loss of silencing activity [40, 43–46]. Similarly, boranophosphate SiRNAs (in which a non-bridging oxygen is substituted for borane moiety) show higher nuclease resistance and potency relative to unmodified SiRNAs [39, 47].

2.2 Sugar Modifications

SiRNAs can also be modified at the ribose ring, and specifically the hydroxyl group at the 2′ carbon atom. The simplest chemical modification is the use of 2′-deoxy (i.e., DNA) bases in the 3′ overhangs (usually TT [2]). This modification has been widely used,

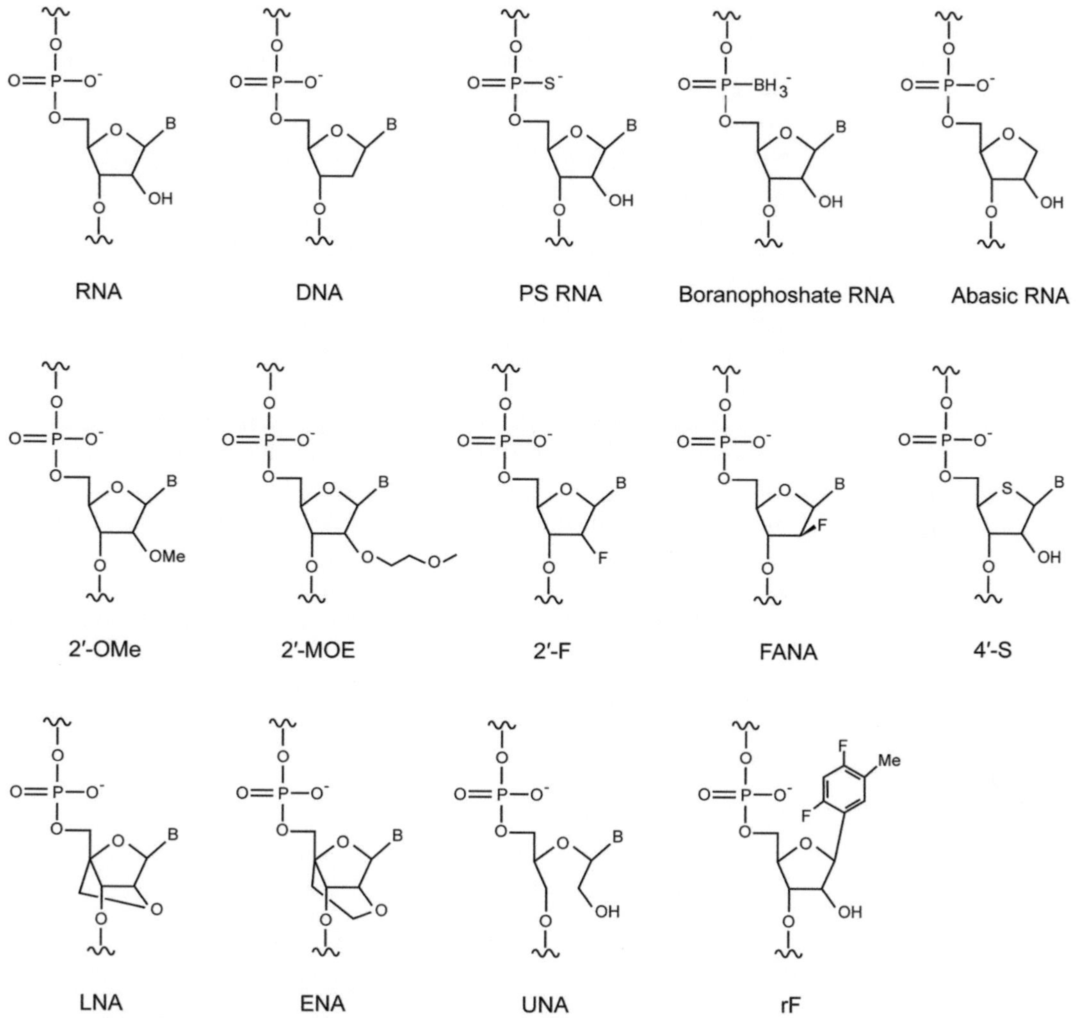

Fig. 1 Examples of nucleic acid chemistries used in modified siRNAs

particularly in the laboratory setting as it was believed to improve the resistance of the duplex to exonucleolytic degradation. However, several studies have shown that this modification can also decrease the potency of the SiRNA and does not offer a significant advantage over unmodified RNA overhangs [48–50]. The inclusion of internal DNA bases has also been shown to reduce off-target effects [51].

Other 2′ substitutions have been evaluated for their effects on SiRNA stability and efficacy. Perhaps the most common are 2′-*O*-methyl (2′-OMe), 2′-*O*-methoxyethyl (2′-MOE) and 2′-fluoro (2′-F) [11, 44–46]. Many of these modifications are well tolerated. For example, SiRNAs which are completely substituted with alternating 2′-fluoro and 2′-OMe nucleotides are RNAi competent and, in some cases, show higher silencing activity than equivalent unmodified duplexes [52]. However, complete substitution at all positions with 2′-OMe nucleotides abolishes silencing activity. Incorporation of 2′-OMe nucleotides in SiRNAs may have additional advantages as substitution at position 2 of the guide strand reduced miRNA-like silencing of seed matched off-target transcripts [53], and similarly, as few as two substitutions at uridine or guanosine bases in one strand was sufficient to abrogate immunostimulatory effects [54].

An alternative sugar modification is to substitute oxygen for sulfur in the ribose ring itself (the so-called 4′-thio or 4′-S modification). 4′-S-SiRNAs have improved nuclease stability and potency relative to unmodified SiRNAs [55, 56].

Replacement of the ribose sugar with arabinose to generate arabinonucleic acids (ANAs) has been another approach to SiRNA modification. A variation on this theme called FANA (2′-deoxy-2′-fluoro-β-D-arabinonucleic acid) has been shown to improve stability and potency with FANA modifications throughout the passenger strand and at the 3′ end of the guide strand [17, 57].

Unlocked nucleic acid (UNA) is an acyclic nucleotide chemistry also known as 2′,3′-seco-RNA [58]. Incorporation of a single UNA base at the 5′ end of the sense strand of an SiRNA abolishes sense strand gene silencing leading to improved potency of on-target silencing [37].

2.3 Bridged Nucleic Acids

Locked nucleic acids (LNA) are bicyclic nucleotides in which the pucker of the ribose ring is "locked" in a 3′-*endo* conformation by a methylene bridge linking the 2′ oxygen and 4′ carbon atoms. The major advantage of the LNA chemistry is that the *Tm* of the oligonucleotide is increased by 5–10 °C per LNA base. As such, incorporation of one, or a few, LNA nucleotides has the effect of improving binding affinity for the target RNA with minimal alteration to the SiRNA chemistry. LNA-SiRNAs are effective gene silencing agents both in cell culture and in vivo [45, 59, 60]. Incorporation of LNA bases also increases serum stability, although modification at certain nucleotide positions reduces silencing activity. Mook et al. have suggested that minimal incorporation of LNA bases at the 3′ terminus of an SiRNA strand provides a good trade-off between stability and efficacy [60]. A chemistry with similar properties 2′-*O*,4′-*C*-ethylene-bridged nucleic acid (ENA) has also been developed [61].

2.4 Nucleobase Modifications

Modifications to the nucleotide base have also been used in modified SiRNAs. For example, an SiRNA containing a single ribo-difluorotoluyl (rF) nucleotide (which is incapable of forming Watson–Crick base pairs) retained gene silencing potential with improved nuclease resistance relative to an equivalent unmodified SiRNA [62]. Conversely, abasic SiRNA substitutions have recently been shown to act as potent RNAi triggers [63]. Abasic nucleotides lack a nucleobase and may have additional backbone/ribose ring modifications. Removal of a nucleobase results in the loss of one Watson–Crick base-pair with a concomitant reduction in *Tm*. Importantly, incorporation of abasic nucleotides in the central region of an SiRNA abolished AGO2-mediated slicing but did not interfere with miRNA-like silencing effects [63, 64].

3 SiRNA Conjugates

Direct covalent conjugation was one of the first approaches employed to enhance the delivery of SiRNA. Given that SiRNAs alone lack intrinsic targeting towards specific cell types, covalent conjugation can be utilized to impart cell-targeting and/or or membrane penetrating properties on SiRNAs. A wide variety of conjugates have been described including SiRNAs covalently linked to antibodies [65], aptamers [66, 67], and other ligands (such as folate) [68]. Here we discuss conjugates with lipophilic moieties, polyethylene glycol (PEG), cell penetrating peptides (CPPs), and GalNAc in more detail. Further information about SiRNA conjugates can be found in reference [69].

3.1 Lipophilic Conjugates

Cholesterol-conjugated SiRNAs (chol-SiRNA) were utilized in the early proof-of-concept studies demonstrating SiRNA-mediated systemic knockdown of genes in mice [18] and in nonhuman primates [70]. In both studies, chol-SiRNA was used to target apolipoprotein B (ApoB) demonstrating efficient gene knockdown in the liver after systemic injection. Other lipophilic conjugates include bile acids and α-tocopherol (vitamin-E) [71, 72]. Bile acid conjugates were shown to mediate SiRNA uptake via interaction with lipoprotein particles such as high-density lipoprotein (HDL), which directs the SiRNA into liver, kidney, and steroidogenic organs, and low-density lipoprotein (LDL) which delivers SiRNA primarily to the liver. Uptake was dependent on receptor-mediated endocytosis via LDL-receptors expression for SiRNA delivery by LDL particles, and the scavenger receptor BI (SR-BI) for SiRNA delivery by HDL particles [71]. Targeting ApoB mRNA in this manner also led to accumulation of lipid droplets in the liver indicative of efficacious gene silencing.

3.2 PEG Conjugates

PEGylation (addition of polyethylene glycol, PEG) has been widely used to enhance the pharmacokinetics of large molecules and nanoparticles [73]. PEG-SiRNA conjugates were shown to have higher levels of stability than naked SiRNA in the presence of serum and lasted up to 16 h without a significant loss of integrity [74]. When incorporated with polycationic delivery vectors such as polyethylenimine (PEI), the resulting micelles displayed much higher stability than naked SiRNA and reduced VEGF expression by up to

96.5 % in prostate carcinoma cells (PC-3) [74]. In vivo, 20 kDa PEG-SiRNA conjugates demonstrated significantly prolonged circulation time after intravenous injection [75]. PEG-SiRNA also exhibited a better biodistribution profile with increased concentration in liver, kidney, spleen and lung and delayed urine excretion without significant degradation up to 24 h post-injection [75].

3.3 Cell Penetrating Peptides

Cell-Penetrating Peptides (CPPs) (also called Protein Transduction Domains, PTDs) are polybasic and/or amphipathic peptides, usually less than 30 amino acids in length, that possess the ability to traverse the plasma membrane [76, 77]. CPPs have attracted much interest in recent years as promising vectors for the delivery of a wide variety of therapeutics ranging from small molecules to nanoparticles and have been shown to enhance SiRNA uptake through covalent conjugation. SiRNAs conjugated to TAT peptides showed efficient RNAi activity and perinuclear SiRNA localization [78]. Furthermore, conjugation of TAT to the guide strand of an SiRNA via a non-cleavable thioether linkage did not interfere with RISC incorporation or RNAi activity. Similarly, SiRNAs coupled to penetratin and transportan via a cleavable disulfide bond showed efficient suppression of target reporter genes (luciferase or green fluorescence protein) in various cell lines with equivalent or better activity than cationic liposomes [79]. Furthermore, penetratin-SiRNA conjugates were shown to mediate efficient SiRNA delivery to entire populations of cultured primary mammalian hippocampal and sympathetic neurons, which are known to be hard to transfect [80]. Despite the promising results in vitro, penetratin-SiRNA conjugates were shown to induce nonspecific immune responses in vivo [81]. Penetratin conjugated SiRNA administered intratracheally led to a significant increase in the expression of TNFα and IL-12p40, as well as elevated release of IFNα, which shows that peptide-associated toxicity must be considered when designing CPP-SiRNA conjugates. It is also important to emphasize that generating pure covalent conjugates between cationic CPPs and anionic SiRNAs is nearly impossible without the problem of extensive charge-induced aggregation.

3.4 GalNAc Conjugates

Recently, *N*-acetylgalactosamine–SiRNA conjugates (GalNAc–SiRNAs) have attracted much interest. GalNAc conjugation has been shown to enhance delivery of liposomes and oligonucleotides to hepatocytes by targeting asialoglycoprotein receptors [82, 83]. Trivalent GalNAc conjugated SiRNA was shown to achieve specific accumulation in the liver with high levels of sustained liver-specific gene knockdown following subcutaneous administration (http://www.alnylam.com/web/wp-content/uploads/2013/10/ALNY-ConjugateUpdate-OTS2013.pdf). GalNAc-conjugated antisense oligonucleotides have demonstrated similar efficacy [84], thereby demonstrating that this is a promising approach for targeting oligonucleotide therapeutics to the liver.

4 Nanoparticle Approaches

Over the last couple of decades, a range of delivery vectors have been developed which condense nucleic acids to form nanoparticles that promote cellular uptake by endocytosis and protect their oligonucleotide cargoes from nuclease digestion. Although initially developed for the purpose of gene delivery, many of formulations have been adapted for the delivery of shorter oligonucleotides (including SiRNAs). These can be broadly divided into cationic polymers and lipid/lipid-like vectors. A central concept underlying the majority of these systems is that they rely, at least partially, on electrostatic interactions between the positively charged carrier and negatively charged nucleic acids to form non-covalent nanoparticles with diverse configurations. In addition to these, other interesting nanoparticle systems have evolved that have an element of covalent interactions and do not rely on charge-charge interactions. For example, Mirkin and colleagues have used small gold nanoparticles that are functionalized with nucleic acids [85–87]. Such particles are efficiently taken up by cells and promote robust gene silencing in vitro and in vivo, despite their overall negative charge.

4.1 Cationic Polymers

Synthetic and naturally occurring cationic polymers represent a large group of nucleic acid carriers. Typically these polymers are linear or branched, and functionalized in various ways. The polycationic nature of these polymers facilitates the condensation of SiRNAs into small particles that can be taken up by endocytosis based on electrostatic interactions, and facilitates disruption of the endosome via the proton sponge effect [88]. In order to further increase escape from endosomes, various fusogenic peptides and other endosome destabilizing agent have been covalently attached or non-covalently included in the nanoparticle formulations. Despite showing potency in vitro, SiRNA nanoparticles based on polyethyleneimine (PEI) and derivatives thereof have limited activity in vivo. This is due in part to vehicle-associated toxicity (PEI is nonbiodegradable) and to unfavorable interactions with various serum proteins [89]. Another approach has been to modify polymers with peptides in order to decrease the effective dose of polymer by either improving targeting or overall cell penetration. For example, combinations of polymers and peptides have been reported for systemic RNAi delivery in tumor models [90]. Wagner and colleagues have tested a range of PEI-peptide conjugates to improve targeting [91]. Similarly, PEGylated-PEI conjugated to the integrin-binding peptide RGD was successfully used for SiRNA delivery to target VEGFR in tumor bearing mice as early as 2004 [92]. Recently, the Wagner lab has stabilized polyplexes via inclusion of tyrosine trimers that assist endosomal escape and improve SiRNA delivery in vivo [93]. In addition to conventional polycations such as PEI, a range of other vectors, including protamine and chitosans, have been previously exploited with varying success, as reviewed in ref. [94]. Despite a myriad of publications on the successful use of PEI and PEI-like materials, the first polymer used in man for RNAi delivery was based on cyclodextrin [95]. In this study, the cyclodextrin-based polymer (CDP) was a short polycation that condenses SiRNA. Inclusion of PEG-moieties improved stabilization and the nanoparticle surface was coated with peptides for cellular targeting. Incorporation of imidazole groups at the polymer termini serves to buffer endocytic pH and promote endosomal escape [96]. Subsequently, several other groups have used similar modified cyclodextrin-based systems for SiRNA delivery [97, 98].

Despite many years of development for polyplexes since the initial discovery of polylysine [99], their utility for SiRNA delivery has been limited. We believe that the main reason for this is their limited tolerability in vivo. In contrast, lipid-based vectors have recently shown much greater potential.

4.2 Lipid Nanoparticles and Lipid-Like Materials

Lipid-based vectors have emerged as leading candidates for SiRNA delivery. These range from traditional cationic liposomes, initially discovered for gene delivery purposes in 1987 by Felgner [100], to more recent lipid-like materials such as the lipidoids. As excellently reviewed recently by Cullis and coworkers [101], the leading compounds for intravenously administered RNAi-based therapeutics are the lipid nanoparticles (LNPs). LNP-SiRNA systems are lipid-based particles with diameters less than 100 nm that are composed of an ionizable lipid, phosphatidylcholine, cholesterol, and a PEGylated coat lipid with defined molar ratios (also called stable nucleic acid lipid nanoparticles, SNALPs [70, 102]). In contrast to conventional liposomes, the core of LNPs is more electron dense, and encapsulation of SiRNA is close to 100 %. LNPs have shown remarkable gene silencing potency in hepatocytes following intravenous delivery at doses as low as 0.005 mg/kg [103, 104]. Similar LNP/SiRNA formulations have recently been used for silencing neuronal gene expression in the brain by intracerebroventricular injection, thereby enabling the development of gene therapies that go beyond liver-related diseases (e.g., neurodegenerative diseases) [105]. Multiple LNP/SiRNA compounds have been tested, or are under assessment, in clinical trials [106].

The other major nanoparticle platform developed for the purpose of SiRNA delivery is the lipid-like nanoparticles or lipidoids. These could be regarded as modified polymers, i.e., polyamines modified with hydrophobic acrylates that form hydrophobically stabilized cores of SiRNA. The synthesis scheme of lipidoids is based on the conjugate addition of alkyl-acrylates or alkyl-acrylamides to primary or secondary amines. Unlike many other lipid synthesis chemistries, lipidoids are generated in reactions without solvents or catalysts, which eliminates the need for protection/deprotection steps, or purification and concentration steps. They can thus be rapidly generated in large quantities. These lipid-like materials were first reported in 2008 when it was shown that lipidoids (formulated with cholesterol and PEG-lipid at defined molar ratios) form stable nanoparticles with SiRNAs and mediated robust hepatic gene silencing in mice, rats, and nonhuman primates [107]. Since then, Anderson and Langer, in collaboration with Alnylam Pharmaceuticals, have performed further screens and identified compounds with improved potency that mediated hepatic gene silencing at similar doses to LNPs [108, 109]. In addition to targeting liver, lipidoids have recently also been exploited for systemic SiRNA delivery to myeloid cells both in rodents and nonhuman primates [110]. Intriguingly, by targeting TNFα in a mouse model of rheumatoid arthritis, the authors observed comparable attenuation of disease progression as with conventional potent biologics. This study emphasizes that lipid-like materials could have utility beyond the targeting of hepatic cells.

Until very recently, few studies had reported SiRNA delivery to cells other than hepatocytes, immune cells and tumor cells. However, an exceptionally important study has now reported the utility of a polymeric nanoparticle system composed of low-molecular-weight polyamines and lipids that confers robust extra-hepatic gene silencing [111]. In contrast to lipids and lipidoids, this new platform efficiently promotes SiRNA delivery to endothelial cells with only minor gene silencing observed in liver.

4.3 CPP-Based Nanoparticles

Using CPPs to deliver oligonucleotides after non-covalent complexation was first demonstrated by the group of Gilles Divita using the MPG peptide [112]. The positively charged MPG peptide was shown to form nanoparticles with negatively charged single- and double-stranded oligonucleotides which were efficiently internalized by cells. This strategy has drastically expanded the range of therapeutic cargos that can be delivered via CPPs as it avoids laborious chemical conjugation and has almost no limitation on the size of the cargo. Since this initial publication, many other CPPs have been developed that can form nanoparticles with nucleic acid cargos and efficiently mediate their delivery in a plethora of in vitro and in vivo settings [113–115].

Chemical modification of CPPs has been used to improve nanoparticle formation and enhance membrane interaction following non-covalent complexation (Table 1). C-terminal cysteamide modification was shown to be crucial for CPP-mediated SiRNA delivery using the MPG, PEP and CADY peptide families by increasing membrane association and particle stability through the formation of peptide dimers [116–119]. Acetylation of these peptides was also required to enhance stability [118]. The addition of hydrophobic moieties to CPPs has been shown to be an efficient

Table 1
Examples of CPPs and chemical modifications compatible with non-covalent delivery of nucleic acids (adapted from ref. [122])

CPPs	Sequence	Modification	References
MPG peptides			
MPG	Ac-GALFLGWLGAAGSTM GAPKKKRKV-Cya	Acetylation, cysteamidation	[112]
MPGΔ^{NLS}	Ac-GALFLAFLAAALSLMGLWS QPKKKRKV-Cya	Acetylation, cysteamidation	[116]
MPG-8	β-AFLGWLGAWGTMGWS PKKKRK-Cya	Cysteamidation	[128]
Chol-MPG-8	chol-β-AFLGWLGAWGTMG WSPKKKRK-Cya	Cholesterol, cysteamidation	[128]
Pep and CADY peptides			
Pep-2	Ac-KETWFETWFTEWSQPKK KRKV-Cya	Acetylation, cysteamidation	[129]
Pep-3	Ac-KWFETWFTEWPKKRK-Cya	Acetylation, cysteamidation	[130]
PEG-Pep-3	PEG- KWFETWFTEWPKKRK-Cya	PEGylation, cysteamidation	[130]
CADY	Ac-GLWRALWRLLRSLW RLLWRA-Cya	Acetylation, cysteamidation	[119]
Polyarginine peptides			
Stearyl-Arg8	Stearyl-RRRRRRRR-NH_2	Stearylation	[131]
Chol-Arg9	Chol-RRRRRRRRR-NH_2	Cholesterol	[120]
Stearyl-$(RxR)_4$	Stearyl-RXRRXRRXRRXR-NH_2	Stearylation	[132]
TAT-DRBD			
TAT-DRBD	TAT fusion protein with double-stranded RNA-binding domain	–	[125]
PepFect peptides			
PF3	Stearyl-AGYLLGKINLKALAAL AKKIL-NH_2	Stearylation	[133]
PF6	Stearyl-AGYLLGK[a]INLKALAAL AKKIL-NH_2	Stearylation	[123]
PF14	Stearyl-AGYLLGKLLOOLAA AALOOLL-NH_2	Stearylation	[134]
NickFect1	Stearyl-AGY(PO_3) LLGKTNLKALAALAKKIL-NH_2	Stearylation, phosphorylation	[135]

[a] Trifluoromethylquinoline

means of increasing nanoparticle delivery efficiency. Similarly, cholesteryl modification of polyarginines and MPG-8 was demonstrated to enhance their activity for delivery of SiRNA in vivo [120]. The group of Shiroh Futaki at Kyoto university was the first to demonstrate that stearylation of polyarginines could enhance their transfection efficiency by 100-fold [121]. In recent years, stearylation and other chemical modifications to CPPs such as TP10 led to the development of the PepFect family of peptides (PFs) [122]. We have shown that PF6 is very efficient at delivering SiRNA to various cell-lines and in vivo after intravenous administration [123]. PF6 is N-terminally modified TP10 with a trifluoromethyl-quinoline moiety attached to one of the lysines in the backbone via a lysine tree. The trifluoromethyl-quinoline moiety enhances the endosomal escape of nanoparticles with a concomitant improvement in RNAi response. Furthermore, we have shown that another peptide (PF14) is very efficient at the delivery of various nucleic acid cargoes, including SiRNAs, both in solution and in solid formulation [124]. PF14-SiRNA solid formulations were as active as those freshly prepared in solution. Interestingly, the complexes were even stable and RNAi-competent after incubation at very low pH simulated gastric conditions. Together these studies demonstrate the feasibility of peptide-SiRNA delivery via multiple routes of administration (including the oral route).

Another strategy to enhance CPP association with SiRNA was to generate TAT fusion protein with double-stranded RNA-binding domain (DRBD) [125]. The DRBD binds SiRNA with high affinity, and thereby masks its negative charge allowing TAT-mediated cellular uptake. This system was shown to be very efficient at delivering SiRNA to primary cell-lines with no cytotoxicity or induction of the innate immune response. The TAT-DRBD SiRNA delivery approach has also been used successfully in vivo [126].

5 Conclusions

RNAi has enormous potential as a novel therapeutic modality. Strategies for delivering SiRNAs have focused on two general approaches (1) chemical modification of the SiRNA molecule itself, and (2) conjugation/formulation with a carrier molecule(s). Alterations in SiRNA chemistry effectively allow for its physicochemical properties to be tuned in order to maximize activity (especially in terms of improving the stability of the SiRNA to nucleases). In certain cases, highly optimized extensive SiRNA modification has even permitted in vivo delivery in the absence of delivery vehicle [8–10]. Thanks in part to advances in antisense oligonucleotide development [127], a wealth of chemical modifications are available for SiRNA incorporation, with a vast array of possible modification combinations possible [56, 57]. Alternatively, direct conjugation of cell-targeting/membrane-penetrating moieties to SiRNAs also improves delivery and can impart cell-specific targeting.

Non-covalent complexation has been extensively used to facilitate the delivery of nucleic acids. Here we have discussed complexes consisting of cationic polymers, lipids, lipidoids, and peptides for SiRNA delivery. In all cases, the carrier and nucleic acid are mixed together at defined molar ratios to form nanoparticle formulations which shield the SiRNAs from nucleolytic degradation, are of sufficient size to promote uptake by endocytosis, and bypass renal clearance. The properties of the nanoparticle complexes (e.g., nanoparticle diameter, charge, and tissue-targeting capabilities) can be further optimized via the addition of further lipid, PEG, or peptide components. Given the multitude of possibilities for SiRNA conjugation/complexation it is likely than many highly functional combinations have yet to be discovered. Whether, the success of SiRNA delivery in animal studies translates into success in clinical trials remains to be seen.

References

1. Fire A, Xu SQ, Montgomery MK, Kostas SA, Driver SE, Mello CC (1998) Potent and specific genetic interference by double-stranded RNA in Caenorhabditis elegans. Nature 391:806–811
2. Elbashir SM, Harborth J, Lendeckel W, Yalcin A, Weber K, Tuschl T (2001) Duplexes of 21-nucleotide RNAs mediate RNA interference in cultured mammalian cells. Nature 411:494–498
3. Bernstein E, Caudy AA, Hammond SM, Hannon GJ (2001) Role for a bidentate ribonuclease in the initiation step of RNA interference. Nature 409:363–366
4. Matranga C, Tomari Y, Shin C, Bartel DP, Zamore PD (2005) Passenger-strand cleavage facilitates assembly of siRNA into Ago2-containing RNAi enzyme complexes. Cell 123:607–620
5. Rand TA, Petersen S, Du FH, Wang XD (2005) Argonaute2 cleaves the anti-guide strand of siRNA during RISC activation. Cell 123:621–629
6. Liu JD, Carmell MA, Rivas FV, Marsden CG, Thomson JM, Song JJ, Hammond SM, Joshua-Tor L, Hannon GJ (2004) Argonaute2 is the catalytic engine of mammalian RNAi. Science 305:1437–1441
7. Kim DH, Behlke MA, Rose SD, Chang MS, Choi S, Rossi JJ (2005) Synthetic dsRNA Dicer substrates enhance RNAi potency and efficacy. Nat Biotechnol 23:222–226
8. Yu DB, Pendergraff H, Liu J, Kordasiewicz HB, Cleveland DW, Swayze EE, Lima WF, Crooke ST, Prakash TP, Corey DR (2012) Single-stranded RNAs use RNAi to potently and allele-selectively inhibit mutant huntingtin expression. Cell 150:895–908
9. Lima WF, Prakash TP, Murray HM, Kinberger GA, Li WY, Chappell AE, Li CS, Murray SF, Gaus H, Seth PP, Swayze EE, Crooke ST (2012) Single-stranded siRNAs activate RNAi in animals. Cell 150:883–894
10. Byrne M, Tzekov R, Wang Y, Rodgers A, Cardia J, Ford G, Holton K, Pandarinathan L, Lapierre J, Stanney W, Bulock K, Shaw S, Libertine L, Fettes K, Khvorova A, Kaushal S, Pavco P (2013) Novel hydrophobically modified asymmetric RNAi compounds (sd-rxRNA) demonstrate robust efficacy in the eye. J Ocul Pharmacol Ther 29:855–864
11. Czauderna F, Fechtner M, Dames S, Aygun H, Klippel A, Pronk GJ, Giese K, Kaufmann J (2003) Structural variations and stabilising modifications of synthetic siRNAs in mammalian cells. Nucleic Acids Res 31:2705–2716
12. Yuan ZP, Wu XL, Liu C, Xu GX, Wu ZW (2012) Asymmetric siRNA: new strategy to improve specificity and reduce off-target gene expression. Hum Gene Ther 23:521–532
13. Bramsen JB, Laursen MB, Damgaard CK, Lena SW, Babu BR, Wengel J, Kjems J (2007) Improved silencing properties using small internally segmented interfering RNAs. Nucleic Acids Res 35:5886–5897
14. Dallas A, Ilves H, Shorenstein J, Judge A, Spitler R, Contag C, Wong SP, Harbottle RP, MacLachlan I, Johnston BH (2013) Minimal-

length Synthetic shRNAs Formulated with Lipid Nanoparticles are Potent Inhibitors of Hepatitis C Virus IRES-linked Gene Expression in Mice. Mol Ther Nucleic Acids 2, e123

15. Gao S, Dagnaes-Hansen F, Nielsen EJB, Wengel J, Besenbacher F, Howard KA, Kjems J (2009) The effect of chemical modification and nanoparticle formulation on stability and biodistribution of siRNA in mice. Mol Ther 17:1225–1233
16. Tsui NBY, Ng EKO, Lo YMD (2002) Stability of endogenous and added RNA in blood specimens, serum, and plasma. Clin Chem 48:1647–1653
17. Dowler T, Bergeron D, Tedeschi AL, Paquet L, Ferrari N, Damha MJ (2006) Improvements in siRNA properties mediated by 2′-deoxy-2′-fluoro-beta-D-arabinonucleic acid (FANA). Nucleic Acids Res 34:1669–1675
18. Soutschek J, Akinc A, Bramlage B, Charisse K, Constien R, Donoghue M, Elbashir S, Geick A, Hadwiger P, Harborth J, John M, Kesavan V, Lavine G, Pandey RK, Racie T, Rajeev KG, Rohl I, Toudjarska I, Wang G, Wuschko S, Bumcrot D, Koteliansky V, Limmer S, Manoharan M, Vornlocher HP (2004) Therapeutic silencing of an endogenous gene by systemic administration of modified siRNAs. Nature 432:173–178
19. Zeng Y, Cullen BR (2002) RNA interference in human cells is restricted to the cytoplasm. RNA 8:855–860
20. Gagnon KT, Li LD, Chu YJ, Janowski BA, Corey DR (2014) RNAi factors are present and active in human cell nuclei. Cell Rep 6:211–221
21. Robb GB, Brown KM, Khurana J, Rana TM (2005) Specific and potent RNAi in the nucleus of human cells. Nat Struct Mol Biol 12:133–137
22. Roberts TC (2014) The MicroRNA biology of the mammalian nucleus. Mol Ther Nucleic Acids 3, e188
23. Sahay G, Querbes W, Alabi C, Eltoukhy A, Sarkar S, Zurenko C, Karagiannis E, Love K, Chen DL, Zoncu R, Buganim Y, Schroeder A, Langer R, Anderson DG (2013) Efficiency of siRNA delivery by lipid nanoparticles is limited by endocytic recycling. Nat Biotechnol 31:653–658
24. Khvorova A, Reynolds A, Jayasena SD (2003) Functional siRNAs and miRNAs exhibit strand bias (vol 115, pg 209, 2003). Cell 115:505
25. Schwarz DS, Hutvagner G, Du T, Xu ZS, Aronin N, Zamore PD (2003) Asymmetry in the assembly of the RNAi enzyme complex. Cell 115:199–208
26. Jackson AL, Burchard J, Schelter J, Chau BN, Cleary M, Lim L, Linsley PS (2006) Widespread siRNA "off-target" transcript silencing mediated by seed region sequence complementarity. RNA 12:1179–1187
27. Scacheri PC, Rozenblatt-Rosen O, Caplen NJ, Wolfsberg TG, Umayam L, Lee JC, Hughes CM, Shanmugam KS, Bhattacharjee A, Meyerson M, Collins FS (2004) Short interfering RNAs can induce unexpected and divergent changes in the levels of untargeted proteins in mammalian cells. Proc Natl Acad Sci U S A 101:1892–1897
28. Persengiev SP, Zhu XC, Green MR (2004) Nonspecific, concentration-dependent stimulation and repression of mammalian gene expression by small interfering RNAs (siRNAs). RNA 10:12–18
29. Hornung V, Guenthner-Biller M, Bourquin C, Ablasser A, Schlee M, Uematsu S, Noronha A, Manoharan M, Akira S, de Fougerolles A, Endres S, Hartmann G (2005) Sequence-specific potent induction of IFN-alpha by short interfering RNA in plasmacytoid dendritic cells through TLR7. Nat Med 11:263–270
30. Judge AD, Sood V, Shaw JR, Fang D, McClintock K, MacLachlan I (2005) Sequence-dependent stimulation of the mammalian innate immune response by synthetic siRNA. Nat Biotechnol 23:457–462
31. Robbins M, Judge A, MacLachlan I (2009) siRNA and innate immunity. Oligonucleotides 19:89–101
32. Heil F, Hemmi H, Hochrein H, Ampenberger F, Kirschning C, Akira S, Lipford G, Wagner H, Bauer S (2004) Species-specific recognition of single-stranded RNA via toll-like receptor 7 and 8. Science 303:1526–1529
33. Doench JG, Petersen CP, Sharp PA (2003) siRNAs can function as miRNAs. Genes Dev 17:438–442
34. Hong J, Qian ZK, Shen SY, Min TS, Tan C, Xu JF, Zhao YC, Huang WD (2005) High doses of siRNAs induce eri-1 and adar-1 gene expression and reduce the efficiency of RNA interference in the mouse. Biochem J 390:675–679
35. Grimm D, Streetz KL, Jopling CL, Storm TA, Pandey K, Davis CR, Marion P, Salazar F, Kay MA (2006) Fatality in mice due to oversaturation of cellular microRNA/short hairpin RNA pathways. Nature 441:537–541
36. John M, Constien R, Akinc A, Goldberg M, Moon YA, Spranger M, Hadwiger P, Soutschek J, Vornlocher HP, Manoharan M, Stoffel M, Langer R, Anderson DG, Horton JD, Koteliansky V, Bumcrot D (2007) Effective RNAi-mediated gene silencing

without interruption of the endogenous microRNA pathway. Nature 449:745–747

37. Snead NM, Escamilla-Powers JR, Rossi JJ, McCaffrey AP (2013) 5′ unlocked nucleic acid modification improves siRNA targeting. Mol Ther Nucleic Acids 2, e103
38. Bramsen JB, Laursen MB, Nielsen AF, Hansen TB, Bus C, Langkjaer N, Babu BR, Hojland T, Abramov M, Van Aerschot A, Odadzic D, Smicius R, Haas J, Andree C, Barman J, Wenska M, Srivastava P, Zhou CZ, Honcharenko D, Hess S, Muller E, Bobkov GV, Mikhailov SN, Fava E, Meyer TF, Chattopadhyaya J, Zerial M, Engels JW, Herdewijn P, Wengel J, Kjems J (2009) A large-scale chemical modification screen identifies design rules to generate siRNAs with high activity, high stability and low toxicity. Nucleic Acids Res 37:2867–2881
39. Choung S, Kim YJ, Kim S, Park HO, Choi YC (2006) Chemical modification of siRNAs to improve serum stability without loss of efficacy. Biochem Biophys Res Commun 342:919–927
40. Harborth J, Elbashir SM, Vandenburgh K, Manninga H, Scaringe SA, Weber K, Tuschl T (2003) Sequence, chemical, and structural variation of small interfering RNAs and short hairpin RNAs and the effect on mammalian gene silencing. Antisense Nucleic Acid Drug Dev 13:83–105
41. Geary RS, Yu RZ, Levin AA (2001) Pharmacokinetics of phosphorothioate antisense oligodeoxynucleotides. Curr Opin Investig Drugs 2:562–573
42. Overhoff M, Sczakiel G (2005) Phosphorothioate-stimulated uptake of short interfering RNA by human cells. EMBO Rep 6:1176–1181
43. Krieg AM, Stein CA (1995) Phosphorothioate oligodeoxynucleotides: antisense or anti-protein? Antisense Res Dev 5:241
44. Chiu YL, Rana TM (2003) siRNA function in RNAi: a chemical modification analysis. RNA 9:1034–1048
45. Braasch DA, Jensen S, Liu Y, Kaur K, Arar K, White MA, Corey DR (2003) RNA interference in mammalian cells by chemically-modified RNA. Biochemistry 42: 7967–7975
46. Amarzguioui M, Holen T, Babaie E, Prydz H (2003) Tolerance for mutations and chemical modifications in a siRNA. Nucleic Acids Res 31:589–595
47. Hall AHS, Wan J, Shaughnessy EE, Shaw BR, Alexander KA (2004) RNA interference using boranophosphate siRNAs: structure-activity relationships. Nucleic Acids Res 32:5991–6000
48. Hohjoh H (2002) RNA interference (RNAi) induction with various types of synthetic oligonucleotide duplexes in cultured human cells. FEBS Lett 521:195–199
49. Boutla A, Delidakis C, Livadaras I, Tabler M (2003) Variations of the 3′ protruding ends in synthetic short interfering RNA (siRNA) tested by microinjection in Drosophila embryos. Oligonucleotides 13:295–301
50. Rettig GR, Behlke MA (2012) Progress toward in vivo use of siRNAs-II. Mol Ther 20:483–512
51. Ui-Tei K, Naito Y, Zenno S, Nishi K, Yamato K, Takahashi F, Juni A, Saigo K (2008) Functional dissection of siRNA sequence by systematic DNA substitution: modified siRNA with a DNA seed arm is a powerful tool for mammalian gene silencing with significantly reduced off-target effect. Nucleic Acids Res 36:2136–2151
52. Allerson CR, Sioufi N, Jarres R, Prakash TP, Naik N, Berdeja A, Wanders L, Griffey RH, Swayze EE, Bhat B (2005) Fully 2′-modified oligonucleotide duplexes with improved in vitro potency and stability compared to unmodified small interfering RNA. J Med Chem 48:901–904
53. Jackson AL, Burchard J, Leake D, Reynolds A, Schelter J, Guo J, Johnson JM, Lim L, Karpilow J, Nichols K, Marshall W, Khvorova A, Linsley PS (2006) Position-specific chemical modification of siRNAs reduces "off-target" transcript silencing. RNA 12:1197–1205
54. Judge AD, Bola G, Lee AC, MacLachlan I (2006) Design of noninflammatory synthetic siRNA mediating potent gene silencing in vivo. Mol Ther 13:494–505
55. Hoshika S, Minakawa N, Kamiya H, Harashima H, Matsuda A (2005) RNA interference induced by siRNAs modified with 4′-thioribonucleosides in cultured mammalian cells. FEBS Lett 579:3115–3118
56. Dande P, Prakash TP, Sioufi N, Gaus H, Jarres R, Berdeja A, Swayze EE, Griffey RH, Bhat B (2006) Improving RNA interference in mammalian cells by 4′-thio-modified small interfering RNA (siRNA): effect on siRNA activity and nuclease stability when used in combination with 2′-O-alkyl modifications. J Med Chem 49:1624–1634
57. Watts JK, Choubdar N, Sadalapure K, Robert F, Wahba AS, Pelletier J, Pinto BM, Damha MJ (2007) 2′-fluoro-4′-thioarabino-modified oligonucleotides: conformational switches linked to siRNA activity. Nucleic Acids Res 35:1441–1451
58. Langkjaer N, Pasternak A, Wengel J (2009) UNA (unlocked nucleic acid): a flexible RNA mimic that allows engineering of nucleic acid duplex stability. Bioorg Med Chem 17:5420–5425

59. Elmen J, Thonberg H, Ljungberg K, Frieden M, Westergaard M, Xu Y, Wahren B, Liang Z, Orum H, Koch T, Wahlestedt C (2005) Locked nucleic acid (LNA) mediated improvements in siRNA stability and functionality. Nucleic Acids Res 33:439–447
60. Mook OR, Baas F, de Wissel MB, Fluiter K (2007) Evaluation of locked nucleic acid-modified small interfering RNA in vitro and in vivo. Mol Cancer Ther 6:833–843
61. Morita K, Hasegawa C, Kaneko M, Tsutsumi S, Sone J, Ishikawa T, Imanishi T, Koizumi M (2002) 2′-O,4′-C-ethylene-bridged nucleic acids (ENA): highly nuclease-resistant and thermodynamically stable oligonucleotides for antisense drug. Bioorg Med Chem Lett 12:73–76
62. Xia J, Noronha A, Toudjarska I, Li F, Akinc A, Braich R, Frank-Kamenetsky M, Rajeev KG, Egli M, Manoharan M (2006) Gene silencing activity of siRNAs with a ribo-difluorotoluyl nucleotide. ACS Chem Biol 1:176–183
63. Liu J, Pendergraff H, Narayanannair KJ, Lackey JG, Kuchimanchi S, Rajeev KG, Manoharan M, Hu J, Corey DR (2013) RNA duplexes with abasic substitutions are potent and allele-selective inhibitors of huntingtin and ataxin-3 expression. Nucleic Acids Res 41:8788–8801
64. Chorn G, Zhao L, Sachs AB, Flanagan WM, Lim LP (2010) Persistence of seed-based activity following segmentation of a microRNA guide strand. RNA 16:2336–2340
65. Song E, Zhu P, Lee SK, Chowdhury D, Kussman S, Dykxhoorn DM, Feng Y, Palliser D, Weiner DB, Shankar P, Marasco WA, Lieberman J (2005) Antibody mediated in vivo delivery of small interfering RNAs via cell-surface receptors. Nat Biotechnol 23:709–717
66. McNamara JO 2nd, Andrechek ER, Wang Y, Viles KD, Rempel RE, Gilboa E, Sullenger BA, Giangrande PH (2006) Cell type-specific delivery of siRNAs with aptamer-siRNA chimeras. Nat Biotechnol 24:1005–1015
67. Dassie JP, Liu XY, Thomas GS, Whitaker RM, Thiel KW, Stockdale KR, Meyerholz DK, McCaffrey AP, McNamara JO 2nd, Giangrande PH (2009) Systemic administration of optimized aptamer-siRNA chimeras promotes regression of PSMA-expressing tumors. Nat Biotechnol 27:839–849
68. Dohmen C, Frohlich T, Lachelt U, Rohl I, Vornlocher HP, Hadwiger P, Wagner E (2012) Defined folate-PEG-siRNA conjugates for receptor-specific gene silencing. Mol Ther Nucleic Acids 1, e7
69. Jeong JH, Mok H, Oh YK, Park TG (2009) siRNA conjugate delivery systems. Bioconjug Chem 20:5–14
70. Zimmermann TS, Lee AC, Akinc A, Bramlage B, Bumcrot D, Fedoruk MN, Harborth J, Heyes JA, Jeffs LB, John M, Judge AD, Lam K, McClintock K, Nechev LV, Palmer LR, Racie T, Rohl I, Seiffert S, Shanmugam S, Sood V, Soutschek J, Toudjarska I, Wheat AJ, Yaworski E, Zedalis W, Koteliansky V, Manoharan M, Vornlocher HP, MacLachlan I (2006) RNAi-mediated gene silencing in non-human primates. Nature 441:111–114
71. Wolfrum C, Shi S, Jayaprakash KN, Jayaraman M, Wang G, Pandey RK, Rajeev KG, Nakayama T, Charrise K, Ndungo EM, Zimmermann T, Koteliansky V, Manoharan M, Stoffel M (2007) Mechanisms and optimization of in vivo delivery of lipophilic siRNAs. Nat Biotechnol 25:1149–1157
72. Nishina K, Unno T, Uno Y, Kubodera T, Kanouchi T, Mizusawa H, Yokota T (2008) Efficient in vivo delivery of siRNA to the liver by conjugation of alpha-tocopherol. Mol Ther 16:734–740
73. Kolate A, Baradia D, Patil S, Vhora I, Kore G, Misra A (2014) PEG—a versatile conjugating ligand for drugs and drug delivery systems. J Control Release 192:67–81
74. Kim SH, Jeong JH, Lee SH, Kim SW, Park TG (2006) PEG conjugated VEGF siRNA for anti-angiogenic gene therapy. J Control Release 116:123–129
75. Iversen F, Yang C, Dagnaes-Hansen F, Schaffert DH, Kjems J, Gao S (2013) Optimized siRNA-PEG conjugates for extended blood circulation and reduced urine excretion in mice. Theranostics 3:201–209
76. Ezzat K, El Andaloussi S, Abdo R, Langel U (2010) Peptide-based matrices as drug delivery vehicles. Curr Pharm Des 16:1167–1178
77. Verdurmen WPR, Brock R (2011) Biological responses towards cationic peptides and drug carriers. Trends Pharmacol Sci 32:116–124
78. Chiu YL, Ali A, Chu CY, Cao H, Rana TM (2004) Visualizing a correlation between siRNA localization, cellular uptake, and RNAi in living cells. Chem Biol 11:1165–1175
79. Muratovska A, Eccles MR (2004) Conjugate for efficient delivery of short interfering RNA (siRNA) into mammalian cells. FEBS Lett 558:63–68
80. Davidson TJ, Harel S, Arboleda VA, Prunell GF, Shelanski ML, Greene LA, Troy CM (2004) Highly efficient small interfering RNA delivery to primary mammalian neurons induces MicroRNA-like effects before mRNA degradation. J Neurosci 24:10040–10046

81. Moschos SA, Jones SW, Perry MM, Williams AE, Erjefalt JS, Turner JJ, Barnes PJ, Sproat BS, Gait MJ, Lindsay MA (2007) Lung delivery studies using siRNA conjugated to TAT(48-60) and penetratin reveal peptide induced reduction in gene expression and induction of innate immunity. Bioconjug Chem 18:1450–1459
82. Rensen PCN, Sliedregt LAJM, Ferns A, Kieviet E, van Rossenberg SMW, van Leeuwen SH, van Berkel TJC, Biessen EAL (2001) Determination of the upper size limit for uptake and processing of ligands by the asialoglycoprotein receptor on hepatocytes in vitro and in vivo. J Biol Chem 276:37577–37584
83. Biessen EAL, Sliedregt-Bol K, Hoen PACT, Prince P, Van der Bilt E, Valentijn ARPM, Meeuwenoord NJ, Princen H, Bijsterbosch MK, Van der Marel GA, Van Boom JH, Van Berkel TJC (2002) Design of a targeted peptide nucleic acid prodrug to inhibit hepatic human microsomal triglyceride transfer protein expression in hepatocytes. Bioconjug Chem 13:295–302
84. Prakash TP, Graham MJ, Yu JH, Carty R, Low A, Chappell A, Schmidt K, Zhao CG, Aghajan M, Murray HF, Riney S, Booten SL, Murray SF, Gaus H, Crosby J, Lima WF, Guo SL, Monia BP, Swayze EE, Seth PP (2014) Targeted delivery of antisense oligonucleotides to hepatocytes using triantennary N-acetyl galactosamine improves potency 10-fold in mice. Nucleic Acids Res 42:8796–8807
85. Jensen SA, Day ES, Ko CH, Hurley LA, Luciano JP, Kouri FM, Merkel TJ, Luthi AJ, Patel PC, Cutler JI, Daniel WL, Scott AW, Rotz MW, Meade TJ, Giljohann DA, Mirkin CA, Stegh AH (2013) Spherical nucleic acid nanoparticle conjugates as an RNAi-based therapy for glioblastoma. Sci Transl Med 5:209ra152
86. Zheng D, Giljohann DA, Chen DL, Massich MD, Wang XQ, Iordanov H, Mirkin CA, Paller AS (2012) Topical delivery of siRNA-based spherical nucleic acid nanoparticle conjugates for gene regulation. Proc Natl Acad Sci U S A 109:11975–11980
87. Giljohann DA, Seferos DS, Prigodich AE, Patel PC, Mirkin CA (2009) Gene regulation with polyvalent siRNA-nanoparticle conjugates. Journal of the American Chemical Society 131:2072
88. Behr JP (1997) The proton sponge: a trick to enter cells the viruses did not exploit. Chimia 51:34–36
89. Merkel OM, Librizzi D, Pfestroff A, Schurrat T, Buyens K, Sanders NN, De Smedt SC, Behe M, Kissel T (2009) Stability of siRNA polyplexes from poly(ethylenimine) and poly(ethylenimine)-g-poly(ethylene glycol) under in vivo conditions: effects on pharmacokinetics and biodistribution measured by Fluorescence Fluctuation Spectroscopy and Single Photon Emission Computed Tomography (SPECT) imaging. J Control Release 138:148–159
90. Wang XL, Xu RZ, Lu ZR (2009) A peptide-targeted delivery system with pH-sensitive amphiphilic cell membrane disruption for efficient receptor-mediated siRNA delivery. J Control Release 134:207–213
91. Wagner E (2012) Polymers for siRNA delivery: inspired by viruses to be targeted, dynamic, and precise. Acc Chem Res 45:1005–1013
92. Schiffelers RM, Ansari A, Xu J, Zhou Q, Tang QQ, Storm G, Molema G, Lu PY, Scaria PV, Woodle MC (2004) Cancer siRNA therapy by tumor selective delivery with ligand-targeted sterically stabilized nanoparticle. Nucleic Acids Res 32
93. Troiber C, Edinger D, Kos P, Schreiner L, Klager R, Herrmann A, Wagner E (2013) Stabilizing effect of tyrosine trimers on pDNA and siRNA polyplexes. Biomaterials 34:1624–1633
94. Castanotto D, Rossi JJ (2009) The promises and pitfalls of RNA-interference-based therapeutics. Nature 457:426–433
95. Davis ME, Zuckerman JE, Choi CHJ, Seligson D, Tolcher A, Alabi CA, Yen Y, Heidel JD, Ribas A (2010) Evidence of RNAi in humans from systemically administered siRNA via targeted nanoparticles. Nature 464:1067–1070
96. Davis ME, Pun SH, Bellocq NC, Reineke TM, Popielarski SR, Mishra S, Heidel JD (2004) Self-assembling nucleic acid delivery vehicles via linear, water-soluble, cyclodextrin-containing polymers. Curr Med Chem 11:179–197
97. Shen JL, Kim HC, Su H, Wang F, Wolfram J, Kirui D, Mai JH, Mu CF, Ji LN, Mao ZW, Shen HF (2014) Cyclodextrin and polyethylenimine functionalized mesoporous silica nanoparticles for delivery of siRNA cancer therapeutics. Theranostics 4:487–497
98. Godinho BMDC, Ogier JR, Quinlan A, Darcy R, Griffin BT, Cryan JF, Driscoll CM (2014) PEGylated cyclodextrins as novel siRNA nanosystems: correlations between polyethylene glycol length and nanoparticle stability. Int J Pharm 473:105–112
99. Meyer M, Dohmen C, Philipp A, Kiener D, Maiwald G, Scheu C, Ogris M, Wagner E (2009) Synthesis and biological evaluation of a bioresponsive and endosomolytic siRNA-polymer conjugate. Mol Pharm 6:752–762
100. Felgner PL, Gadek TR, Holm M, Roman R, Chan HW, Wenz M, Northrop JP, Ringold GM, Danielsen M (1987) Lipofection—a

highly efficient, lipid-mediated DNA-transfection procedure. Proc Natl Acad Sci U S A 84:7413–7417

101. Tam YY, Chen S, Cullis PR (2013) Advances in lipid nanoparticles for siRNA delivery. Pharmaceutics 5:498–507

102. Morrissey DV, Lockridge JA, Shaw L, Blanchard K, Jensen K, Breen W, Hartsough K, Machemer L, Radka S, Jadhav V, Vaish N, Zinnen S, Vargeese C, Bowman K, Shaffer CS, Jeffs LB, Judge A, MacLachlan I, Polisky B (2005) Potent and persistent in vivo anti-HBV activity of chemically modified siRNAs. Nat Biotechnol 23:1002–1007

103. Semple SC, Akinc A, Chen JX, Sandhu AP, Mui BL, Cho CK, Sah DWY, Stebbing D, Crosley EJ, Yaworski E, Hafez IM, Dorkin JR, Qin J, Lam K, Rajeev KG, Wong KF, Jeffs LB, Nechev L, Eisenhardt ML, Jayaraman M, Kazem M, Maier MA, Srinivasulu M, Weinstein MJ, Chen QM, Alvarez R, Barros SA, De S, Klimuk SK, Borland T, Kosovrasti V, Cantley WL, Tam YK, Manoharan M, Ciufolini MA, Tracy MA, de Fougerolles A, MacLachlan I, Cullis PR, Madden TD, Hope MJ (2010) Rational design of cationic lipids for siRNA delivery. Nat Biotechnol 28:172–176

104. Jayaraman M, Ansell SM, Mui BL, Tam YK, Chen JX, Du XY, Butler D, Eltepu L, Matsuda S, Narayanannair JK, Rajeev KG, Hafez IM, Akinc A, Maier MA, Tracy MA, Cullis PR, Madden TD, Manoharan M, Hope MJ (2012) Maximizing the potency of siRNA lipid nanoparticles for hepatic gene silencing in vivo. Angew Chem Int Ed Engl 51:8529–8533

105. Rungta RL, Choi HB, Lin PJ, Ko RW, Ashby D, Nair J, Manoharan M, Cullis PR, Macvicar BA (2013) Lipid nanoparticle delivery of siRNA to silence neuronal gene expression in the brain. Mol Ther Nucleic Acids 2, e136

106. Kanasty R, Dorkin JR, Vegas A, Anderson D (2013) Delivery materials for siRNA therapeutics. Nat Mater 12:967–977

107. Akinc A, Zumbuehl A, Goldberg M, Leshchiner ES, Busini V, Hossain N, Bacallado SA, Nguyen DN, Fuller J, Alvarez R, Borodovsky A, Borland T, Constien R, de Fougerolles A, Dorkin JR, Narayanannair Jayaprakash K, Jayaraman M, John M, Koteliansky V, Manoharan M, Nechev L, Qin J, Racie T, Raitcheva D, Rajeev KG, Sah DW, Soutschek J, Toudjarska I, Vornlocher HP, Zimmermann TS, Langer R, Anderson DG (2008) A combinatorial library of lipid-like materials for delivery of RNAi therapeutics. Nat Biotechnol 26:561–569

108. Akinc A, Goldberg M, Qin J, Dorkin JR, Gamba-Vitalo C, Maier M, Jayaprakash KN, Jayaraman M, Rajeev KG, Manoharan M, Koteliansky V, Rohl I, Leshchiner ES, Langer R, Anderson DG (2009) Development of lipidoid-siRNA formulations for systemic delivery to the liver. Mol Ther 17:872–879

109. Love KT, Mahon KP, Levins CG, Whitehead KA, Querbes W, Dorkin JR, Qin J, Cantley W, Qin LL, Racie T, Frank-Kamenetsky M, Yip KN, Alvarez R, Sah DW, de Fougerolles A, Fitzgerald K, Koteliansky V, Akinc A, Langer R, Anderson DG (2010) Lipid-like materials for low-dose, in vivo gene silencing. Proc Natl Acad Sci U S A 107:1864–1869

110. Novobrantseva TI, Borodovsky A, Wong J, Klebanov B, Zafari M, Yucius K, Querbes W, Ge P, Ruda VM, Milstein S, Speciner L, Duncan R, Barros S, Basha G, Cullis P, Akinc A, Donahoe JS, Narayanannair Jayaprakash K, Jayaraman M, Bogorad RL, Love K, Whitehead K, Levins C, Manoharan M, Swirski FK, Weissleder R, Langer R, Anderson DG, de Fougerolles A, Nahrendorf M, Koteliansky V (2012) Systemic RNAi-mediated gene silencing in nonhuman primate and rodent myeloid cells. Mol Ther Nucleic Acids 1, e4

111. Dahlman JE, Barnes C, Khan OF, Thiriot A, Jhunjunwala S, Shaw TE, Xing YP, Sager HB, Sahay G, Speciner L, Bader A, Bogorad RL, Yin H, Racie T, Dong YZ, Jiang S, Seedorf D, Dave A, Sandhu KS, Webber MJ, Novobrantseva T, Ruda VM, Lytton-Jean AKR, Levins CG, Kalish B, Mudge DK, Perez M, Abezgauz L, Dutta P, Smith L, Charisse K, Kieran MW, Fitzgerald K, Nahrendorf M, Danino D, Tuder RM, von Andrian UH, Akinc A, Panigrahy D, Schroeder A, Koteliansky V, Langer R, Anderson DG (2014) In vivo endothelial siRNA delivery using polymeric nanoparticles with low molecular weight. Nat Nanotechnol 9:648–655

112. Morris MC, Vidal P, Chaloin L, Heitz F, Divita G (1997) A new peptide vector for efficient delivery of oligonucleotides into mammalian cells. Nucleic Acids Res 25:2730–2736

113. Mae M, El Andaloussi S, Lehto T, Langel U (2009) Chemically modified cell-penetrating peptides for the delivery of nucleic acids. Expert Opin Drug Deliv 6:1195–1205

114. Heitz F, Morris MC, Divita G (2009) Twenty years of cell-penetrating peptides: from molecular mechanisms to therapeutics. Br J Pharmacol 157:195–206

115. Deshayes S, Morris M, Heitz F, Divita G (2008) Delivery of proteins and nucleic acids using a non-covalent peptide-based strategy. Adv Drug Deliv Rev 60:537–547

116. Simeoni F, Morris MC, Heitz F, Divita G (2003) Insight into the mechanism of the peptide-based gene delivery system MPG: implications for delivery of siRNA into

mammalian cells. Nucleic Acids Res 31:2717–2724

117. Weller K, Lauber S, Lerch M, Renaud A, Merkle HP, Zerbe O (2005) Biophysical and biological studies of end-group-modified derivatives of Pep-1. Biochemistry 44:15799–15811
118. Gros E, Deshayes S, Morris MC, Aldrian-Herrada G, Depollier J, Heitz F, Divita G (2006) A non-covalent peptide-based strategy for protein and peptide nucleic acid transduction. Biochim Biophys Acta 1758:384–393
119. Crombez L, Aldrian-Herrada G, Konate K, Nguyen QN, McMaster GK, Brasseur R, Heitz F, Divita G (2009) A new potent secondary amphipathic cell-penetrating peptide for siRNA delivery into mammalian cells. Mol Ther 17:95–103
120. Kim WL, Christensen LV, Jo S, Yockman JW, Jeong JH, Kim YH, Kim SW (2006) Cholesteryl oligoarginine delivering vascular endothelial growth factor siRNA effectively inhibits tumor growth in colon adenocarcinoma. Mol Ther 14:343–350
121. Futaki S, Ohashi W, Suzuki T, Niwa M, Tanaka S, Ueda K, Harashima H, Sugiura Y (2001) Stearylated arginine-rich peptides: a new class of transfection systems. Bioconjug Chem 12:1005–1011
122. Lehto T, Ezzat K, Langel U (2011) Peptide nanoparticles for oligonucleotide delivery. Prog Mol Biol Transl Sci 104:397–426
123. Andaloussi SEL, Lehto T, Mager I, Rosenthal-Aizman K, Oprea II, Simonson OE, Sork H, Ezzat K, Copolovici DM, Kurrikoff K, Viola JR, Zaghloul EM, Sillard R, Johansson HJ, Hassane FS, Guterstam P, Suhorutsenko J, Moreno PMD, Oskolkov N, Halldin J, Tedebark U, Metspalu A, Lebleu B, Lehtio J, Smith CIE, Langel U (2011) Design of a peptide-based vector, PepFect6, for efficient delivery of siRNA in cell culture and systemically in vivo. Nucleic Acids Res 39:3972–3987
124. Ezzat K, Zaghloul EM, Andaloussi SEL, Lehto T, El-Sayed R, Magdy T, Smith CIE, Langel U (2012) Solid formulation of cell-penetrating peptide nanocomplexes with siRNA and their stability in simulated gastric conditions. J Control Release 162:1–8
125. Eguchi A, Meade BR, Chang YC, Fredrickson CT, Willert K, Puri N, Dowdy SF (2009) Efficient siRNA delivery into primary cells by a peptide transduction domain-dsRNA binding domain fusion protein. Nat Biotechnol 27:567–571
126. Michiue H, Eguchi A, Scadeng M, Dowdy SF (2009) Induction of in vivo synthetic lethal RNAi responses to treat glioblastoma. Cancer Biol Ther 8:2306–2313
127. Corey DR (2007) RNA learns from antisense. Nat Chem Biol 3:8–11
128. Crombez L, Morris MC, Dufort S, Aldrian-Herrada G, Nguyen Q, Mc Master G, Coll JL, Heitz F, Divita G (2009) Targeting cyclin B1 through peptide-based delivery of siRNA prevents tumour growth. Nucleic Acids Res 37:4559–4569
129. Morris MC, Chaloin L, Choob M, Archdeacon J, Heitz F, Divita G (2004) Combination of a new generation of PNAs with a peptide-based carrier enables efficient targeting of cell cycle progression. Gene Ther 11:757–764
130. Morris MC, Gros E, Aldrian-Herrada G, Choob M, Archdeacon J, Heitz F, Divita G (2007) A non-covalent peptide-based carrier for in vivo delivery of DNA mimics. Nucleic Acids Res 35, e49
131. Khalil IA, Futaki S, Niwa M, Baba Y, Kaji N, Kamiya H, Harashima H (2004) Mechanism of improved gene transfer by the N-terminal stearylation of octaarginine: enhanced cellular association by hydrophobic core formation. Gene Ther 11:636–644
132. Lehto T, Abes R, Oskolkov N, Suhorutsenko J, Copolovici DM, Mager I, Viola JR, Simonson OE, Ezzat K, Guterstam P, Eriste E, Smith CIE, Lebleu B, El Andaloussi S, Langel U (2010) Delivery of nucleic acids with a stearylated (RxR)(4) peptide using a non-covalent co-incubation strategy. J Control Release 141:42–51
133. Lehto T, Simonson OE, Mager I, Ezzat K, Sork H, Copolovici DM, Viola JR, Zaghloul EM, Lundin P, Moreno PMD, Mae M, Oskolkov N, Suhorutsenko J, Smith CIE, Andaloussi SEL (2011) A peptide-based vector for efficient gene transfer in vitro and in vivo. Mol Ther 19:1457–1467
134. Ezzat K, El Andaloussi S, Zaghloul EM, Lehto T, Lindberg S, Moreno PMD, Viola JR, Magdy T, Abdo R, Guterstam P, Sillard R, Hammond SM, Wood MJA, Arzumanov AA, Gait MJ, Smith CIE, Hallbrink M, Langel U (2011) PepFect 14, a novel cell-penetrating peptide for oligonucleotide delivery in solution and as solid formulation. Nucleic Acids Res 39:5284–5298
135. Oskolkov N, Arukuusk P, Copolovici DM, Lindberg S, Margus H, Padari K, Pooga M, Langel U (2011) NickFects, phosphorylated derivatives of transportan 10 for cellular delivery of oligonucleotides. Int J Pept Res Ther 17:147–157

Index

Kato Shum and John Rossi (eds.), *SiRNA Delivery Methods: Methods and Protocols*, Methods in Molecular Biology, vol. 1364, DOI 10.1007/978-1-4939-3112-5, © Springer Science+Business Media New York 2016

MIX
Papier aus verantwortungsvollen Quellen
Paper from responsible sources
FSC® C105338

If you have any concerns about our products, you can contact us on
ProductSafety@springernature.com

In case Publisher is established outside the EU, the EU authorized representative is:
Springer Nature Customer Service Center GmbH
Europaplatz 3, 69115 Heidelberg, Germany

Printed by Libri Plureos GmbH
in Hamburg, Germany